PRELUDE TO GALILEO

BOSTON STUDIES IN THE PHILOSOPHY OF SCIENCE

EDITED BY ROBERT S. COHEN AND MARX W. WARTOFSKY

VOLUME 62

WILLIAM A. WALLACE

School of Philosophy, The Catholic University of America, Washington, D.C.

PRELUDE TO GALILEO

Essays on Medieval and Sixteenth-Century Sources of Galileo's Thought

D. REIDEL PUBLISHING COMPANY

DORDRECHT : HOLLAND / BOSTON : U.S.A.

LONDON : ENGLAND

Library of Congress Cataloging in Publication Data

CIP

Wallace, William A.
Prelude to Galileo.

(Boston studies in the philosophy of science ; v. 62)
Bibliography: p.
Includes indexes.
1. Astronomy, Medieval–Addresses, essays, lectures.
2. Astronomy–History–Addresses, essays, lectures.
3. Galilei, Galileo, 1564-1642–Addresses, essays, lectures.
I. Title. II. Series.
Q174.B67 vol. 62 [QB23] 501s [520'.9'02] 81-364
ISBN-13: 978-94-009-8406-6 e-ISBN-13: 978-94-009-8404-2
DOI: 10.1007/978-94-009-8404-2

Published by D. Reidel Publishing Company,
P.O. Box 17, 3300 AA Dordrecht, Holland.

Sold and distributed in the U.S.A. and Canada
by Kluwer Boston Inc.,
190 Old Derby Street, Hingham, MA 02043, U.S.A.

In all other countries, sold and distributed
by Kluwer Academic Publishers Group,
P.O. Box 322, 3300 AH Dordrecht, Holland.

D. Reidel Publishing Company is a member of the Kluwer Group.

EDITORIAL PREFACE

Can it be true that Galilean studies will be without end, without conclusion, that each interpreter will find his own Galileo? William A. Wallace seems to have a historical grasp which will have to be matched by any further workers: he sees directly into Galileo's primary epoch of intellectual formation, the sixteenth century. In this volume, Wallace provides the companion to his splendid annotated translation of *Galileo's Early Notebooks: The Physical Questions* (University of Notre Dame Press, 1977), pointing to the 'realist' sources, mainly unearthed by the author himself during the past two decades. Explicit controversy arises, for the issues are serious: nominalism and realism, two early rivals for the foundation of knowledge, contend at the birth of modern science, or better yet, contend in our modern efforts to understand that birth. Related to this, continuity and discontinuity, so opposed to each other, are interwoven in the interpretive writings ever since those striking works of Duhem in the first years of this century, and the later studies of Annaliese Maier, Alexandre Koyré and E. A. Moody. Historiographer as well as philosopher, Wallace has critically supported the continuity of scientific development without abandoning the revolutionary transformative achievement of Galileo's labors. That continuity had its contemporary as well as developmental quality; and we note that William Wallace's *Prelude* studies are complementary to Maurice A. Finocchiaro's sensitive study of *Galileo and the Art of Reasoning* (*Boston Studies* 61, 1980), wherein the actuality of rhetoric and logic comes to the fore.

William A. Wallace, O.P. was among the founders of our Boston Colloquium for the Philosophy of Science, twenty years ago, when he was among the scholar-priests at the Dominican House of Studies, then at Dover, Massachusetts. We were grateful for his help then, and for his participation in so many ways, including his lecture some years later on 'Elementarity and Reality in Particle Physics' (*Boston Studies* 3, 1967). Wallace moves easily and with justified confidence among medieval, renaissance, classical and modern physics, as philosopher, historian, and philologist. We are grateful to him once more for all these qualities, evident in the essays of this book.

Center for the Philosophy
and History of Science
Boston University

ROBERT S. COHEN
MARX W. WARTOFSKY
January 1981

TABLE OF CONTENTS

EDITORIAL PREFACE v

PREFACE ix

ACKNOWLEDGEMENTS xv

PART I: MEDIEVAL PROLOGUE

1. The Philosophical Setting of Medieval Science 3
2. The Medieval Accomplishment in Mechanics and Optics 29

PART II: THE SIXTEENTH-CENTURY ACHIEVEMENT

3. The Development of Mechanics to the Sixteenth Century 51
4. The Concept of Motion in the Sixteenth Century 64
5. The *Calculatores* in the Sixteenth Century 78
6. The Enigma of Domingo de Soto 91
7. Causes and Forces at the Collegio Romano 110

PART III: GALILEO IN THE SIXTEENTH-CENTURY CONTEXT

8. Galileo and Reasoning *Ex suppositione* 129
9. Galileo and the Thomists 160
10. Galileo and the *Doctores Parisienses* 192
11. Galileo and the Scotists 253
12. Galileo and Albertus Magnus 264
13. Galileo and the Causality of Nature 286

PART IV: FROM MEDIEVAL TO EARLY MODERN SCIENCE

14. Pierre Duhem: Galileo and the Science of Motion 303

15. Anneliese Maier: Galileo and Theories of Impetus 320
16. Ernest Moody: Galileo and Nominalism 341

BIBLIOGRAPHY 349

INDEX OF NAMES 359

SUBJECT INDEX 365

PREFACE

Those acquainted with the literature on Galileo will recognize the kinship of the studies contained in this volume, in style and content, with the collections produced by Pierre Duhem, Annaliese Maier, Alexandre Koyré, Ernest Moody, and Stillman Drake. They represent the fruit of the author's labors over a period of some fifteen years, much of which has appeared in journals and proceedings that are not readily available to students of the history and philosophy of science. In their elaboration, and particularly in the essays written during the past several years (which make up the bulk of the volume), the author has come to the conviction that Galileo will never be understood, either historically or philosophically, when viewed in isolation from the intellectual background out of which his scientific work emerged. Galileo's classics were written in the first part of the seventeenth century, but his early studies and indeed his seminal work date from the last two decades of the sixteenth century. The latter period has never been adequately treated by historians of science; *a fortiori* it has not been seen in its proper measure of continuity with the centuries that preceded, extending back to the High Middle Ages. The essays reproduced here aim to fill this lacuna. Some have been completely rewritten and all have been reworked with this end in view, so that effectively they present a unified thesis about the medieval and sixteenth-century sources of early modern science. They are titled simply *Prelude to Galileo*, since the author does not wish to make exorbitant claims about Galileo's "precursors," and yet wishes to signal the importance of the studies for an understanding of Galileo's early, as well as his mature, contributions.

The author's thesis departs in significant respects from those advanced by the authorities already mentioned. With regard to Drake and to a lesser extent Koyré, he believes that these writers, like most others who have written about Galileo, have treated him *a parte post* and thus have evaluated him more from hindsight than from a purview of the context in which his work developed. Like the others mentioned – Duhem, Maier, and Moody – the author has adopted the opposite perspective, one that is distinctively *a parte ante*. What is particularly original about his approach is that it focuses on hitherto unknown manuscript sources of Galileo's late sixteenth-century notebooks, and traces their lineage back to thirteenth-century thinkers, mainly scholastic commentators on Aristotle. The purpose of this tactic is not to make Galileo a medieval, a scholastic, or an Aristotelian, but rather to flesh out the heritage we now know to be his, one that he shared with few other professors at Italian universities in the Renaissance. The author's thesis differs from those of Duhem, Maier, and Moody mainly in that it does not accord fourteenth-century nominalism the key role in the genesis of modern science that they do, but stresses instead the importance of more realist movements that flourished from the thirteenth to the sixteenth centuries, viz, Thomism, Scotism, Averroism, and Renaissance Aristotelianism generally. Some specifics that serve to distinguish the author's position from those of Duhem and Maier are set out in Essays 10, 14, and 15, while his critique of Moody's collected essays is included as Essay 16.

The present work is further intended to serve as an interpretative volume that draws out the historiographical and philosophical implications of the author's *Galileo's Early Notebooks: The Physical Questions* (Notre Dame: The University of Notre Dame Press, 1977). Several of the essays in Parts I and II were written while research for that book was in progress, and all of those in Parts III and IV supply technical details and textual information that authenticate and further explicate the materials contained in it. Where feasible, these essays have been revised and keyed to the *Notebooks* so as to facilitate the reader's reference to that work.

The studies in Part I, entitled 'Medieval Prologue,' set the stage for appreciating the materials in the other parts. The first essay, 'The Philosophical Setting of Medieval Science,' is adapted from the author's chapter of the same title in a collection of essays edited by David C. Lindberg, *Science in the Middle Ages* (Chicago: University of Chicago Press, 1978), pp. 91–119. The second, 'The Medieval Accomplishment in Mechanics and Optics,' appeared in slightly different form in the *Dictionary of the History of Ideas*, ed. P. P. Wiener (New York: Charles Scribner's Sons, 1973), Vol. 2, pp. 196–205, under the title 'Experimental Science and Mechanics in the Middle Ages.' These essays document the main philosophical positions and scientific contributions of the Middle Ages that are discussed in what follows, and thus are essential for an understanding of the sixteenth-century development that is central to the book's thesis.

Part II, 'The Sixteenth-Century Achievement,' contains, among others, several papers written while the author was pursuing studies of Domingo de Soto. The first, 'The Development of Mechanics to the Sixteenth Century,' is adapted from an article that appeared originally as 'Mechanics from

Bradwardine to Galileo' in the *Journal of the History of Ideas* **32** (1971), pp. 15–28, and serves to establish continuity with several of the themes developed in Part I. The second essay, 'The Concept of Motion in the Sixteenth Century,' reproduces a study of the same title that appeared in the *Proceedings of the American Catholic Philosophical Association* **41** (1967), pp. 184–195. The third, 'The *Calculatores* in the Sixteenth Century,' is revised slightly from 'The 'Calculatores' in Early Sixteenth-Century Physics,' *The British Journal for the History of Science* **4** (1969), pp. 221–232. The fourth, 'The Enigma of Domingo de Soto,' is adapted from an article with the longer title, 'The Enigma of Domingo de Soto: *Uniformiter difformis* and Falling Bodies in Late Medieval Physics,' originally published in *Isis* **59** (1968), pp. 384–401. The last essay in this section, 'Causes and Forces at the Collegio Romano,' is a somewhat different version of 'Causes and Forces in Sixteenth-Century Physics,' *Isis* **69** (1978), pp. 400–412. This is the most recent of the essays in this part and supplies material that is transitional to Parts III and IV, delineating the contents of Jesuit teachings on motion that show affinities with Galileo's early drafts on this subject, usually referred to as the *De motu antiquiora*.

Part III, 'Galileo in the Sixteenth-Century Context,' contains several of the longest essays and is central to the thesis of the volume as a whole. Its first essay, 'Galileo and Reasoning *Ex suppositione*,' appeared originally in the *Proceedings of the 1974 Biennial Meeting of the Philosophy of Science Association*, ed. R. S. Cohen et al. (Dordrecht–Boston: D. Reidel Publishing Co., 1976), pp. 79–104, as 'Galileo and Reasoning *Ex suppositione*: The Methodology of the *Two New Sciences*.' In view of the discussion this paper has generated since its first publication, the author has added an appendix to it for this volume, clarifying his positions on its key issues. The second essay, 'Galileo and the Thomists,' is heavily revised from the original version of the same title, which was published in *St. Thomas Aquinas Commemorative Studies, 1274–1974* (Toronto: Pontifical Institute for Mediaeval Studies, 1974), Vol. 2, pp. 293–330. The third, 'Galileo and the *Doctores Parisienses*,' is a reprint of an article of similar title that appeared in *New Perspectives on Galileo*, eds. R. Butts and J. Pitt, Western Ontario Series in the Philosophy of Science (Dordrecht: D. Reidel Publishing Co., 1978), pp. 87–138; it supplies much of the textual documentation on which *Galileo's Early Notebooks* is based. An appendix has also been added to this essay analyzing additional texts in which the *Parisienses* are cited, and thus further clarifying the sources of Galileo's knowledge of this fourteenth-century nominalist school. The fourth study, 'Galileo and the Scotists,' represents a fuller elaboration and development of a contribution to the Fourth International Scotist Congress, Padua 1976, which was published with the title, 'Galileo's Knowledge of the Scotistic Tradition,' *Regnum Hominis et Regnum Dei*, ed. C. Bérubé (Rome: Societas Internationalis Scotistica, 1978), Vol. 2, pp. 313–320; it too documents and supplements material contained in the *Notebooks*. The fifth essay, 'Galileo and Albertus Magnus,' was published under the title, 'Galileo's Citations of Albert the Great,' in *Albert the Great: Commemorative Essays*, eds. F. J. Kovach and R. M. Shahan, Norman: University of Oklahoma Press, 1980, pp. 261–283; it supplies detailed information that assists in establishing the date of composition of the *Notebooks* as around 1590. The final essay, 'Galileo and the Causality of Nature,' is new with this volume, although it was developed from a paper entitled 'Some Sixteenth-Century Views of Nature and Its Causality' and read at the Twelfth Annual Conference of the Center for Medieval and Early Renaissance Studies, Binghamton 1976, whose theme was Nature in the Middle Ages.

Finally, Part IV, 'From Medieval to Early Modern Science,' contains three essays, each dealing with attempts by well-known medievalists to show elements of continuity between medieval and modern science. The first, entitled 'Pierre Duhem: Galileo and the Science of Motion,' is essentially the keynote address delivered to the Second Mid-Atlantic Conference on Patristic, Medieval, and Renaissance Studies, Villanova 1977, and printed in the *Proceedings* of that conference (Villanova: Augustinian Historical Institute, 1979), Vol. 2, pp. 1–17. It recapitulates much of the argumentation developed in the earlier essays of the volume, but focuses them as a critique of Duhem's continuity thesis, which was elaborated in the early part of the twentieth century when none of the research reported in the essays was yet available. The essay also includes new material unearthed since the publication of *Galileo's Early Notebooks* and relating to possible Jesuit sources of Galileo's treatises on motion; for this reason it complements the essays contained in Part III, and may be regarded as an extension of them. The second essay, 'Anneliese Maier: Galileo and Theories of Impetus,' was written for a commemorative volume marking the tenth anniversary of Maier's death on December 2, 1971. It builds on the critique of Duhem in the preceding essay and shows how medieval and scholastic theories of impetus influenced Galileo's formulations of his *De motu antiquiora* along lines that were suspected by Maier, although she lacked the manuscript evidence since uncovered by the author to properly document her thesis. The final essay, 'Ernest Moody: Galileo and Nominalism,' is the author's review of Moody's collected papers, which appeared originally as 'Buridan, Ockham, Aquinas: Science in the Middle Ages,' *The*

Thomist **40** (1976), pp. 475–483. All three essays serve collectively to delineate the author's position vis-à-vis those developed by these distinguished scholars. At the same time they acknowledge his overall agreement with them, as well as his indebtedness to them for their lasting contributions to his field of study.

Washington, D.C.
April 25, 1980

WILLIAM A. WALLACE

ACKNOWLEDGEMENTS

The research on which the essays in this book are based began in 1965 and has continued almost uninterruptedly to the present. During this period the author has worked in many libraries and archives in the United States and Europe. In his search for relevant materials he has been generously assisted by librarians, archivists, and curators far too numerous to mention, who invariably have also facilitated the acquisition of microfilms of the more important manuscripts and early printed books on which his studies are based. To acknowledge this debt in a minimal way he lists here, by country, the collections from which he has benefited the most, either by way of direct use or through microfilms subsequently acquired:

In the United States: the libraries of Harvard University, Princeton University, the University of Chicago, Yale University, Columbia University, and the Institute for Advanced Study, Princeton; the New York Public Library; and the Henry E. Huntington Library, San Marino, California.

In Italy: at Rome, the Biblioteca Nazionale Centrale Vittorio Emanuele, the Biblioteca Casanatense, the Biblioteca Vallicelliana, and the libraries of the Pontificia Università Gregoriana and the Pontificia Università di San Tommaso d'Aquino; at the Vatican, the Biblioteca Apostolica Vaticana; at Padua, the Biblioteca Antoniana, the Biblioteca Universitaria, and the Archivo Antico of the University; at Milan, the Biblioteca di Brera and the Biblioteca Ambrosiana; at Florence, the Biblioteca Nazionale Centrale and the Biblioteca Laurenziana; at Pisa, the Domus Galileana and the Archivio di Stato; at Bologna, the Biblioteca Comunale dell'Archiginnasio; at Venice, the Biblioteca Nazionale Marciana; at Parma, the Biblioteca Palatina; and at Pistoia, the Biblioteca Forteguerriana.

On the Iberian Peninsula: at Madrid, the Biblioteca Nacional; at the Escorial, the library of the Monasterio de El Escorial; at Salamanca, the Biblioteca Universitaria and the library of the Convento de San Esteban; at Valladolid, the Biblioteca Universitaria and the Archivo General de Simancas; at Seville, the Biblioteca Colombina; at Lisbon, the Biblioteca Nacional; and at Coimbra, the Biblioteca Universitaria.

Elsewhere in Europe: at Munich, the Bayerischer Staatsbibliothek; at Bamberg, the Staatsbibliothek; at Ueberlingen (Bodensee), the Leopold

Sophien Bibliothek; at Vienna, the Oesterreichische Nationalbibliothek; at Cracow, the Biblioteka Jagiellonska; at Paris, the Bibliothèque Nationale; at London, the British Library and the library of the Warburg Institute; and at Oxford, the Bodleian Library.

For copyright permissions to reprint essays in whole or in part that have appeared in previous publications, the author acknowledges with thanks:

The University of Chicago Press, for Essay 1.

Charles Scribner's Sons, New York, for Essay 2.

The *Journal of the History of Ideas* for Essay 3.

The American Catholic Philosophical Association for Essay 4.

The British Society for the History of Science for Essay 5.

The History of Science Society for Essays 6 and 7.

D. Reidel Publishing Company, Dordrecht and Boston, and Boston Studies in the Philosophy of Science for Essay 8.

The Pontifical Institute of Mediaeval Studies, Toronto, for Essay 9.

D. Reidel Publishing Company, Dordrecht and Boston, and the University of Western Ontario Series in Philosophy of Science for Essay 10.

The Societas Internationalis Scotistica, Rome, for Essay 11.

The University of Oklahoma Press for Essay 12.

The Augustinian Historical Institute, Villanova, Pennsylvania, for Essay 14.

The Edizioni di Storia e Letteratura, Rome, for Essay 15.

The Thomist Press, Washington, D.C., for Essay 16.

For financial and academic support the author finally wishes to express his gratitude to the National Science Foundation, for research grants in the years 1965–1967, 1972–1974, and 1975–1977; to Harvard University, for appointment as a Research Associate in the History of Science in 1965–1967; to the Institute for Advanced Study, Princeton, for membership in 1976–1977; and to The Catholic University of America, for a sabbatical and other leaves of absence in the intervening years that enabled him to bring these researches to completion.

In addition to the foregoing institutions, the author expresses his special thanks to Professor Jean Dietz Moss for her continued encouragement and assistance, without which these essays could not have been assembled in the form in which they now appear.

W. A. W.

PART ONE

MEDIEVAL PROLOGUE

1. THE PHILOSOPHICAL SETTING OF MEDIEVAL SCIENCE

Unlike the science of the present day, which is frequently set in opposition to philosophy, the science of the Middle Ages was an integral part of a philosophical outlook. The field of vision for this outlook was that originally defined by Aristotle, but enlarged in some cases, restricted in others, by insights deriving from religious beliefs – Jewish, Islamic, and Christian. The factors that framed this outlook came to be operative at different times and places, and they influenced individual thinkers in a variety of ways. As a consequence, there was never, at any period of the Middle Ages, a uniform philosophical setting from which scientific thought, as we now know it, emerged. Rather, medieval philosophy, itself neither monolithic, authoriitarian, nor benighted – its common characterizations until several decades ago – underwent an extensive development that can be articulated into many movements and schools spanning recognizable chronological periods. Not all of this development, it turns out, is of equal interest to the historian of science. The movement that invites his special attention is a variety of Aristotelianism known as high scholasticism, which flourished in the century roughly between 1250 and 1350, and which provides the proximate philosophical setting for an understanding of high and late medieval science. Explaining such a setting will be the burden of this essay: how it came into being, why it stimulated the activity that interests historians of science, and how it ultimately dissolved, giving way in the process to the rise of the modern era.[1]

The scope of medieval philosophy was quite broad, encompassing everything that can be known speculatively about the universe by reason alone, unaided by any special revelation. Theology, or sacred doctrine, thus fell outside the scope of philosophy, and so did practical arts and disciplines such as grammar, mechanics, and medicine. Ethics, then as now, pertained to philosophical discourse, as did logic, natural philosophy, and metaphysics. Epistemology as we know it did not yet exist, although the problem of human knowledge, its objects and its limits, interested many thinkers during this period. Psychology, the study of the soul, was regarded as a branch of natural philosophy, as were all the disciplines we now view as sciences, that is, astronomy, physics, chemistry, and biology. Even mathematics was seen as a part of philosophy, broadly speaking, but there was no general agreement on the way in which mathematical reasoning was related to natural philosophy, and disputes over this led ultimately to the mathematical physics of Galileo and Newton, which has become paradigmatic for much of modern science.[2]

I. THE PERIOD BEFORE THE RECEPTION OF ARISTOTLE

By the year 1250 the works of Aristotle were well diffused and understood in the Latin West, and the materials from which medieval science would take its distinctive form were then ready at hand. It would be a mistake, however, to think that there was no philosophy in the Middle Ages before the thirteenth century. Much of this, it is true, had developed in a theological context, when the teachings of Greek philosophers and the teachings of the Scriptures were juxtaposed. Some of the early Church Fathers were openly hostile to Greek philosophy; Tertullian saw only error and delusion in secular learning and asked, on this account, "What has Athens to do with Jerusalem?" Others, such as Clement of Alexandria and Gregory of Nyssa, having been trained in rhetoric and philosophy, saw in Christianity the answers to questions raised in those disciplines and so proposed to use them, at least as preparatory studies, in the service of revealed truth (Gilson, 1938, pp. 3–33; Copleston, 1972, pp. 17–26). Generally the thought of Plato was seen as best approximating Christian wisdom, mainly because of the creation account in the *Timaeus* and the teachings on the soul and its immortality in the *Phaedo*.[3] The most complete blending of Neoplatonic and Christian doctrines that emerged from such syncretism appeared in the early sixth century in a series of works ascribed to Dionysius the Areopagite, the philosopher recorded as being baptized by St. Paul in Athens.[4] Because of this ascription, false though it was, these works (*On the Divine Names, On the Celestial and Ecclesiastical Hierarchy*, and *On Mystical Theology*) were accorded unprecedented authority in the later Middle Ages (Copleston, 1972, pp. 50–56).

Among the Latin Fathers, the writer who gave most systematic expression to Neoplatonic thought was St. Augustine, bishop of Hippo (*DSB* 1: 333–338). Trained in rhetoric and Latin letters, Augustine was first attracted to the sect of the Manichees because of the solution they offered to the problem of evil in the world. His early philosophical leanings were toward skepticism, but he turned from this to Neoplatonism after reading some 'Platonic treatises' (probably the *Enneads* of Plotinus) that had been translated into Latin by Marius Victorinus. This experience prepared for his conversion to

Christianity by convincing him of the existence of a spiritual reality, by disclosing the nature of evil as a privation and not something positive, and by showing how evil did not rule out the creation of the world by a God who is all good. Convinced that philosophy is essentially the search for wisdom, itself to be found only in the knowledge and love of God, Augustine wrote his *Confessions*, his *City of God*, and many smaller treatises wherein reason and faith are closely intertwined but which, nonetheless, have a recognizable philosophical content.

Augustine's early encounter with skepticism led him to anticipate Descartes' *cogito* with the similar dictum *si fallor sum* – that is, if he could be deceived, he must exist. He was aware of the limitations of sense knowledge but did not think that man's senses err; rather error arises from the way in which his soul judges about the appearances that are presented to it. A world of eternal truths exists, and the soul can grasp these because it is illumined by God to see them and to judge all things in their light. Augustine's theory of knowledge thus utilized the themes of light and illumination and was readily adaptable to Plotinus's view of creation as an emanation analogous to the diffusion of light from a unitary source. For Augustine eternal ideas are in the mind of God, where they serve as exemplars for creation and also, through the process of illumination, exercise a regulative action on the human mind; they enable it to judge correctly and according to changeless standards without themselves being seen. With regard to the created universe Augustine held that God had placed *rationes seminales* or germinal forms in matter at the beginning and that these actualize their latent potentialities in the course of time. One of Augustine's best examples of philosophizing is, indeed, his treatment of the nature of time and the paradoxes that are presented by discourse about its existence prior to the creation of the world (Copleston, 1972, pp. 27–49).

Plato was thus the first of the Greek philosophers to be baptized and to enter the mainstream of medieval thought under the patronage of Augustine, the pseudo-Dionysius, and other Neoplatonic writers. Aristotle, by contrast, exerted little influence in the early Middle Ages. Some of his logical writings (the "old logic," the *Categories* and *On Interpretation*) were made available to the Latin West through the translations of Boethius (*DSB* 2: 228–236), however, and an interest was thereby whetted in his thought. In many ways Boethius was a mediator between ancient culture and scholasticism, introducing once again the liberal arts (the *trivium* and the *quadrivium*), promoting the use of logic in rational inquiry, and setting the stage for the controversy over universals that was to recur in later centuries. He commented on the

Isagoge of Porphyry and wrote a number of theological tractates that were influential for their views on the division of philosophy. He also proposed to translate into Latin all the writings of Aristotle and Plato, aiming to show the basic agreement of these thinkers in matters philosophical. The lines of his reconciliation are seen in his discussion of general ideas or universals as these were treated by Porphyry; pondering the question whether genus and species are real or simply conceptions of the mind, Boethius leaned toward the Platonic view that they not only are conceived separately from bodies but actually exist apart from them (Knowles, 1962, pp. 107–115).

The problem of universals posed by Boethius assumed considerable importance for Peter Abelard (*DSB* 1: 1–4), who insisted that universality must be attributed to names, not things, thus anticipating nominalism, though not in the precise form in which this movement achieved prominence in the fourteenth century. As Abelard saw it, there are no universal entities or things, for all existents are singular. So, if man has exact and vivid representations of objects, these apply to individuals alone; only his weak and confused impressions can fit the members of an entire class. Man's confused grasp of natures, moreover, can never match God's universal concepts of the substances that he alone creates. Such knowledge, therefore, must be associated with a word that has at best a significative or pragmatic import, seeing that it does not strictly denominate anything in the order of existents. An expert dialectician, Abelard also produced a work entitled *Sic et non*, variously translatable as "Yes and No," "For and Against," "This Way and That." Based on the conviction that controversies can often be solved by showing how various authors used the same words in different senses, Abelard's work gave powerful stimulus to the development of scholastic method. The stylized proposal of questions, with arguments for and against, that resulted was used extensively in the interpretation of biblical, canonical, and philosophical texts, and is best illustrated in the *Decretum* of Gratian and the *Sentences* of Peter Lombard, which became highly successful textbooks in canon law and theology, respectively, during the later Middle Ages.[5]

II. ARISTOTLE AND HIGH ARISTOTELIANISM

The early scholasticism of Abelard and others, such as St. Anselm of Canterbury, was developed in the Parisian School of Saint-Victor and in the cathedral school of Chartres and prepared the way for the full flowering of scholasticism at Paris and Oxford in the thirteenth century. Around 1200 such schools were organized as the guild or "university of the masters and scholars of Paris," and that city quickly became the prestigious center of medieval European learning. Hitherto unknown works of Aristotle, together with commentaries and treatises composed by Arab and Jewish thinkers, became available there in Latin translation, and the schoolmen were suddenly confronted with vast branches of new learning that had to be assimilated into their existing syntheses. A like situation developed at Oxford, where a university was legally constituted in 1214, at which time Robert Grosseteste (*DSB* 5: 548–554), who was among the first to set himself to making Aristotle Catholic, assumed its leadership (Leff, 1968, pp. 75–115).

Grosseteste is a convenient figure with which to begin discussion of high scholasticism, for his Aristotelianism was more allied to Augustinian ways of thought than the Aristotelianism that developed at Paris, and at the same time it had fruitful implications for the growth of medieval science. Unlike many scholastics Grosseteste knew Greek, translated many works from that language into Latin, including the *Nicomachean Ethics* and the *De caelo* of Aristotle and the major treatises of the pseudo-Dionysius, and commented on parts of Aristotle's "new logic" (the *Posterior Analytics* and *On Sophistical Refutations*) and the *Physics*. One could say that he pioneered in introducing Aristotelian learning to Oxford, even though his interpretations were strongly influenced by Neoplatonism. Augustine's theme of illumination pervades Grosseteste's writings; and, indeed, the resulting philoosphy has been aptly described as a "metaphysics of light." In Grosseteste's view *lux* is the first form to come to primary or primordial matter, and it multiplies its own likeness or species in all directions, thereby constituting corporeal dimensionality and the entire universe according to determined laws of mathematical proportionality. Even the human soul is a special manifestation of light, although to possess knowledge it must also be illumined by God, the source of all light. Moreover, since the multiplication of forms or species follows geometrical patterns, the world of nature has a mathematical substructure, and so the key to natural science will be found in mathematics. The science of geometrical optics provides an ideal illustration of the required methodology: observation and experience (*experimentum*) can provide the facts (*quia*), but mathematics is necessary to see the reason for the facts (*propter quid*). Grosseteste's ideas were taken up by the Oxford Franciscans, especially Roger Bacon (*DSB* 1: 377–385) and John Pecham (*DSB* 10: 473–476), who did important work in optics; his emphasis on mathematics also stimulated the fourteenth-century development of physics at Merton College in Oxford, to be discussed later (Lindberg, 1976, pp. 94–102).

Grosseteste was a secular master, but with close connections to the

Franciscans, a mendicant order that had come into being along with the Dominicans early in the thirteenth century. Both orders of mendicants were welcomed at Paris by another secular master, William of Auvergne (*DSB* 14: 388–389), a somewhat eclectic theologian who had been made bishop of that "city of books and learning." The Paris Franciscans, whose luminaries included Alexander of Hales and St. Bonaventure, had some knowledge of Aristotle, but they preferred to work within the older Augustinian tradition and were not particularly receptive to the newly available scientific learning. The same cannot be said of the Dominican masters who assumed chairs at the University of Paris, especially St. Albert the Great and St. Thomas Aquinas. These men, as it turned out, were the architects of a new Aristotelianism at Paris that put secular learning on an almost equal footing with revealed truth and laid firm foundations for the growth of medieval science (Gilson, 1938, pp. 69–99).

Albert the Great (*DSB* 1: 99–103) was apparently the first to realize how Greco-Arabic science could best serve Christian faith by granting it proper autonomy in its own sphere. He was quite willing to accord Augustine primacy in matters of faith and morals, but in medicine, as he said, he would much rather follow Galen or Hippocrates, and in physics Aristotle or some other expert on nature.[6] Remarkable for his range of interests and for his prodigious scholarly activity, Albert was called "the Great" in his own lifetime and was commonly given the title of Universal Doctor. Much of his fame derived from his encyclopedic literary activity; he made available in Latin, for example, a paraphrase of the entire Aristotelian *corpus* ranging from metaphysics through all of the specialized sciences. Himself an indefatigable observer and cataloger of nature, he added to these accounts and generated interest and enthusiasm in his students for their further development.

Among Albert's students was the young Italian Dominican, Thomas Aquinas (*DSB* 1: 196–200), destined to become the greatest theologian in the High Middle Ages. From Albert, Aquinas derived his inspiration to christianize Aristotle, and subsequently became so proficient in Aristotelian methodology that his ultimate theological synthesis can almost be seen as an Aristotelianization of Christianity. Aquinas's basic metaphysical insight consisted in a thorough grasp of the Aristotelian principles of potentiality and actuality, which he first used to refine the distinction between essence and existence, and then applied the resulting doctrine in a novel way to a whole range of problems, from those relating to God and creation to that of the human soul and its activities. While treating Augustine respectfully, Aquinas preferred to speak of the potency of matter rather than of *rationes seminales*; he saw the human soul as the unique entelechy, or active principle, in man; and he substituted a theory of abstraction, effected through each man's *intellectus agens*, for Augustine's theory of divine illumination. While best known in the present day for his metaphysical and theological innovations, Aquinas was regarded in his own day as a competent logician and natural philosopher also, and his commentaries on Aristotle's *Posterior Analytics, Physics, De caelo, De generatione et corruptione*, and *Meteorology* rank among the best produced within the medieval period (Weisheipl, 1974a).

For both Albert and Aquinas the natural philosophy contained in Aristotle's *Physics* was important for laying the foundations of metaphysics and theology, but it was even more important for the general theory it provided for the scientific study of the entire world of nature. In their view physics was prior to metaphysics, and they were concerned to preserve the autonomy of physics from the more abstract disciplines of mathematics and metaphysics. The difference between these disciplines is set out by Aquinas in his commentary on Boethius's *De trinitate* in terms of his theory of intellectual abstraction. Physics is the least abstract of the speculative or theoretical sciences, in that it always considers material objects that have sensible matter as part of their definition; mathematics is more abstract, in that it leaves aside such sensible matter to construct numbers and figures in the imagination out of a matter that is pure extension – called "intelligible matter" because bereft of sensible qualities; and metaphysics is most abstract, in that it separates its objects from matter entirely and considers them purely under the aspect of being. Apart from these three disciplines Aquinas also allowed for sciences intermediate between them, which he referred to as *scientiae mediae*. Astronomy and optics would be examples of these, being situated between mathematics and physics and so using mathematical principles to attain an understanding of physical objects. Apart from such mathematical principles, of course, there would also be physical principles that are proper to natural philosophy, and these would guarantee the autonomy of natural philosophy from mathematics and metaphysics. In insisting on this autonomy, Albert and Aquinas were consciously at variance with Grosseteste and the Oxford school, not sharing in their "light metaphysics" and the mathematicism this implied (Weisheipl, 1959, pp. 48–62).

For the medievals, as has been mentioned, physics was a speculative science, but this did not mean that it consisted in haphazard and groundless speculation, as some have caricatured it. For Aquinas it took its roots from experience, which means that it had to be based on sense knowledge. His epistemology was realist in this regard, for when Aquinas spoke of sensible matter, he meant matter as possessed of qualities that are directly apprehended by the senses. Such qualities were the attributes or accidents of the individual substances in which they were perceived. Another type of accident, for him, and this in the case of corporeal substances, would be their quantitative extension, and since a quality such as heat would require bodily extension to be present in a substance, such extension could also indirectly quantify the quality and, thus, be the basis for the latter's quantification or measurement.[7] Thus, Aquinas did not rule out the possibility of a quantitative physics, even though the sense experience to which he referred was primarily qualitative, and the natural philosophy he elaborated, following Aristotle, shared in this characteristic. But the methodology he advocated was still basically empirical, since he had rejected Augustine's theory of divine illumination and held that no natural knowledge could come to man's mind without first originating in his senses (Pedersen, 1953, pp. 98–100).

For Aquinas, again, natural philosophy was a science, a *scientia*, and as such could yield true and certain knowledge of the material universe. It would have to do so, of course, through principles that could be discovered in experience by rational inference and that would have a self-evident character, thus being able to serve as premises in a strict proof or demonstration.[8] The experience on which he relied was that of ordinary sense observation, wherein man was a passive spectator and not an active interrogator in an experimental way. Although sometimes Aquinas used the Latin word *experimentum* (as did other medievals) to refer to such experience, and even spoke of the resulting knowledge as *scientia experimentalis*, these expressions should not be taken to connote systematic and controlled experimentation in the modern sense. It should be noted also that Aquinas was acquainted with hypothetico-deductive methodology, and pointed out the conjectural character of arguments that had not been so recognized by Aristotle. Yet he identified such examples mainly in the *scientiae mediae*, such as Ptolemaic astronomy, which he recognized as merely "saving the appearances" and not as demonstrating, for example, the reality of eccentrics and epicycles (Pedersen, 1953, pp. 90–98).

The key problem for the natural philosopher, in Aquinas's view, was that of understanding motion in terms of its causes, taking motion in the widest possible sense to mean any change perceptible in sense experience.[9] Since Aristotle had classified causes into four kinds – material, formal, efficient, and final – the investigation of the causes of motion amounted to identifying the material subject that underlies it, its formal definition, the agents that produce it, and the purposes or ends that it serves. Because of Aristotle's inclusion of final causality in this classification, his physics is generally labeled as teleological. For Aquinas, however, natural processes exhibit only an immanent teleology, in the sense that they terminate, for the most part, in forms that are perfective of the subject that undergoes the change. This he would differentiate from an extrinsic teleology, which would be some further goal or end to which the process or its product could be put. Thus, the fully grown olive tree (not the use to which olives may be put by man) is the final cause of the plant's growth from a seed, and a body's attainment of its natural place is the final cause of its movement under the influence of gravity or levity. Since natural processes are radically contingent and can always be impeded by defects in either matter or agent, Aquinas held that the natural philosopher is usually restricted to demonstrating *ex suppositione finis*, that is, on the supposition of an effect's attainment. Thus, for example, supposing that a perfect olive tree is to be generated, he can reason to all the causes that would be required to bring the generative process to completion, and this counts as scientific knowledge through causes, even though such causes might not be actually effective in the individual case.[10]

When considering the material cause of motion, we come immediately to Aristotle's hylomorphism, or matter-form theory, which pervaded all of scholastic science.[11] In explaining the coming to be and passing away of substances by the natural processes of generation and corruption, themselves examples of substantial change, Aristotle was led to maintain that in all such changes something perdures or is conserved, whereas something else changes. The enduring substrate he called *hulē* or matter, and the changing but determining factor he called *morphē* or form; the word *hylomorphic* merely transliterates these terms and so refers to the matter-form composition of corporeal substances. It is important to note here that neither matter nor form as so conceived is itself an existing thing or substance; rather, they are substantial principles, that is, factors that enter into the composition of a substance and so are parts of its nature or essence. For this reason, and to differentiate *hulē* from the matter of ordinary experience, the scholastics referred to it as primary matter (*materia prima*). Similarly, they called the form that gave primary matter its substantial identity and made it a substance of a particular kind the substantial form (*forma substantialis*). Primary matter, for them, is a material principle; it is undetermined, passive, and, being the same in all bodies, can serve to explain such common features as extension and mobility. Substantial form, by contrast, is a formal principle; as determining and actualizing, it can account for specific properties that serve to differentiate one kind of body from another.

For Aquinas, whose analysis of existence (*esse*) was considerably more refined than Aristotle's, *materia prima* had to be a purely potential principle, bereft of all actuality, including existence, and incapable of existing by itself even through God's creative action. Since God himself is Pure Actuality, he is at the opposite pole of being from primary matter, which is pure potentiality. Unlike Bonaventure, who followed in this the Jewish thinker Ibn Gebirol, Aquinas did not countenance the existence of a spiritual matter of which angels and human souls would be composed. Matter for him was the basic substrate of changeable or corporeal being, the proper subject of physical investigation. The presence of this qualityless protomatter that is conserved in all natural change, moreover, did not rule out a composition of integral parts; corporeal substances could also have elemental components, and to the extent that these might be separated out and themselves made to undergo substantial change, even such elements were essentially composed of primary matter and substantial form.

In Aquinas's view, therefore, as in that of other scholastics, the four elements of Greek natural philosophy were also considered as a type of material cause (Pedersen, 1953, pp. 102–103). These correspond roughly to the states of matter that fall under sense observation: fire (flame), air (gas), water (liquid), and earth (solid). To each of these could be assigned pairs of qualities that seemed most obvious and pervasive throughout the world of nature, that is, hot and cold, wet and dry. So fire was thought to be hot and dry; air, hot and wet; water, cold and wet; and earth, cold and dry. The elements could be readily transmuted into one another, moreover, by conserving one quality and varying the other: so fire could be changed into air by conserving its hotness and converting its dryness to wetness; air could be changed into water by conserving its wetness and converting its hotness to coldness; and so on. Such elements, for Aquinas and for Aristotelians generally, did not exist in a pure state; all of the substances that come under sense experience are composites (*mixta*) of elements in varying proportions. The combination of the elements' primary qualities thus gave rise to the entire range of secondary qualities that are observed in nature. Because Aquinas held to the unicity of the substantial form in each natural substance, this raised for him the problem of how the forms of elements could be present in compounds, and he was led to propose that they are present not actually but only virtually – a position that was widely accepted in the later Middle Ages (Grant, 1974, pp. 603–605). Like most medieval philosophers, Aquinas believed that material substances are continuous, and thus rejected Democritus's theory of matter being composed of atoms with interstitial voids. He did believe in *minima naturalia*, however, maintaining that minimum quantitative dimensions are required for the existence of most natural substances, and it is noteworthy that Albert the Great identified such *minima* with the atoms of Democritus.[12]

With regard to the formal definition of motion, Aquinas followed Aristotle in defining motion as an imperfect actuality or act, that is, the actuality of a being whose potentiality is actualized while still remaining in potency to further actualization.[13] This definition is not easy to comprehend, and, indeed, it raises many questions about the reality and existence of motion (as formulated, for example, in Zeno's paradoxes): Is motion anything more than the actuality or terminus that is momentarily attained by it at any instant, and should it be conceived as a *forma fluens* or as a *fluxus formae*? The latter question was adumbrated by Albert the Great and answers to it divided the nominalists and the realists of the fourteenth century. Ockham, for example, thought of motion as nothing more than the forms successively acquired by a subject, and so he defined motion from the nominalist viewpoint as merely a *forma fluens*. By contrast, Walter Burley, while admitting that motion could be viewed in the Ockhamist way, made the realist claim that motion is also a flux, a *fluxus formae*, that is, an actual transformation by which these new termini or forms are being successively acquired.[14] Burley saw motion also as a successive quantity which is continuous in the same way as corporeal substances appear to be; whereas bodies were static continua, however, motions (and likewise time) came to be regarded as "flowing" continua. The ways in which these various entities could be said to be constituted of quantitative parts quickly gave rise to all the problems of the continuum, many of which could not be solved before the invention of the infinitesimal calculus and modern theories of infinity.[15]

The definition of motion in terms of actuality and potentiality also had profound implications for investigating the agent causes that produce change. Since a thing could not be in both actuality and potentiality to the same terminus at the same time, it seemed obvious that no object undergoing change could be the active source of its own motion: rather, it would have to be moved by an agent that already possessed the actuality it itself lacked. Water, for example, could not be the active cause of its own heating, whereas fire could be such a cause, since fire was actually hot and could reduce the water's potentiality for heat from potency to act. This insight led medieval thinkers into an imaginative search for the movers behind all motions observed in nature, especially when such movers were not patently observable.

Typical queries would be: What causes the continued motion of a projectile after it has left the hand of the thrower? What causes the fall of heavy bodies? and What causes hot water to cool when left standing by itself? (Weisheipl, 1965, pp. 26–45).

Many of these problems, of course, had already been broached by Aristotle, but they acquired new interest for medievals in view of the arguments Aristotle had presented in the last part of his *Physics* to prove the existence of a First Unmoved Mover. For Aquinas, not only did Aristotle's principles open up the possibility of rendering intelligible all of nature's operations, but they could even lead one to a knowledge of the Author of nature by purely rational means. In a word, they made available reasonable proofs for the existence of a mover that was incorporeal, immaterial, of infinite power, and eternal in duration, who could be identified with the God of revelation. Small wonder, then, that for him Aristotelian physics held the greatest of promise. It allowed one to reassert the autonomy of reasoning based on sense experience, it explained the magnificent hierarchy of beings from the pure potentiality of primary matter through all the higher degrees of actuality, and it even provided access to the Pure Actuality, God himself, *Ipsum Esse Subsistens*, who had revealed the details of his inner being to all who accepted on faith his divine revelation (Weisheipl, 1959, pp. 31–48).

This brief sketch of the natural philosophy of the thirteenth century may help to explain the enthusiasm with which Aristotle's *libri naturales* came to be accepted at Paris and at similar centers in Christendom. Topics other than those already indicated occupied the attention of philosophers and theologians alike, and these cumulatively constituted the subject matter that would later become modern science. Among such topics were the nature of space and time, the existence of a vacuum and the possibility of motion through it, the kinematical and dynamical aspects of local motion, the various forces and resistances that determine a body's movement, the intensification of qualities, and a variety of problems associated with the structure of the cosmos and the relationships between celestial and terrestrial motions.[16] Such topics were approached anew with great confidence, for Greek learning now appeared to be buttressed by Catholic faith, and an all-knowing God seemed to be beckoning men, as it were, to uncover the rationality and intelligibility hitherto concealed in his material creation. Historians have seen in this situation the basic charter that underlies the whole scientific enterprise (Whitehead, 1925). Whether this be true or not, there seems little doubt that between Oxford and Paris some fundamental contributions had already been made, such as highlighting the problem of the role of mathematics in physical science, asserting the primacy of empirical investigation in studying the world of nature, and granting physics its autonomy from metaphysics and theology as a source of valid knowledge concerning the cosmos.

III. THE CRISIS OF AVERROISM

Mention has been made of Islamic influences on the development of natural philosophy, and these now need to be examined in some detail. The problem of the relationship between reason and belief came to a head earlier in Islam than it did in Latin Christendom, with results that very often asserted the primacy of reason over faith, rather than the other way around, as accepted without question by the Latins during the long period from Augustine to Aquinas. The main inspiration behind the Arab position was Averroës (*DSB* 12: 1–9), whose thought has been characterized as a twelfth-century rationalism similar in some respects to the later movement in modern Europe. Apart from Averroës' polemical writings against Arab divines such as Algazel, however, what is most significant is that he commented in detail on all the works of Aristotle, with such skill that when his writings were made available in Latin they earned for him the undisputed title of "Commentator," the interpreter of Aristotle *par excellence*. The Aristotle he interpreted, moreover, was somewhat Neoplatonized, but he had not been baptized and, thus, not contaminated with elements surreptitiously derived from Christian doctrine. Where Aquinas, therefore, had made room for a rational understanding of the universe in the light of Aristotle but in a general thought context provided by faith, Averroës pushed the claims of unaided reason even further. And where Aquinas felt compelled to question Aristotle's authority and even to modify his teachings as the occasion demanded, Averroës was under no such constraint; indeed, he regarded Aristotle as a god, the summit of all rational understanding, an infallible guide to knowledge of the world of nature (Gilson, 1938, pp. 35–66; Copleston, 1972, pp. 104–124).

Averroës's contributions, of course, were made in an Islamic milieu and were critical of the thought of other Arabs, such as Avicenna, rather than that of the early scholastics. Among his distinctive theses was the teaching that there is only one intellect for all men. Avicenna (*DSB* 15: 494–501) had taught that there was a single *intellectus agens*, or active intellect, but that each individual possessed his own *intellectus possibilis*, or passive intellect, wherein he would have his own ideas. Averroës disagreed with this teaching, holding that both the passive and the agent intellects were a separated substance, one and the same for all mankind, and so denying the accepted basis for man's personal immortality. Another of Averroës' theses was the eternity of the universe, for he followed Aristotle literally in maintaining that the heavens, motion, time, and primary matter had no temporal beginning or end; so they were not created, nor would they ever cease to exist. He disagreed too with Avicenna's teaching on essence and existence, and this affected his understanding of the relationships between God and the universe, effectively necessitating God's action in ways that would turn out to be contrary to Christian teaching.[17]

Averroës' commentaries on Aristotle were known to William of Auvergne, Albert the Great, and others who first advanced the cause of Aristotelianism at the University of Paris. The fact that Averroës' teachings could be inimical to Catholic belief was not immediately recognized, but as Averroës' influence increased along with the reception of Aristotle, his distinctive interpretations gradually got a hearing. Not only that, but they soon attracted adherents within the faculty of arts. Thus, a movement got underway that has been characterized as Latin Averroism or heterodox Aristotelianism, whose chief proponents were Siger of Brabant and Boethius of Sweden. The Latin Averroists taught the oneness of the passive intellect for all men, the eternity of the universe and of all its species (with the result that there would be no first man), and the necessity of God's causality in the world, although they admitted contingency in the sublunary regions because of matter's presence there. These teachings had ramifications that were opposed to Christian faith, and it is noteworthy that the Latin Averroists never denied this faith, although they held that their philosophical conclusions were probable, or even necessary.[18]

The theologians at the University of Paris reacted, predictably, to such teachings, and Bonaventure, Albert the Great, Aquinas, Giles of Rome, and others all wrote polemical treatises against Siger and his followers. Alarmed at the growth of naturalistic rationalism, ecclesiastical authorities also attempted to halt heterodox Aristotelianism with a series of condemnations. In 1270 the bishop of Paris, Etienne Tempier, reinforcing some earlier prohibitions, condemned thirteen propositions that contained Averroist Aristotelian teachings, namely, the oneness of the intellect, the eternity of the world, the mortality of the human soul, the denial of God's freedom and providence, and the necessitating influence of the heavenly bodies in the sublunary world. Tempier followed this on March 7, 1277, with the condemnation of 219 propositions, all linked with philosophical naturalism, including some theses upheld by Aquinas. In the prologue to the condemnation, Tempier accused the Averroists of saying that what was true according to philosophy was not true according to the Catholic faith, "as if there could be two contrary truths," thus giving rise to what has been called a theory of double truth, although such a theory was not explicitly advocated by Averroës, Siger, or anybody else in the movement.[19]

It is undeniable that the condemnation of 1277 had an effect on the development of medieval science, although not as profound as was maintained by Pierre Duhem, who actually proposed 1277 as the birthdate of modern science. Duhem's argument was that among the condemned propositions were two that bore on subject matters later of interest to scientists: one, denying that God had the power to move the universe with a rectilinear motion for the reason that a vacuum would result; the other, denying God's power to create more than one world. By thus opening up the possibility that theses rejected by Aristotle were true, and asserting God's omnipotence as a factor that would henceforth have to be taken into account when deciding cases of possibility or impossibility in the cosmological order, Tempier stimulated the scientific imagination, as Duhem saw it, and so opened the way to a full consideration of various alternatives to an Aristotelian cosmology (Duhem, 1909, p. 412).

It is generally agreed that Duhem's thesis is extreme, for there is no indication of a spurt in scientific thought or activity following 1277, and it is doubtful whether any authoritarian restriction on cosmological teachings could have stimulated the free spirit of inquiry that is generally seen as characteristic of modern science. There is no doubt, however, that the condemnation of 1277 did have profound consequences on the development of natural philosophy in the decades that followed, and particularly on the relationships between philosophical and theological thought, as will be discussed presently (Grant, 1971, pp. 27–35, 84–90).

While on the subject of Averroism, it should be noted that Averroës had a number of distinctive teachings relating to natural philosophy that were incorporated into commentaries and questionaries on Aristotle's *Physics*, thus becoming part of the Aristotelian tradition that would be taught in the universities to the end of the sixteenth century. Among these were the theses that substantial form is prior to and more knowable than the substance of which it is a part, which takes its entire essence or quiddity from the form; that motion through a vacuum would be instantaneous (which has the effect of ruling out motion in a vacuum, since no motion can take place in an instant); that the substantial form is the principal mover as an active principle in the natural motion of heavy and light bodies, and that the gravity and levity of such bodies are secondary movers as instruments of the substantial form; that a projectile is moved by the surrounding medium, and so there can be no projectile motion in a vacuum; that mover and moved must always be in contact, with the result that action at a distance is impossible; and that curvilinear motion cannot be compared to rectilinear motion because the two are incommensurable. Many of these teachings were propounded by John of Jandun at Paris in the early fourteenth century, and they were commonly accepted in northern Italy, especially at Padua, from the fourteenth through the sixteenth century. Not all commentators, of course, subscribed to them. Thomists, for example, basing their interpretations of Aristotle on the many writings of Aquinas, taught that the essence or quiddity of a natural substance must include matter as well as form; that motion through a vacuum would not be instantaneous; that the principal mover in natural motion is the generator of the moved object (that which brought it into being) and that internal forms such as gravity serve only as passive principles of such motion; that a projectile can be moved by an internal form, or impetus, analogous to the form of gravity for falling motion; and that actual physical contact between mover and moved is not essential, but that in some cases a virtual contact (*secundum virtutem*) suffices. These few examples may serve to show that even within Aristotelianism there was no uniform body of doctrine, and that different schools, such as Averroists and Thomists, had their own distinctive teachings in natural philosophy, despite the fact that they took their basic inspiration from Aristotle.[20]

IV. THE CRITICAL AND SKEPTICAL REACTION

To come now to other schools that figured in late medieval science, we must return again to the Franciscans, who took the ascendancy in the late thirteenth and early fourteenth centuries through the writings of John Duns Scotus and William of Ockham, themselves Franciscan friars. Both the Scotistic and Ockhamist movements were part of a critical and skeptical reaction in philosophy, primarily motivated by theological interests, and following on the condemnation of 1277. The high scholasticism of the thirteenth century had seen Aristotle welcomed enthusiastically as "the master of all who know." Now the theologians had pointed out what disastrous consequences could attend the uncritical acceptance of Aristotle's teachings in matters that touched on their discipline. There was no doubt, then, that Aristotle had erred in matters theological; might it not be the case that he likewise erred in matters philosophical? Both Scotus and Ockham implicitly answered this question in the affirmative, and in so doing set scholasticism on a different course from that which it had been following under the inspiration of Albert the Great and Thomas Aquinas (Vignaux, 1959, pp. 146–213; Copleston, 1972, pp. 213–256).

Duns Scotus (*DSB* 4: 254–256) is known as the Subtle Doctor, and his writings abound in fine distinctions and closely reasoned arguments that put him very much in the scholastic mold. He was a critical thinker, moreover, and quite concerned with elaborating a systematic metaphysics, which, in turn, had important consequences for his natural philosophy. He understood Arabic thought quite well, consistently favoring Avicenna over Averroës. Again, though he denied the theory of divine illumination, he was much indebted to Augustine, Bonaventure, and the Oxford tradition within Franciscan thought. Unlike Aquinas, he was not so much interested in assimilating philosophy into theology as he was in preserving the autonomy, and, indeed, the very possibility, of theology against the encroachments of a naturalistic rationalism.

Scotus attempted to do this by developing his own theory of knowledge, which focused on being as a univocal concept; by stressing the primacy of the will, to assure God's absolute freedom and also the preeminence of freedom in man; and by viewing God under the aspect of infinity as his essential characteristic. He held to primary matter but did not conceive this as pure potentiality, as had traditional Aristotelians; rather, he saw it as a positive reality and actuality capable of receiving further perfection. Moreover, apart from matter and form, Scotus held that every concrete reality also has metaphysical components of universality and particularity, thereby reopening the debate over universals. Every being, in his view, contains a common nature (*natura communis*) that is itself indifferent to such universality and particularity but that is rendered particular by an individuating principle, which he referred to as "thisness" (*haecceitas*). Scotus developed also a complex system of distinctions, including a novel one, the formal distinction, which he proposed as intermediate between the real distinction and the distinction of reason generally invoked by the scholastics.[21] He denied the necessary validity of the principle, "Whatever is moved is moved by another," saying that this is true of violent but not of natural motions, and, thus, he did not use this principle, as Aquinas had, to prove the existence of a First Unmoved Mover. His own proof for the existence of God was distinctive, being based on an analysis of the notion of possibility, and showing how this entails the existence of a necessary uncaused being (Effler, 1962; Wolter, 1962).

Even from this brief summary it can be seen that Scotus was more the metaphysician than the natural philosopher, and that in some ways his desire to guarantee the possibility of theology as a science led him to negate the gains made by Aquinas and Albert the Great in their attempts to maintain the autonomy of reason against the Augustinians. Moreover, though Scotus and his followers did not neglect natural philosophy entirely, their importance for medieval science derives less from their positive contributions to that discipline than from the skeptical reaction they provoked from William of Ockham. Ockham (*DSB* 10: 171–175) had been exposed to Scotistic teaching during his years of training in the Franciscan Order, and he developed his own thought in conscious opposition to that of the Subtle Doctor. Like that of Scotus, however, Ockham's intent was theological from the beginning, and although his critique of Aristotle was indeed philosophical, its motivation is directly traceable to the condemnation of 1277. Again, like Scotus, Ockham stressed the traditional Franciscan themes of divine omnipotence and divine freedom and was concerned to eliminate any element of necessity from God's action, as this had been found in Neoplatonic emanationism and in Arab thought generally (Leff, 1975).

In working out his own philosophical position, Ockham consistently invoked two main theses. The first was that God has the power to do anything whose accomplishment does not involve a contradiction. The net effect of this teaching was to admit that, in the order of nature, whatever is not self-contradictory is possible; thus, there is no *a priori* necessity in nature's operation, and whatever is the case must be ascertained from experience alone. Ockham's second thesis was a principle of parsimony, referred to as "Ockham's razor" and commonly expressed in the maxim "Beings are not to be multiplied without necessity." The application of this principle led Ockham to formulate a new logic, similar to the nominalism of Abelard, wherein he no longer sought to find real counterparts in the universe for all the categories, as most Aristotelians after Boethius had done.[22] Concepts or universals for Ockham became simply words, and the only real existents were "absolute things" (*res absolutae*), which he conceived of as individual substances and their qualities. All other categories were to be regarded as abstract nouns, used for the sake of brevity in discourse but having no real referent other than substance and quality. Much of Ockham's polemic was, in fact, directed against Scotus's "common nature" and his formal distinction between such a nature and its individuating principle. Ockham also denied, however, Aquinas's real distinction between essence and existence, and most other metaphysical distinctions that had come to play a dominant role in thirteenth-century scholasticism (Weisheipl, 1959, pp. 63–69; Pedersen, 1953, pp. 120–121).

Fourteenth-century nominalism, as fathered by Ockham and quickly taken up by others, thus incorporated a view of the universe that was radically contingent in its being, where the effect of any secondary cause could be dispensed with and immediately replaced by God's direct causality. The theory of knowledge on which it was based was empiricist, and the problems it addressed were mainly those of the philosophy of language. While scholastic in setting, Ockham's philosophy was thoroughly modern in orientation. Referred to as the *via moderna*, in opposition to the *via antiqua* of the earlier scholastics, it has been seen as a forerunner of the modern age of analysis – indeed, as a fourteenth-century attempt to unite logic and ontology in ways that had to await the twentieth century for their more rigorous formulation (Moody, 1975, pp. 300–302, 316–319).

In natural philosophy, following Scotus, Ockham accorded actual existence to primary matter and saw form as providing geometrical extension, more as *figura* than as *forma substantialis*, which in the Thomistic understanding confers actual existence on the composite. Motion, for Ockham, could not be an absolute entity, and, thus, it was not a reality distinct from the body that is in motion. Most of the difficulties in prior attempts to define motion, as Ockham saw it, arose from the inaccurate use of language, from speaking of motion as if it were something different from the body moved and the terminus it attains. In effectively rejecting Aristotle's definition of motion as the actualization of the potential *qua* potential, Ockham also had to dispense with the need for a mover to produce local motion, whether this be located in the medium through which the body moves or within the moving body itself. Some have seen in this rejection of motor causality an adumbration of the concept of inertia, but this seems unwarranted, as Ockham dispensed with a moving cause not merely in the case of uniform motion in a straight line, but in all instances of local movement (Dijksterhuis, 1961, pp. 175–176; Whittaker, 1946, pp. 45–47, 139–143). And, in the long run, Ockham's analysis of motion was not very profound – its logic led to the rejection of the very reality it was devised to explain – although it did have some consequences for the development of medieval science at Merton College in Oxford and at Paris, as will be explained later (Pedersen, 1953, pp. 121–128).

Ockham's philosophy may be viewed as the first consistent attempt to renovate Aristotelian scholasticism, but it was neither the most critical nor the most radical. Other thinkers who reacted yet more skeptically include John of Mirecourt and Nicholas of Autrecourt. John of Mirecourt insisted on the merely probable character of most human knowledge – this because he had extreme views as to what might guarantee certitude and because he, like Ockham, wished to make due allowance for the unlimited freedom of God. In philosophy, for John, there is little hope of reaching anything better than probability, since sense knowledge is deceptive, truths can rarely be reduced to the principle of non-contradiction, and God can always intervene miraculously to produce a different result. A similar strain runs through the writings of Nicholas of Autrecourt, who likewise held that all knowledge arises from sensation and that there can be only one valid criterion of certitude, again the principle of non-contradiction. Since the senses deceive and since it is difficult to resolve any arguments to non-contradictory assertions, human knowledge must be essentially limited to probabilities. Applying these principles to Aristotelian physics, and particularly to its teaching on causality, Nicholas argued that the atomism that was rejected by Aristotle is just as likely as his theory of matter-form composition, and that even if causality exists in nature it can never be demonstrated. Because of such skeptical views Nicholas has been seen as the "medieval Hume," the forerunner of modern empiricism (Copleston, 1972, pp. 260–266; Pedersen, 1953, pp. 128–134).

By the middle of the fourteenth century, then, under the critical and skeptical attacks of philosophers as diverse as Scotus, Ockham, and Nicholas of Autrecourt, the scholastic program that had been initiated with such enthusiasm by Grosseteste, Albert, and Aquinas effectively came to an end. The newer, more critical movements did not negate the basic Aristotelian insights that had enlivened the theology of the High Middle Ages, and, indeed, a new philosophy of nature was about to emerge that would be eclectic in many particulars, but still would contain the seeds from which modern science could arise in the early seventeenth century. By this time, however, high scholasticism had already peaked and had begun to disintegrate, fragmented into many schools and opposing factions, and given over to endless subtleties of disputation that were to be caricatured by Renaissance humanists and moderns alike. The final impression would be that scholastic Aristotelianism had failed, initially because reason had claimed too little, ultimately because it had claimed too much, as it competed with the Christian faith in its efforts to seek an understanding of the world of nature.

V. THE NEW NATURAL PHILOSOPHY

At this stage the elements were at hand for the forging of new and innovative patterns of scientific thought, which would broaden natural philosophy beyond the bounds set for it by Aristotle. The critical and skeptical reactions to Aristotle were the immediate forerunners, but they had little direct influence on the new natural philosophy that was making its way through the tangle of traditional ideas. They did reinforce, however, the general impression already produced by the condemnation of 1277, namely, that Aristotle's views had to be examined critically, corrected, reformulated, and sometimes rejected entirely – and this when they came into conflict not only with divine revelation but also with the manifest data of experience. The principal innovations that resulted in the natural philosophy of the mid-fourteenth century, when this finally assumed recognizable contours, were the emergence of new mathematical methods for use in physical investigations and the introduction of kinematical and dynamical concepts that would finally raise the mechanics of moving bodies to the status of a science. Neither of these innovations was especially fostered by Scotus or Ockham, whose philosophies had little need of mechanical concepts and who assigned to mathematics a role in physics no larger than that given it by Albert and Aquinas and far smaller than that given it by Grosseteste. As it turned out, however, the latter's mathematical tradition was still alive at Oxford, and it was there, at Merton College, that the first innovations were made. These quickly passed to the University of Paris, where they merged with new mechanical concepts, and thence were transmitted to other centers of learning in western Europe (Copleston, 1972, pp. 270–275).

A fuller elaboration of the resulting conceptual development is taken up in subsequent essays of this volume. Here only a few general observations need be made about this development as it relates to matters already discussed. At Oxford the principal contributors to the new natural philosophy were Walter Burley, Thomas Bradwardine, William of Heytesbury, and Richard Swineshead. Burley (*DSB* 2: 608–612), as we have seen, is noteworthy for his "realist" reaction to Ockham's nominalism and for reopening most of the problems relating to the causal agents involved in local motion that Ockham had sought to bypass. Whereas Burley thus initiated a traditionalist revival at Oxford, Bradwardine (*DSB* 2: 390–397), on the other hand, was more the innovator. His own interests were heavily mathematical, and he set himself the problem of resolving some of the internal contradictions that were detectable in Aristotle's so-called dynamical laws. In Bradwardine's day many Arabic and Latin commentators were interpreting Aristotle's statements, especially those in the fourth and seventh books of the *Physics*, to imply precise quantitative relationships between motive force, resistance, and velocity of movement. In such a context Bradwardine proposed an ingenious interpretation of Aristotle that used a relatively complex mathematical function to render consistent the various ratios mentioned in his writings. This formulation, while incorrect from the viewpoint of Newtonian mechanics, introduced the concept of instantaneous velocity and adumbrated some of the computational apparatus of the infinitesimal calculus. Bradwardine was followed at Merton College by Heytesbury (*DSB* 6: 376–380) and Swineshead (*DSB* 13: 184–213), who presupposed the validity of his dynamic analysis and extended it to a fuller examination of the comparability of all types of motions or changes. In so doing they discussed the intension and remission of forms, and spoke of the "latitude of forms," regarding even qualitative changes as traversing a distance (*latitudo*) and, thus, as quantifiable. They also employed a letter calculus that lent itself to the discussion of logical subtleties – the *sophismata calculatoria* soon to be decried by humanists. Such "calculations" were of unequal value from the viewpoint of natural philosophy, but they did suggest new techniques for dealing with the problems of infinity; they also led to a sophisticated terminology for describing rates of change that would have important applications in mechanics (Pedersen, 1953, pp. 134–142).

Mertonians such as Bradwardine, Heytesbury, and Swineshead were sympathetic to nominalism, and, as a consequence, they did not have the realist concerns that were to become influential at Paris in the mid-fourteenth century. They were highly imaginative in their treatment of kinematical problems, but did so in an abstract mathematical way, generally without reference to the motions actually found in nature. By contrast, a group of thinkers at the University of Paris devoted themselves rather consistently to investigating the physical causes of motion, introducing the concept of impetus and quantifying, in ways suggested by the Mertonians, the forces and resistance involved in the natural movements of bodies. Foremost among these were the Scotist, Franciscus de Marchia, and Jean Buridan, Albert of Saxony, and Nicole Oresme, the last three generally referred to as "Paris terminists." Terminism is sometimes equated with nominalism, and it is true that these thinkers were all nominalist in their logic, making extensive use of Ockham's *logica moderna*, but in natural philosophy they rejected the nominalist analysis of motion and developed realist views of their own.

Jean Buridan is regarded as the leader of the Paris group, playing a role there similar to that of Bradwardine at Oxford, and being best known for his development of the concept of impetus as a cause of projectile motion and of the acceleration of falling bodies. He also defined, against Nicholas of Autrecourt, the character of natural philosophy as a science *secundum quid* (that is, in a qualified sense) because based on evidence *ex suppositione*, thus using Aquinas's methodological expression; his concern was not merely with the contingency found in nature, however, but rather with the possibility of nature's order being set aside miraculously through divine intervention (Wallace, 1976b). Buridan's pupils, Albert of Saxony and Nicole Oresme, showed greater competence than he in mathematics and applied Mertonian techniques to the discussion of both terrestrial and celestial motions. Oresme pioneered, in fact, in the development of geometrical methods of summing series and integrating linear functions, and adumbrated some of the concepts of analytical geometry. The writings of the Paris terminists were widely diffused throughout western Europe and were much discussed in commentaries and questionaries on the *Physics* because of their obvious relevance to the problems of natural philosophy. It is for this reason that pioneer historians of medieval science, such as Duhem, spoke of them as the "Parisian precursors of Galileo" (1913, p. 583).

By the end of the fourteenth century, therefore, a considerable body of new knowledge had become available that was basically Aristotelian and yet had been enriched by mathematical and dynamical concepts showing considerable affinity with those of modern science. The history of the diffusion of this new natural philosophy throughout the fifteenth and sixteenth centuries is quite complex, and will be sketched in more detail in Part II of this volume. In general outline, however, one can say that the diversity of schools and movements continued, although with a noticeable relaxation of the fierce partisan loyalties that had characterized debates in the late thirteenth and early fourteenth centuries. A tendency toward eclecticism began to manifest itself, with most commentators picking and choosing theses that suited their purposes and seemed most consistent with their own experiences. In such an atmosphere, full-length commentaries on the *Physics* gave way to shorter tracts on various subjects, and treatises entitled *De motu* appeared in increasing numbers. Some of these were nominalist, others Thomist or Scotist, yet others Averroist in inspiration, but all covered essentially the same subject matter – motion, its definition, its causes, its quantitative aspects. Such treatises were taught in the universities when Galileo and the other founders of modern science pursued their formal studies, and they provided the proximate background for the emergence of the "new science" of the seventeenth century.[23]

This, then, completes our account of the philosophical setting of medieval science. The ideas discussed were developed over a span of a thousand years, from Augustine to Oresme, though they had received their initial formulation in Greek antiquity, in the writings of Plato and Aristotle. What we now call medieval science, as it was understood over most of this period, was actually identical with natural philosophy, or *scientia naturalis*, except for the ancillary role played by the *scientiae mediae* in the development of mathematical methodology (Murdoch, 1974). It is perhaps noteworthy that most of the problems of natural philosophy, and particularly those formulated by Aristotle, still resist definitive solution in the present day, and in the main they have passed into a related discipline known as the philosophy of science, where realists and nominalists (now called positivists) continue to be divided over the basic issues (Wallace, 1968c). How seventeenth-century science succeeded in disentangling itself from philosophy and in defining its own limits in apparent independence of philosophical thought still awaits adequate treatment. As far as the science of the Middle Ages is concerned, however, this problem did not present itself. It was part and parcel of man's attempt to comprehend the world of nature with the light of unaided reason. Once this is understood, the Middle Ages can no longer be regarded by historians of science as the Dark Ages, but, rather, must be seen as a period of gradual enlightenment, culminating in the thirteenth and fourteenth centuries, when recognizable foundations were laid for the modern scientific era.

NOTES

[1] A good general introduction to philosophy in the Middle Ages is Frederick Copleston's *A History of Medieval Philosophy* (London, 1972), which contains an extensive and up-to-date bibliography relating to all philosophers and schools mentioned in this essay. Also basic as a reference work, although it accents metaphysics to the neglect of natural philosophy, is Etienne Gilson's *History of Christian Philosophy in the Middle Ages* (New York, 1955). General surveys of intellectual history of the Middle Ages will be found in David Knowles, *The Evolution of Medieval Thought* (New York, 1962), and in Gordon Leff, *Medieval Thought: St. Augustine to Ockham* (Baltimore, 1958). Shorter treatments that are practically classics in the field are Etienne Gilson's *Reason and Revelation in the Middle Ages* (New York, 1938), which focuses on Augustine, Anselm, Averroës, and Aquinas; Paul Vignaux, *Philosophy in the Middle Ages: An Introduction*, trans. by E. C. Hall (New York, 1959), which explains well the period from Anselm to Ockham; and Fernand Van Steenberghen, *Aristotle in the West: The Origins of Latin Aristotelianism*, trans. by Leonard Johnston (Louvain, 1955), which covers in detail the period from 1200 to 1277. For individuals, concepts, and movements, brief but informative summaries are to be found in specialized encyclopedias, especially the *New Catholic Encyclopedia*, ed. W. McDonald, 15 vols. (New York, 1967) (hereafter *NCE*), which gives extensive coverage to medieval philosophy and its relation to science; see also the supplement, vol. 16 (Washington, 1974). The *Encyclopedia of Philosophy*, ed. Paul Edwards (New York, 1967), and the *Dictionary of Scientific Biography*, ed. Charles C. Gillispie (New York, 1970–) (hereafter *DSB*), likewise include up-to-date articles on the more important personages; in the latter see, for example, G. E. L. Owen et al., 'Aristotle,' 1: 250–281.

[2] See David C. Lindberg, ed., *Science in the Middle Ages* (Chicago, 1978), chapter 14. For a comprehensive survey of natural philosophy in the Middle Ages and its relation to medieval science, see James A. Weisheipl, *The Development of Physical Theory in the Middle Ages* (New York, 1959), which contains a guide to further reading. Also helpful is the article by Olaf Pedersen, 'The Development of Natural Philosophy, 1250–1350,' *Classica et Mediaevalia* **14** (1953):86–155. The collected papers of Ernest A. Moody, published under the title *Studies in Medieval Philosophy, Science, and Logic* (Berkeley, 1975), contain considerable material relating to the themes of this essay, and are particularly good on William of Ockham. William A. Wallace, *Causality and Scientific Explanation*, vol. 1. *Medieval and Early Classical Science* (Ann Arbor, 1972), stresses elements of methodological and epistemological continuity from the thirteenth to the seventeenth century. More detailed studies are to be found in Anneliese Maier, *Studien zur Naturphilosophie der Spätscholastik*, 5 vols. (Rome, 1949–58).

[3] *Timaeus* 52D–57C; *Phaedo* in its entirety; see also D. J. Allan, 'Plato,' *DSB* 11:22–31.

[4] Acts 17:34; see F. X. Murphy, 'Pseudo-Dionysius,' *NCE*, 11:943–944.

[5] For details, see E. A. Synan, 'Universals,' *NCE*, 14:452–454, I. C. Brady, 'Scholasticism,' *NCE*, 12:1153–58; and J. A. Weisheipl, 'Scholastic Method,' ibid., pp. 1145–1146.

[6] *Librum II sententiarum*, dist. 13, art. 2, in *Omnia opera*, ed. A. Borgnet, **27** (Paris, 1894), p. 247a.

[7] There are two resulting types of measurement of a quality such as heat, one based more directly on the quantitative extension of the body, called quantity of heat, and the other based on the qualitative intensity itself, called degree of heat (the modern notion of temperature). See R. F. O'Neill, 'Quality,' *NCE*, 12:2–5; C. F. Weiher, 'Extension,' ibid., 5:766–767; and W. A. Wallace, 'Measurement,' ibid., 9:528–529.

[8] For a fuller explanation, see W. A. Wallace, 'Science (*Scientia*),' *NCE*, 12:1190–1193; M. A. Glutz, 'Demonstration,' ibid., 4:757–760; and E. Trépanier, 'First Principles,' ibid., 5:937–940.

[9] See M. A. Glutz, 'Change,' *NCE*, 3:448–449.

[10] See *infra*, pp. 132–134.

[11] These concepts are explained in W. A. Wallace, 'Hylomorphism,' *NCE*, 7:284–285; V. E. Smith, 'Matter and Form,' ibid., 9:484–490; and A. Robinson, 'Substantial Change,' ibid., 13:771–772.

[12] *Librum I de generatione*, tract. 1, cap. 12, in *Omnia opera*, ed. Borgnet, **4** (Paris, 1890), p. 354b; for the relation of *minima* to atomism, see W. A. Wallace, 'Atomism,' *NCE*, 1:1020–1024.

[13] See M. A. Glutz, 'Motion,' *NCE*, 10:24–27.

[14] Note that on p. 100b of my article on Albert in *DSB* 1:99–103, these positions are incorrectly reversed. On *forma fluens and fluxus formae*, see Lindberg, ed., *Science in the Middle Ages*, chapter 7.

[15] For a survey, see the articles by John Murdoch and Edith Sylla on Walter Burley and Richard Swineshead, *DSB*, 2:608–612, and 13:184–213.

[16] See Lindberg, ed., *Science in the Middle Ages*, chapters 6–8; also Pedersen, 'Development of Natural Philosophy,' pp. 107–114. For translations of specific texts dealing with motion and the vacuum, and with cosmological questions generally, see Grant, 1974, pp. 211–367 and 442–568.

[17] See B. H. Zedler, 'Arabian Philosophy,' *NCE*, 1:722–726, and 'Averroës,' ibid., pp. 1125–1127; L. Gardet, 'Avicenna,' ibid., pp. 1131–1132.

[18] A. Maurer, 'Latin Averroism,' *NCE*, 1:1127–1129.

[19] 'The Condemnation of 1277,' in Grant, 1974, pp. 45–50; B. H. Zedler, 'Theory of Double Truth,' *NCE*, 4:1022–1023.

[20] These observations are based on my studies of sixteenth-century commentaries and questionaries on Aristotle's *Physics*, as yet unpublished; some background information is provided by J. A. Weisheipl, 1965 and 1974b. See also Essay 13, *infra.*

[21] A real distinction exists between things when they are non-identical in their own right, apart from any insight of human reason; thus, there is a real distinction between a dog and a man, and between Peter and John. A modal distinction is also considered a real distinction, though weaker than that between one thing and another; it is the difference between a thing and its mode, for example, between a line segment and the point that terminates the segment, or between a stick and its ends. A distinction of reason, by contrast, originates in the mind that understands or reasons about things, and formulates a proposition such as "Man is man" or attributes to Peter predicates such as "body" and "living." The distinction between "man" as it is the subject of the proposition and "man" as it is the predicate is called a distinction of reason reasoning (*rationis ratiocinantis*) because it arises *only* in the mind formulating the proposition. The distinction between "body" and "living" as said of Peter, on the other hand, is said to be a distinction of reason reasoned about (*rationis ratiocinatae*) because, though "body" and "living" are both really the same as Peter, there is an objective foundation in Peter, that is, in the thing reasoned about, that gives rise to the diverse predicates. Now Scotus's formal distinction is said to be midway between the modal real distinction and the distinction of reason reasoned about; on this account it is called the "intermediate distinction," that is, intermediate between the lesser of the real distinctions and the greater of the distinctions of reason. According to Scotus, the formal distinction is what differentiates the individuating principle, "thisness," from the common nature. The fineness of the distinction perhaps gives some indication as to why Scotus is referred to as the Subtle Doctor. See J. J. Glanville, 'Kinds of Distinction,' *NCE*, 4:908–911.

[22] According to Aristotle, the categories are the ten different classes of predicates that represent the ultimate ways of speaking about things, that is, as substance or as the nine different types of accident, for example, quantity, quality, relation, action, and so on. In addition to these being modes of predicating (hence, likewise, called predicaments), they were also commonly regarded as modes of being, and, thus, every existent entity would have to be located in one way or another within the categories; again, to each of the categories there would have to correspond some type of entity in the real order. See R. M. McInerny, 'Categories of Being,' *NCE*, 3:241–244.

[23] See the essays in Parts II and III of this volume.

2. THE MEDIEVAL ACCOMPLISHMENT IN MECHANICS AND OPTICS

The scientific revolution of the seventeenth century, as is commonly acknowledged, had its remote antecedents in Greek and early medieval thought. In the period from the thirteenth to the sixteenth centuries this heritage gradually took shape in a series of methods and ideas that formed the background for the emergence of modern science. The methods adumbrated were mainly those of experimentation and mathematical analysis, while the concepts were primarily, though not exclusively, those of the developing sciences of mechanics and optics. The history of their evolution may be divided conveniently on the basis of centuries: (1) the thirteenth, a period of beginnings and reformulation; (2) the fourteenth, a period of development and culmination; and (3) the fifteenth and sixteenth, a period of dissemination and transition. It is the purpose of this essay to sketch the essentials of this achievement in mechanics and optics by relating it to the philosophical setting sketched in the previous essay, and so to provide a framework in which the essays constituting the remaining parts of the volume can conveniently be located.

I. THIRTEENTH-CENTURY BEGINNINGS

Experimental science owes its beginnings in Western Europe to the influx of treatises from the Near East, by way of translations from Greek and Arabic, which gradually acquainted the schoolmen with the entire Aristotelian corpus and with the computational techniques of antiquity. The new knowledge merged with an Augustinian tradition prevalent in the universities, notably at Oxford and at Paris, deriving from the Church Fathers. This tradition, as we have seen, owed much to Platonism and Neo-Platonism, and already was favorably disposed toward a mathematical view of reality. The empirical orientation and systematization of Aristotle were welcomed for their value in organizing the natural history and observational data that had survived the Dark Ages through the efforts of encyclopedists, while the new methods of calculation found a ready reception among those with mathematical interests. The result was the appearance of works, first at Oxford and then at Paris, which heralded the beginnings of modern science in the Middle Ages (Crombie, 1959, Vol. 1; Dales, 1973).[1]

1. *Origins at Oxford*

Aristotle's science and his methodology could not be appreciated until his *Physics* and *Posterior Analytics* had been read and understood in the universities. Among the earliest Latin commentators to make the works of Aristotle thus available was Robert Grosseteste (*DSB* 5: 548–554), who composed the first full-length exposition of the *Posterior Analytics* shortly after 1200 (Crombie, 1953, pp. 44–60; Wallace, 1972, pp. 27–53). This work, plus a briefer commentary on the *Physics* and the series of opuscula on such topics as light and the rainbow, served as the stimulus for other scientific writings at Oxford. Taken collectively, their authors formed a school whose philosophical orientation we have characterized as the "metaphysics of light," but which did not preclude their doing pioneer work in experimental methodology.

The basis for the theory of science that developed in the Oxford school under Grosseteste's inspiration was Aristotle's distinction between knowledge of the fact (*quia*) and knowledge of the reason for the fact (*propter quid*). In attempting to make the passage from the one to the other type of knowledge, these writers, implicitly at least, touched on three methodological techniques that have come to typify modern science, namely the inductive, the experimental, and the mathematical.

Grosseteste, for example, treated induction as a discovery of causes from the study of effects, which are presented to the senses as particular physical facts. The inductive process became, for him, one of resolving the composite objects of sense perception into their principles, or elements, or causes – essentially an abstractive process. A scientific explanation would result from this when one could recompose the abstracted factors to show their causal connection with the observed facts. The complete process was referred to as "resolution and composition," a methodological expression that was to be employed in schools such as Padua until the time of Galileo (Crombie, 1953, pp. 61–81).

Grosseteste further was aware that one might not be able to follow such an orderly procedure and then would have to resort to intuition or conjecture to provide a scientific explanation. This gave rise to the problem of how to discern a true from a false theory. It was in this context that the Oxford school worked out primitive experiments, particularly in optics, designed to falsify theories. They also employed observational procedures for verification and falsification when treating of comets and heavenly phenomena that could not be subjected to human control (Crombie, 1953, pp. 82–90).[2]

The mathematical component of this school's methodology was inspired by its metaphysics of light. Convinced, as already noted, that light (*lux*) was the first form that came to primary matter at creation, and that the entire structure of the universe resulted from the propagation of luminous *species* according to geometrical laws, they sought *propter quid* explanations for physical phenomena in mathematics, and mainly in classical geometry. Thus they focused interest on mathematics as well as on experimentation, although they themselves contributed little to the development of new methods of analysis (Grant, 1974, pp. 384–435).

2. *Science on the Continent*

The mathematicist orientation of the Oxford school foreshadowed in some ways the Neo-Pythagoreanism and rationalism of the seventeenth century. This aspect of their thought was generally rejected, however, by their contemporaries at the University of Paris, especially Albertus Magnus and Thomas Aquinas. Both of the latter likewise composed lengthy commentaries on the *Posterior Analytics* and on the physical works of Aristotle, primarily to put the Stagirite's thought at the service of Christian theology, but also to aid their students in uncovering nature's secrets (Wallace, 1972, pp. 65–88). Not convinced of an underlying mathematical structure of reality, they placed more stress on the empirical component of their scientific methodology than on the mathematical (Weisheipl, 1959, pp. 58–62).

Albertus Magnus (*DSB* 1: 99–103) is particularly noteworthy for his skill at observation and systematic classification. He was an assiduous student of nature, intent on ascertaining the facts, and not infrequently certifying observations with his *Fui et vide experiri* ("I was there and saw it for myself"). He recognized the difficulty of accurate observation and experimentation, and urged repetition under a variety of conditions to ensure accuracy. He was painfully aware of and remonstrated against the common failing of the schoolmen, i.e., their uncritical reliance on authority, including that of Aristotle. Among his own contributions were experiments on the thermal effects of sunlight, which employed the method of agreement and difference later to be formulated by J. S. Mill; the classification of some hundred minerals, with notes on the properties of each; a detailed comparative study of plants, with digressions that show a remarkable sense of morphology and ecology; and studies in embryology and reproduction, which show that he experimented with insects and the lower animals (Crombie, 1953, pp. 195–196). Albert also had theoretical and mathematical interests, stimulating later thinkers such as William of Ockham and Walter Burley with his analysis of motion, and doing much to advance the Ptolemaic conception of the structure of the universe over the more orthodox Aristotelian views of his contemporaries.[3]

The best experimental contribution of this period, however, was that of Peter Peregrinus of Maricourt (*DSB* 10: 532–540), whose *Epistola de magnete* (1269) reveals a sound empirical knowledge of magnetic phenomena. Peter explained how to differentiate the magnet's north pole from its south, stated the rule for the attraction and repulsion of poles, knew the fundamentals of magnetic induction, and discussed the possibility of breaking magnets into smaller pieces that would become magnets in turn. He understood the workings of the magnetic compass, viewing magnetism as a cosmic force somewhat as Kepler was later to do. His work seems to be the basis for Roger Bacon's extolling the experimental method, and it was praised by William Gilbert as "a pretty erudite book considering the time."[4]

3. *Use of Calculation*

Mathematical analysis was not entirely lacking from scientific investigation in the thirteenth century. One unexpected source came at the end of the century in the work of Arnald of Villanova (*DSB* 1: 289–291), who combined alchemical pursuits with those of pharmacy and medicine. Arnald was interested in quantifying the qualitative effects of compound medicines, and refined and clarified a proposal of the Arabian philosopher Alkindi that linked a geometric increase in the number of parts of a quality to an arithmetic increase in its sensed effect. The exponential function this implies has been seen by some as of the function later used by Thomas Bradwardine in his dynamic analysis of local motion (McVaugh, 1967).

A more noteworthy mathematical contribution was found, however, in earlier work on mechanics, particularly in statics and kinematics, that definitely came to fruition in the fourteenth century. Jordanus Nemorarius (*DSB* 7: 171–179) and his school took up and developed (though not from original sources) the mechanical teachings of antiquity, exemplified by Aristotle's justification of the lever principle, by Archimedes' axiomatic treatment of the lever and the center of gravity, and by Hero's study of simple machines. They formulated the concept of "positional gravity" (*gravitas secundum situm*), with its implied component forces, and used a principle analogous to that of virtual displacements or of virtual work to prove the law of the lever (Moody & Clagett, 1952). Gerard of Brussels (*DSB* 5: 360) was similarly heir to the kinematics of antiquity. In his *De motu* he attempted to reduce various possible curvilinear velocities of lines, surfaces, and solids to the uniform rectilinear velocity of a moving point. In the process he anticipated the "mean-speed theorem" later used by the Mertonians, successfully equating the varying rotational motion of a circle's radius with a uniform translational motion of its midpoint.[5]

Other conceptual work in the study of motive powers and resistances, made in the context of Aristotle's rules for the comparison of motions, laid the groundwork for the gradual substitution of the notion of force (as exemplified by *vis insita* and *vis impressa*) for that of cause, thereby preparing for later more sophisticated analyses of gravitational and projectile motion.[6]

II. FOURTEENTH-CENTURY DEVELOPMENT

The more valuable scientific contributions of the thirteenth century were in most instances those of isolated individuals, who reformulated the science of antiquity and made new beginnings in both experimentation and mathematical analysis. The fourteenth century saw a fuller development along these same lines, culminating in important schools at both Oxford and Paris whose members are commonly regarded as the forerunners of modern science.

1. *Theory and Experiment*

These "precursors", to use Duhem's term, worked primarily in the area of mechanics, concentrating on logical and mathematical analyses that led to somewhat abstract formulations, only much later put to experimental test. They never reached the stage of active interchange between theory and experiment that characterizes twentieth-century science, and that could only be begun in earnest with the mechanical investigations of Galileo and Newton. In another area of study, however, a beginning was made even in this type of methodology; the area, predictably enough, was optics, which from antiquity had been emerging, along with mechanics, as an independent branch of physics.

The reasons for the privileged position enjoyed by optics in the late thirteen and early fourteenth centuries are many. One was the eminence it earlier had come to enjoy among the Greeks and the Arabs. Another was its easy assimilation within the theological context of "Let there be light" (*Fiat lux*) and the philosophical context of the "metaphysics of light" already alluded to. Yet other reasons can be traced in the striking appearances of spectra, rainbows, halos, and other optical phenomena in the upper atmosphere, in the perplexity aroused by optical delusions or by an awareness of their possibility, and above all in the applicability of a simple geometry toward the solution of optical problems (Dijksterhuis, 1961, pp. 145–152).

Whatever the reasons, the fact is that considerable progress had already been made in both catoptrics, the study of reflected light, and dioptrics, the study of refraction. In the former, the works of Euclid, Ptolemy, and Alhazen (*DSB* 6: 189–210) had shown that the angles of incidence and reflection from plane surfaces are equal; they also explained how images are formed in plane mirrors and, in the case of Alhazen, gave exhaustive and accurate analyses of reflection from spherical and parabolic mirrors. Similarly in dioptrics Ptolemy and Alhazen had measured angles of incidence and refraction, and knew in a qualitative way the difference between refraction away from, and refraction toward, the normal, depending on the media through which the light ray passed. Grosseteste even attempted a quantitative description of the phenomenon, proposing that the angle of refraction equals half the angle of incidence, which is, of course, erroneous.[7] In this way, however, the stage was gradually set for more substantial advances in optics by Witelo (*DSB* 14: 457–462) and Theodoric of Freiberg. Perhaps the most remarkable was Theodoric's work on the rainbow (*De iride*), composed shortly after 1304, wherein he explained the production of the bow through the refraction and reflection of light rays (Crombie, 1953, pp. 233–259).

Theodoric's treatise is lengthy and shows considerable expertise in both experimentation and theory, as well as the ability to relate the two (Wallace, 1959, 1974c). On the experimental side Theodoric passed light rays through a wide variety of prisms and crystalline spheres to study the production of spectra. He traced their paths through flasks filled with water, using opaque surfaces to block out unwanted rays, and obtained knowledge of angles of refraction at the various surfaces on which the rays in which he was interested were incident, as well as the mechanics of their internal reflection within the flask. Using such techniques he worked out the first essentially correct explanation of the formation of the primary and secondary rainbows (Figures 1 and 2). The theoretical insight that lay behind this work, and that had escaped all of his predecessors, was that a globe of water could be thought of – not as a diminutive watery cloud, as others viewed it – but as a magnified raindrop. This, plus the recognition that the bow is actually the cumulative effect of radiation from many drops, provided the principles basic to his solution. Theodoric's experimental genius enabled him to utilize these principles in a striking way: the first to immobilize the raindrop, in magnified form, in what would later be called a "laboratory" situation, he was able to

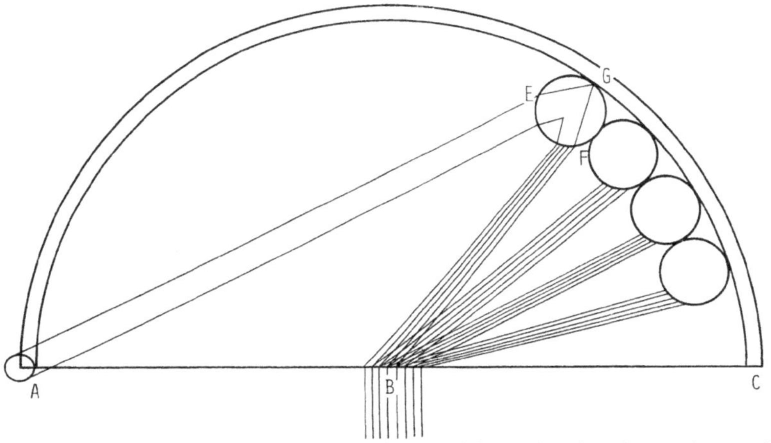

Fig. 1. The formation of the primary or lower rainbow, showing the much magnified drops (or collection of drops) that produce the four colors Theodoric held were present in the bow. The sun is at *A*, the observer at *B*, and a point directly in front of the observer on the horizon at *C*. Rays from the sun enter the uppermost drop (or drops) at *E*, are refracted there, then are internally reflected within the drop at *G*, and finally are refracted again at *F* and transmitted to the eye of the observer. Each drop (or group of drops) reflects a different color at the eye position.

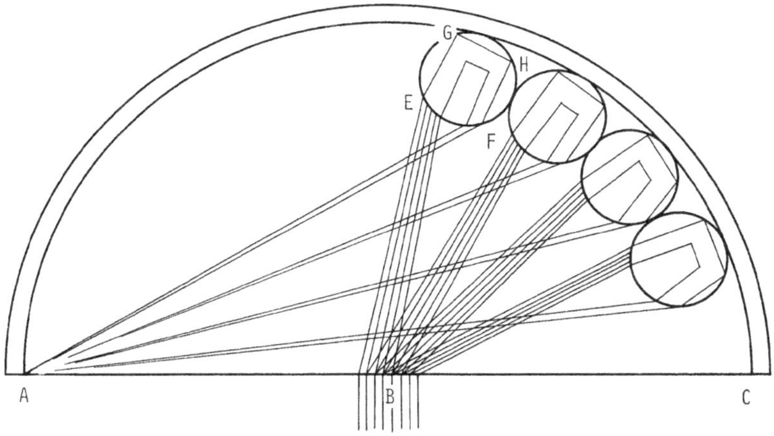

Fig. 2. The formation of the secondary or upper rainbow, showing the four drops (or collection of drops) that produce the four colors Theodoric held were present in this bow also. *A*, *B*, and *C* are as in Figure 1. Rays from the sun enter the uppermost drop (or drops) at *F*, are refracted there, then are internally reflected within the drop twice, at *H* and *G*, before being finally refracted at *E* and transmitted to the eye of the observer. These drops reflect the same colors, but in the reverse order to those in Figure 1.

examine leisurely and at length the various components involved in the rainbow's production (Boyer, 1959, pp. 110–124).

Theodoric proposed the foregoing methodology as an application of Aristotle's *Posterior Analytics* wherein he identified the causes of the bow and demonstrated its properties using a process of resolution and composition. In attempting to explain the origin and ordering of the bow's colors, however, he engaged in a far more hypothetical type of reasoning, and coupled this with experiments designed to verify and falsify his alternative hypotheses. This work, while closer methodologically to that of modern science, was not successful. There were errors too in his geometry, and in some of his measurements; these were corrected in succeeding centuries, mainly by Descartes and Newton. Theodoric's contribution, withal, was truly monumental, and represents the best interplay between theory and experiment known in the high Middle Ages (Wallace, 1959).

2. *Nominalism and Its Influence*

Most historians are agreed that some break with Aristotle was necessary before the transition could be made from natural philosophy to science in the classical sense. One step toward such a break came with the condemnation of 1277, already mentioned in our first essay. Another was the rise of nominalism or terminism in the universities, a phenomenon not unconnected with the condemnation. Under the auspices of William of Ockham and his school, this movement developed in an Aristotelian thought context but quickly led to distinctive views in logic and natural philosophy. Its theory of supposition, as already remarked, questioned the reality of universals or "common natures," generally admitted by Aristotelians, and restricted the ascription of reality to individual "absolute things" (*res absolutae*), which could be only particular substances or qualities. Quantity, in Ockham's system, is merely an abstract noun: it cannot exist by itself; it can increase or decrease without affecting the substance, as is seen in the phenomena of rarefaction and condensation; and by God's absolute power it can even be made to disappear entirely, as is known from the mystery of the Eucharist. Thus, with Ockham, quantity became a problem more of language than of physical science; his followers soon were involved in all manner of linguistic analyses relating to quantity, but not infrequently the physical problems involved got lost in a maze of logical subtleties. This notwithstanding, however, their analyses prepared the way for sophisticated, if highly imaginative, calculations of spatiotemporal relationships between motions with various velocities.

These calculations opened the path to considerable advances in kinematics, soon to be made at Merton College in Oxford.

Nominalism quickly spread from Oxford to the universities on the Continent, where it merged its thought patterns with both "orthodox" and "heterodox" (from the viewpoint of the Christian faith) schools of Aristotelianism. From this amalgam came a renewed interest in the problems of physical science, a considerably revised conceptual structure for their solution, and a growing tolerance of skepticism and eclecticism. Most of the fruits were borne in mechanics and astronomy, but some were seen in new solutions to the problems of the continuum and of infinity. Nicholas of Autrecourt is worthy of mention for his advocacy of atomism – at a time when Democritus' thought was otherwise consistently rejected – and for his holding a particulate theory of light.[8]

3. *Merton College and Kinematics*

One of the most significant contributors to the mathematical preparation for the modern science of mechanics was Thomas Bradwardine (*DSB* 2: 390–397), fellow of Merton College and theologian of sufficient renown to be mentioned by Chaucer in his *Nun's Priest's Tale*. While at Oxford Bradwardine composed treatises on speculative arithmetic and geometry wherein he not only summarized the works of Boethius and Euclid, but expanded their treatments of ratios (*proportiones*) and proportions (*proportionalitates*) to include new materials from the Arabs Thâbit and Ahmad ibn Yusuf. He then applied this teaching to a problem in dynamics in his *Treatise on the ratios of velocities in motions* (*Tractatus de proportionibus velocitatum in motibus*) composed in 1328 (Crosby, 1955). By this time various Arab and Latin writers had been interpreting Aristotle's statements (mostly in Books 4 and 7 of the *Physics*) relating to the comparability of motions to mean that the velocity V of a motion is directly proportional to the weight or force F causing it and inversely proportional to the resistance R of the medium impeding it. This posed a problem when taken in conjunction with another Aristotelian statement to the effect that no motion should result when an applied force F is equal to or less than the resistance R encountered. In modern notation, V should equal 0 when $F \leqslant R$, and this is clearly not the case if $V \propto F/R$, since V becomes finite for all cases except $F = 0$ and $R = \infty$.

In an ingenious attempt to formulate a mathematical relationship that would remove this inconsistency, Bradwardine equivalently proposed an exponential law of motion that may be written

$$\left(\frac{F_2}{R_2}\right) = \left(\frac{F_1}{R_1}\right)^{V_2/V_1}$$

Referred to as the "ratio of ratios" (*proportio proportionum*), Bradwardine's law came to be widely accepted among schoolmen up to the sixteenth century. It never was put to experimental test, although it is easily shown to be false from Newtonian dynamics. Its significance lies in its representing, in a moderately complex function, instantaneous changes rather than completed changes (as hitherto had been done), thereby preparing the way for the concepts of the infinitesimal calculus (Clagett, 1959, pp. 421–444, 629–671).

Bradwardine composed also a treatise on the continuum (*Tractatus de continuo*) which contains a detailed discussion of geometrical refutations of mathematical atomism. Again, in a theological work he analyzed the concept of infinity, using a type of one-to-one correspondence to show that a part of an infinite set is itself infinite; the context of this analysis is a proof showing that the world cannot be eternal. In such ways Bradwardine made use of mathematics in physics and theology, and stimulated later thinkers to make similar applications (Murdoch, 1962).

Although occasioned by a problem in dynamics, Bradwardine's treatise on ratios actually resulted in more substantial contributions to kinematics by other Oxonians, many of whom were fellows of Merton College in the generation after him. Principal among these were William of Heytesbury (*DSB* 6: 376–380), John of Dumbleton (*DSB* 7: 116–117), and Richard Swineshead (*DSB* 13: 184–213). All writing towards the middle of the fourteenth century, they presupposed the validity of Bradwardine's dynamic function and turned their attention to a fuller examination of the comparability of all types of motions, or changes, in its light. They did this in the context of discussions on the "intension and remission of forms" or the "latitude of forms," conceiving all changes (qualitative as well as quantitative) as traversing a distance or "latitude" which is readily quantifiable. They generally employed a "letter-calculus" wherein letters of the alphabet represented ideas (not magnitudes), which lent itself to subtle logical arguments, referred to as "calculatory sophisms." These were later decried by humanists and more traditional scholastics, who found the arguments incomprehensible, partly, at least, because of their mathematical complexity.

One problem to which these Mertonians addressed themselves was how to "denominate" or reckon the degree of heat of a body whose parts are heated not uniformly but to varying degrees. Swineshead devoted a section of his *Book of Calculations* (*Liber calculationum*) to solve this problem for a body

A which has greater and greater heat, increasing arithmetically by units to infinity, in its decreasing proportional parts (Figure 3). He was able to show

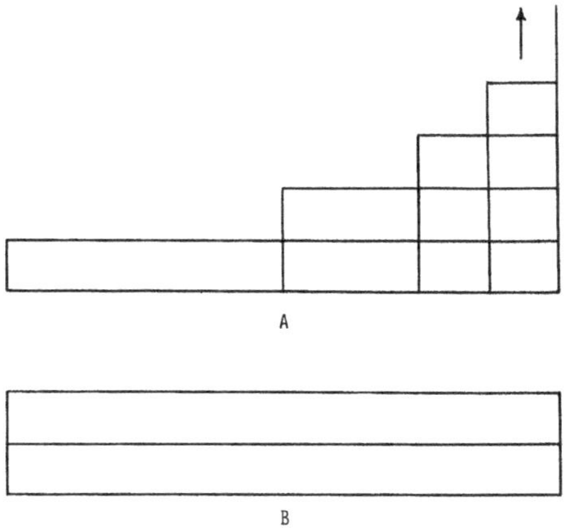

Fig. 3. A schematic representation showing how a non-uniform heat in body *A*, with one degree in the first half of its length, two degrees in the next quarter, three degrees in the next eighth, etc., may be reckoned to have the same heat as body *B* of equal length which is uniformly heated to two degrees throughout. Similar diagrams appear in the margin of a fourteenth-century manuscript of Swineshead's *Calculationes*, Paris BN Lat. 9558, fol. 6r, and the person who drew them was apparently familiar with Oresme's configurational geometry (see Figure 4).

that *A* should be denominated as having the same heat as another body *B* which is heated to two degrees throughout its entire length, thus equivalently demonstrating that the sum of the series $1 + 1/2 + 1/4 + 1/8 \ldots$ converges to the value 2. Swineshead considerably advanced Bradwardine's analysis relating to instantaneous velocity and other concepts necessary for the calculus; significantly his work was known to Leibniz, who wished to have it republished (Boyer, 1949, pp. 74–88).

Motion was regarded by these thinkers as merely another quality whose latitude or mean degree could be calculated. This type of consideration led Heytesbury to formulate one of the most important kinematical rules to come out of the fourteenth century, a rule that has since come to be known as the Mertonian "mean-speed theorem" (Grant, 1974, pp. 237–243). The theorem states that a uniformly accelerated motion is equivalent, so far as the space traversed in a given time is concerned, to a uniform motion whose velocity is equal throughout to the instantaneous velocity of the uniformly accelerating body at the middle instant of the period of its acceleration. The theorem was formulated during the early 1330's, and at least four attempts to prove it arithmetically were detailed at Oxford before 1350. As in the previous case of Bradwardine's function, no attempt was made at an experimental proof, nor was it seen (so far as is known) that the rule could be applied to the case of falling bodies. The "Calculatores," as these writers are called, restricted their attention to imaginative cases conceived in abstract terms: they spoke of magnitudes and moving points, and various types of resistive media, but usually in a mathematical way and without reference to nature or the physical universe (Wilson, 1960). When they discussed falling bodies, as did Swineshead in his chapter "On the Place of an Element" (*De loco elementi*), it was primarily to show that mathematical techniques are inapplicable to natural motions of this type (Hoskin and Molland, 1966).

A final development among the Mertonians that is worthy of mention for its later importance is their attempts at clarifying the expression "quantity of matter" (*quantitas materiae*), which seems to be genetically related to the Newtonian concept of mass. Swineshead took up the question of the "latitude" of rarity and density, and in so doing answered implicitly how one could go about determining the meaning of "amount of matter" or "quantity of matter." His definition of *quantitas materiae*, it has been argued, is not significantly different from Newton's "the measure of the same arising from its density and magnitude conjointly" (Weisheipl, 1963).

4. *Paris and the Growth of Dynamics*

As in the thirteenth century an interest in science with emphasis on the mathematical began at Oxford, to be followed by a similar interest with emphasis on the physical at Paris, so in the fourteenth century an analogous pattern appeared. The works of the English "Calculatores" were read and understood on the Continent shortly after the mid-fourteenth century by such thinkers as John of Holland at the University of Prague and Albert of Saxony at the University of Paris. Under less pronounced nominalist influence than the Mertonians, and generally convinced of the reality of motion, the Continental philosophers again took up the problems of the causes and effects of local motion. Particularly at Paris, in a setting where both Aristotelian and terminist views were tolerated, "calculatory" techniques were applied to natural and violent motions and new advances were made in both terrestrial and celestial dynamics.

The first concept of significance to emerge from this was that of impetus, which has been seen by historians of medieval science, such as Duhem, as a forerunner of the modern concept of inertia. The idea of impetus was not completely new on the fourteenth-century scene; the term had been used in biblical and Roman literature in the general sense of a thrust toward some goal, and John Philoponus (*DSB* 7: 134–139), a Greek commentator on Aristotle, had written in the sixth century of an "incorporeal kinetic force" impressed on a projectile as the cause of its motion. Again Arabs such as Avicenna and Abū'l-Barakāt (*DSB* 1: 26–28) had used equivalent Arabic terminology to express the same idea, and thirteenth-century scholastics took note of impetus as a possible explanation (which they rejected) of violent motion. What was new about the fourteenth-century development was the technical significance given to the concept in contexts that more closely approximate later discussions of inertial and gravitational motion (Clagett, 1959, pp. 505–519).

The first to speak of impetus in such a context seems to have been the Italian Scotist Franciscus de Marchia (*DSB* 5: 113–115). While discussing the causality of the Sacraments in a commentary on the *Sentences* (1323), Franciscus employed impetus to explain how both projectiles and the Sacraments produced effects through a certain power resident within them; in the former case, the projector leaves a force in the projectile that is the principal continuer of its motion, although it also leaves a force in the medium that helps the motion along. The principal mover is the "force left behind" (*virtus derelicta*) in the projectile – not a permanent quality, but something temporary ("for a time"), like heat induced in a body by fire, and this even apart from external retarding influences. The nature of the movement is determined by the *virtus*: in one case it can maintain an upward motion, in another a sideways motion, and in yet another a circular motion. The last case allowed Franciscus to explain the motion of the celestial spheres in terms of an impetus impressed in them by their "intelligences" – an important innovation in that it bridged the peripatetic gap between the earthly and the heavenly, and prepared for a mechanics that could embrace both terrestrial and celestial phenomena (Clagett, 1959, pp. 520, 526–531).

A more systematic elaborator of the impetus concept was John Buridan (*DSB* 2: 603–608), rector of the University of Paris and founder of a school there that soon rivaled in importance the school of Bradwardine at Oxford. Buridan, perhaps independently of Franciscus de Marchia, saw the necessity of some type of motive force within the projectile; he regarded it as a permanent quality, however, and gave it a rudimentary quantification in terms of the primary matter of the projectile and the velocity imparted to it. Although he offered no formal discussion of its mathematical properties, Buridan thought that the impetus would vary directly as the velocity imparted and as the quantity of matter put in motion; in this respect, at least, his concept was similar to Galileo's *impeto* and to Newton's "quantity of motion." The permanence of the impetus, in Buridan's view, was such that it was really distinct from the motion produced and would last indefinitely (*ad infinitum*) if not diminished by contrary influences. Buridan also explained the movement of the heavens by the imposition of impetus on them by God at the time of the world's creation. Again, and in this he was anticipated by Abū'l-Barakāt, Buridan used his impetus concept to explain the acceleration of falling bodies: continued acceleration results because the gravity of the body impresses more and more impetus (Clagett, 1959, pp. 521–525, 532–540).

Despite some similarities between impetus and inertia, critical historians have warned against too facile an identification. Buridan's concept, for example, was proposed as a further development of Aristotle's theory of motion, wherein the distinction between natural and violent (compulsory) still obtained. A much greater conceptual revolution was required before this distinction would be abandoned and the principle of inertia, in its classical understanding, would become accepted among physicists (Maier, 1949, pp. 132–154).

Buridan's students, Albert of Saxony (*DSB* 1: 93–95) and Marsilius of Inghen (*DSB* 9: 136–138), popularized his theory and continued to speak of impetus as an "accidental and extrinsic force," thereby preserving the Aristotelian notions of nature and violence. Albert is important for his statements regarding the free fall of bodies, wherein he speculates that the velocity of fall could increase in direct proportion to the distance of fall or to the time of fall, without seemingly recognizing that the alternatives are mutually exclusive. (This confusion was to continue in later authors such as Leonardo da Vinci and the young Galileo.) Albert himself seems to have favored distance as the independent variable, and thus cannot be regarded as a precursor of the correct "law of falling bodies" (Clagett, 1959, pp. 565–569).

Perhaps the most original thinker of the Paris school was Nicole Oresme (*DSB* 10: 223–230). Examples of his novel approach are his explanation of the motion of the heavens using the metaphor of a mechanical clock, and his speculations concerning the possible existence of a plurality of worlds. An ardent opponent of astrology, he developed Bradwardine's doctrine on ratios to include irrational fractional exponents relating pairs of whole-number ratios, and proceeded to argue that the ratio of any two unknown celestial ratios is probably irrational. This probability, in his view, rendered all astrological prediction fallacious in principle. Oresme held that impetus is not permanent, but is self-expending in its very production of motion; he apparently associated impetus with acceleration, moreover, and not with sustaining a uniform velocity. In discussing falling bodies, he seems to suggest that the speed of fall is directly proportional to the time (and not the distance) of fall, but he did not apply the Mertonian mean-speed theorem to this case, although he knew the theorem and in fact gave the first geometrical proof for it. Further he conceived the imaginary situation of the earth's being pierced all the way through; a falling body would then acquire an *impétuosité* that would carry it beyond the center, and thereafter would oscillate in gradually decreasing amplitudes until it came to rest. A final and extremely important contribution was Oresme's use of a two-dimensional figure to plot a distribution of the intensity of a quality in a subject or of velocity variation with time (Figure 4).

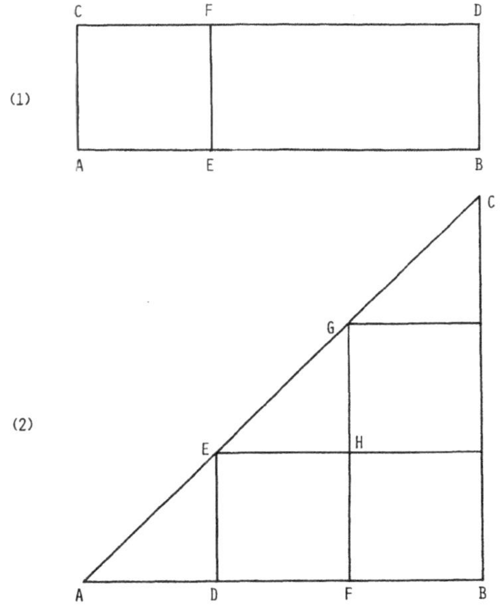

Fig. 4. In Oresme's system, the rectangle (1) and the right triangle (2) above measure the quantity of some quality (or motion). Line *AB* in each case represents the extension of the quality in the subject, whereas perpendiculars erected to this base line, e.g., *AC, EF,* and *BD* in (1) and *DE, FG,* and *BC* in (2), represent the intensity of the quality at a particular point. Oresme designated the limiting line *CD* in (1) and *AC* in (2) as the "line of the summit" or the "line of intensity." This is comparable to a "curve" in modern analytic geometry, while the figures themselves are comparable to the "areas under curves."

Possibly this method of graphical representation was anticipated by the Italian Franciscan Giovanni di Casali, but Oresme perfected it considerably, and on this account is commonly regarded as a precursor of Descartes' analytic geometry (Grant, 1966; Clagett, 1968; Menut & Denomy, 1968).

III. DISSEMINATION AND TRANSITION

The fourteenth century marked the high point in optical experimentation and in the conceptual development of mechanics during the late Middle Ages. The fifteenth and sixteenth centuries served mainly as periods of transition, where the underlying ideas were diffused throughout Europe, entered into combination with those of other cultures, and provided the proximate setting for the emergence of classical science. Much of this interplay took place in Italy, although France and Spain also figured in it to a limited extent.

1. *Italy and Renaissance Influences*

The tradition perhaps most opposed to scholasticism was that of humanism, with its interest in classical antiquity, its emphasis on the arts, and its general preference for Plato over Aristotle. Writers such as Marsilio Ficino and Erasmus ridiculed, respectively, the Paduan schoolmen and the "calculatory sophisms" of their Parisian counterparts. Their overriding interest in philology, moreover, led humanists to make much of original texts, and, even in the case of Aristotle, to confer unprecedented force on arguments from the authority of the classical author. Yet they did make available, in Greek and in accurate translation, the mathematical and mechanical treatises of Euclid, Archimedes, Apollonius, Pappus, Diophantus, and Ptolemy – works that perforce had a salutary effect in preparing for the new scientific mentality.

The writings of particular authors also contributed in different ways to the coming revolution. Nicholas of Cusa (*DSB* 3: 512–516) is important for his use of mathematical ideas in elaborating his metaphysics, which prepared for the transition, in Koyré's apt expression, "from the closed world to the infinite universe" (1957). He also placed great emphasis on measurement, and preserved elements of the medieval experimental tradition in his treatise on 'Experiments with Scales' (*De statics experimentis*) – this despite the fact that most of his experiments are purely fictitious and not one mentions a numerical result. Leonardo da Vinci (*DSB* 8: 192–245) is perhaps overrated for his contributions to science, since his was more the mentality of the engineer; his notebooks are neither systematic nor lucid expositions of physical concepts. Yet he too supplied an important ingredient, wrestling as he did with practical problems of mechanics with great genius and technical ability. He brought alive again the tradition of Jordanus Nemorarius and Albert of Saxony, and his speculations on kinematics and dynamics, if inconclusive, reveal how difficult and elusive were the conceptual foundations of mechanics for its early practitioners. Giordano Bruno (*DSB* 2: 539–544) may also be mentioned as a supporter and successor of Nicholas of Cusa; his works abound in Neo-Platonism and mysticism, and show a heavy reliance on Renaissance magic and the Hermetic-Cabalist tradition. Of little importance for mechanics, his ideas are significant mainly for the support they gave to Copernicanism and to the concept of an infinite universe (Yates, 1964).

Of more direct influence, on the other hand, was work done at the University of Padua under Averroist and terminist influences. Aristotelianism flourished there long after it had gone into eclipse at Oxford and Paris, not so much in subordination to theology as it was among Thomists, but rather under the patronage of the Arab Averroës or of Alexander of Aphrodisias, a Greek commentator on Aristotle. The Averroists were Neo-Platonic in their interpretation of Aristotle, whereas the Alexandrists placed emphasis instead on his original text. Again, at Padua the arts faculty was complemented not by the theology faculty but by the medical faculty; in this more secularized atmosphere the scientific writings of Aristotle could be studied closely in relation to medical problems and with much aid from Arab commentators (Randall, 1961; Schmitt, 1971).

The result was the formation of a new body of ideas within the Aristotelian framework that fostered, rather than impeded, the scientific revival soon to be pioneered by the Paduan professor, Galileo Galilei. Among these ideas some were methodological. They derived from extended discussions of what Galileo would refer to as the "method of analysis" (*metodo risolutivo*) and the "method of synthesis" (*metodo compositivo*). Writers such as Jacopo Zabarella systematized these results, showing how they could be applied to detailed problems in physical science, thereby bringing to perfection the methodology outlined by Grosseteste, which has already been discussed (Gilbert, 1960; Schmitt, 1969).

More than a century before Zabarella, Paul of Venice (*DSB* 10: 419–421), who had studied at Oxford in the late fourteenth century, returned to Padua and propagated Mertonian ideas among his students. A number of these wrote commentaries on Heytesbury that were published and widely disseminated throughout Europe. Noteworthy is the commentary of Gaetano da Thiene, who illustrated much of Heytesbury's abstract reasoning on uniform and difform motions with examples drawn from nature and from artifacts that might be constructed from materials close at hand (Valsanzibio, 1949). As far as is known this fifteenth-century group performed no experiments or measurements, but they took a step closer to their realization by showing how "calculatory" techniques were relevant in physical and medical investigations.[9]

2. *Paris and the Spanish Universities*

The Paduan school exerted considerable influence throughout northern Italy; it also stimulated a renewed interest in Mertonian ideas at the University of Paris at the beginning of the sixteenth century. The group in which this renewal took place centered around John Major or Maior (*DSB* 9: 32–33), the Scottish nominalist, who numbered among his students Jean Dullaert of Ghent, Alvaro Thomaz, and Juan de Celaya. Dullaert (*DSB* 4: 237–238) edited many of the works of Paul of Venice, while he and the others were generally familiar with the "calculatory" writings of Paul's students. Major's group was eclectic in its philosophy, and saw no inconsistency in making a fusion of nominalist and realist currents, the former embracing Oxonian and Parisian terminist thought and the latter including Thomist and Scotist as well as Averroist views. The Spaniard Gaspar Lax (*DSB* 8: 100) and the Portuguese Alvaro Thomaz (*DSB* 13: 349–350) supplied the mathematical expertise necessary to understand Bradwardine's, Swineshead's, and Oresme's more technical writings. Several good physics texts came out of this group; especially noteworthy is that of Juan de Celaya (*DSB* 3: 171–172), who inserted lengthy excerpts from the Mertonians and Paduans, seemingly as organized and systematized by Thomaz, into his exposition of Aristotle's *Physics* (1517). Celaya treated both dynamical and kinematical questions, as by then had become the custom, and thus transmitted much of the late medieval development in mechanics (statics excluded) to sixteenth-century scholars (Wallace, 1969).

Celaya was but one of many Spanish professors at Paris in this period; these attracted large numbers of Spanish students, who later returned to Spain and were influential in modeling Spanish universities such as Alcalá and Salamanca after the University of Paris. An edition of Swineshead's *Liber calculationum* was edited by Juan Martinez Silíceo and published at Salamanca in 1520; this was followed by a number of texts written (some poorly) in the "calculatory" tradition. Theologians who were attempting to build their lectures around Thomist, Scotist, and nominalist concepts soon complained over their students' lack of adequate preparation in logic and natural philosophy. It was such a situation that led Domingo de Soto (*DSB* 12: 547–548), a Dominican theologian and political theorist who had studied under Celaya at Paris as a layman, to prepare a series of textbooks for use at the University of Salamanca. Among these were a commentary and a "questionary" on Aristotle's *Physics*; the latter, appearing in its first complete edition in 1551, was a much simplified and abridged version of the type of physics text that was used at Paris in the first decades of the sixteenth century. It reflected the same concern for both realist and "calculatory" interests, but with changes of emphasis dictated by Soto's pedagogical aims.

One innovation in Soto's work has claimed the attention of historians of science. In furnishing examples of motions that are "uniformly difform" (i.e., uniformly accelerated) with respect to time, Soto explicitly mentions that freely falling bodies accelerate uniformly as they fall and that projectiles (presumably thrown upward) undergo a uniform deceleration; thus he saw the distance in both cases to be a function of the time of travel. He includes numerical examples that show he applied the Mertonian "mean-speed theorem" to the case of free fall, and on this basis, at the present state of knowledge, he is the first to have adumbrated the correct law of falling bodies (Wallace, 1968a). As far as is known, Soto performed no measurements, although he did discuss what later thinkers have called "thought experiments," particularly relating to the vacuum (Schmitt, 1967). An extensive survey of all physics books known to be in use in France and Spain at the time has failed to uncover similar instance of this type, and one can only speculate as to the source of Soto's examples.

3. *Italy Again: Galileo*

With Soto, the conceptual development of medieval mechanics reached its term. What was needed was an explicit concern with measurement and experimentation to complement the mathematical reasoning that had been developed along "calculatory" and Archimedean lines. This final development took place in northern Italy, again mainly at Padua, while Galileo was teaching there. The stage was set by works of considerable mathematical sophistication, under the inspiration of Archimedes, by sixteenth-century authors such as Girolamo Cardano (*DSB* 3: 64–67), Niccolò Tartaglia (*DSB* 13: 258–262), and Giovanni Battista Benedetti (*DSB* 1: 604–609). Also the technical arts had gradually been perfected, and materials were at hand from which instruments and experimental apparatus could be constructed (Drake & Drabkin, 1969).

The person of Galileo provided the catalyst and the genius to coordinate these elements and educe from them a new kind of synthesis that would reach perfection with Isaac Newton. How Galileo knew of these components and how he shaped from them his *nuova scienze* of motion has long been a subject of speculation among historians of science. It is to the elucidation of this development, and particularly the role therein of the medieval and scholastic authors discussed in this study, that the essays in Parts II and III of this volume are directed.

NOTES

[1] A sampling of the types of materials thus made available to the Latin West, all in English translation, is given in Grant, *Source Book*, 1974, pp. 3–41.

[2] A particularly complete documentation of the work in optics, again in English translation and with abundant annotation, is to be found in Grant, *Source Book*, pp. 376–435. The principal contributions detailed, apart from Grosseteste's, are those of Alhazen, Witelo, Roger Bacon, and John Pecham.

[3] For references to specific texts in Albertus's vast literary output that justify these assertions, see Crombie, 1961, Vol. 1, pp. 147–162; Dijksterhuis, 1961, pp. 133–134, 508; and Wallace, 1972, pp. 70–71, 226. English translations of selections from Albertus's treatises on minerals, plants, and animals are to be found in Grant, *Source Book*, pp. 586–603, 624–629, 654–657, and 681–700.

[4] On Gilbert and his relation to the medieval tradition, see Wallace, 1974d, pp. 241–253. Peter's letter on the magnet is translated in its entirety and annotated in Grant, *Source Book*, pp. 368–376.

[5] Representative selections in English from these authors are in Grant, *Source Book*, pp. 211–227 and 234–237.

[6] The principal contribution in this field derived from Averroës's discussion of Aristotle's rules and of the interpretation given them by another Arab philosopher, Avempace or Ibn Bājja (*DSB* 1:408–410). This passage in Averroës, which is translated in Grant, *Source Book*, pp. 253–262, became the subject of extensive commentary in the thirteenth century (Moody, 1951). See also Essay 7 in this volume for more on the transition from causes to forces in the development of modern mechanics.

[7] For a rather extensive series of texts discussing the problems of reflection and refraction, see Grant, *Source Book*, pp. 385–391 and 410–435.

[8] Grant has translated and annotated selections from Nicholas's writings affirming the probable existence of indivisible atoms and interparticulate vacua within bodies in his *Source Book*, pp. 352–359.

[9] For examples and a fuller analysis, see Essay 3 in this volume.

PART TWO

THE SIXTEENTH-CENTURY ACHIEVEMENT

3. THE DEVELOPMENT OF MECHANICS TO THE SIXTEENTH CENTURY

It is difficult to understand any movement, as the schoolmen would say, without knowing its *terminus ad quem.*[1] The movement that gave rise to modern science is no exception. Assuming that science in its classical form arrived in Western Europe during the seventeenth century, to investigate how medieval and scholastic thought may have accelerated its arrival one must have clearly in mind what is to be understood by science. It will not do, for instance, to equate this simply with the modern interplay between theory and experiment, as though the seventeenth-century scientist was intent on elaborating a hypothetical-deductive system that would entail empirically verifiable consequents. Such a view of science had to wait at least two more centuries; it belongs to a different thought-context than that of the seventeenth century.

Actually the *terminus ad quem* of the movement giving rise to modern science is much simpler than that. At the risk of oversimplifying it, in this study the concept of science will be narrowed to that of physics, and only the part of physics known as mechanics will be considered. More particularly still, discussion will be centered only on the part of mechanics that deals with what the schoolmen called *motus localis*, or local motion, comprising the present-day disciplines of kinematics and dynamics.[2] From such a restricted viewpoint, seventeenth-century mechanics may be characterized by its attempts at a precise mathematical formulation of laws that regulate such natural phenomena as falling bodies, and by parallel attempts to measure and determine experimentally how well such a formulation corresponds to reality. In sum, the developing science of mechanics, the *terminus ad quem* of late medieval thought, may be seen as made up of two elements: (1) the mathematical analysis of motion, and (2) its experimental verification. The problem that this conception of the science of mechanics suggests to the historian of science is this: To what extent did medieval and scholastic physics contribute to either or both of these elements, and thus influence the development of mechanics in its seventeenth-century understanding?

I. BRADWARDINE AND THE MERTONIANS

As to the first element, that of mathematical analysis, the main lines of the medieval contribution are fairly well established, and will be dealt with in some detail in subsequent essays of this volume. In broad outline, with the appearance in 1328 of Thomas Bradwardine's *Tractatus de proportionibus* (Treatise on Ratios), a new and distinctive mathematical approach to the study of motion was inaugurated. This was developed and refined by Bradwardine's successors for the next two centuries. Out of the development came the concept of instantaneous velocity, the use of fairly complex mathematical functions to correlate factors affecting motions, procedures for calculating distances traversed in uniformly accelerated motions, and the rudimentary notions of analytical geometry and the calculus. Historians of medieval science such as Maier (1949), Clagett (1959), Crosby (1955), Wilson (1960), and Grant (1966) have sufficiently documented the extent of this contribution. One may argue on points of detail, but there is a general consensus that the late Middle Ages contributed substantially to the mathematical foundation on which seventeenth-century mechanics was built. What is more, most of this contribution was already implicit in Bradwardine's treatise or was contained in the extensions of this treatise by Heytesbury,[3] Dumbleton,[4] and Swineshead.[5] Thus the fourteenth-century Mertonians were definite contributors to the mathematical component that made classical mechanics possible.

With regard to the second component, experimentation and measurement, the historical origins are not so clear. Certainly the role of medieval Aristotelians in its development has not been emphasized, and there has been a tendency to look elsewhere for its historical antecedents. The remainder of this essay will address itself to this problem of experimental origins, and will attempt to trace some factors contributing to its solution that seem to derive from Bradwardine and his successors. The thesis to be defended is that such factors were present, and that at least they set the stage, or established the climate of opinion, wherein experimentation would be sought as a natural complement to the mathematical formulation of laws of motion. Unlike the mathematical component, this experimental component (if one may call it such) was not clearly present in the work of the Mertonians, but it did evolve gradually, over two centuries, as their ideas came to be diffused on the Continent.

Why it was that a mathematical basis for seventeenth-century mechanics was apparent to the Mertonians whereas an experimental basis was not, poses an interesting question. The answer seems to lie in a certain ambivalence that was latent within the Mertonian analysis of motion. The tension that should have resulted from this ambivalence was not sensed immediately; had it been, perhaps the experimental component would have gotten off to as good a start as the mathematical. Yet there is evidence for maintaining that this latent tension did come to be recognized as the "calculatory" analyses developed at Merton College were propagated in France, Italy, and Spain during the fourteenth, fifteenth, and sixteenth centuries. The tension was gradually resolved, and in its resolution the way was prepared for an experimental investigation of nature that would complement a mathematical formulation of its laws.

To argue the thesis, one must be more explicit about the ambivalence present in the treatises of Bradwardine, Heytesbury, and Swineshead. At first approximation, this may be identified as the merging there of both Aristotelian and non-Aristotelian elements, as a heterogeneous blending of what was called the *via antiqua* with the *via moderna.*[6] Possibly this ambivalence can be concretized by examining the concept of motion contained in the Mertonian treatises, focusing in particular on the reality of motion and on the causality involved in its production.

Bradwardine would undoubtedly have identified himself as an Aristotelian, for the problem to which he set himself was to save the rules given by Aristotle for comparing motions and deciding on their commensurability. Yet in defining motion he and the other fellows of Merton College implicitly abandoned Aristotle's analysis in favor of that furnished by William of Ockham. A close study of their writings shows that, rather than conceive motion as the act of a being in potency precisely as it is in potency (*actus entis in potentia inquantum huiusmodi*), as Aristotle had done, the Mertonians regarded motion essentially as a ratio.[7] For them, the formal cause of motion was velocity, or the ratio of space traversed to the time elapsed.[8] Following Ockham, they even denied the independent physical reality of motion; in their analysis motion became nothing more than the object moved.[9] And, although they cited extensively the *loci* in Aristotle where he spoke of ratios or velocities of motion, in general they were insensitive to what for Aristotle was an important distinction, namely, that between motions that are natural, proceeding from a source within the body, and those that are violent, resulting from some type of externally applied force. The rules they formulated applied indifferently to both.

Despite the conceptual changes that these emphases implied, however, the Mertonians continued to speak of motion as having causes and effects. In fact, some writers have been intent on showing that the distinction between dynamics, which ostensibly studies motion from the point of view of the causes or factors producing it, and kinematics, which studies motion in terms of its effects or its spatiotemporal characteristics, was already known to the

Oxford school.[10] It is here that the ambivalence of the Mertonian position lies. If motion is not something real, as Ockham himself was quick to point out, then there is actually no point in seeking out its causes or its effects.[11] Causal terminology becomes meaningless in such a context; a *flatum vocis* or an *ens rationis* is really no basis for differentiating dynamics from kinematics. One may speak of a ratio being the "cause" of another ratio, but what one really means by this is that the ratios are functionally related. And, as the writings of the Mertonians so abundantly show, their interest was ultimately in kinematics. They discussed all types of imaginary motions generally without reference to nature or even to artifacts; they spoke of abstractly conceived and mathematicized motive powers and resistances and examined every type of functionality to be found between them.[12] This explains how they could lay the mathematical foundation on which modern mechanics was to be based. But it also explains why they failed to lay any foundation for the experimental component of this science. The experimenter, according to the classical analysis at least, attempts to cause motions, and to study the effects of what he himself causes, in an attempt to duplicate nature's operation in a laboratory situation.[13] The Mertonians may have paid lip service to the causes and effects of local motion, but for all practical purposes they did not believe in them, and so they lacked an important requisite for the experimentalist's mentality.

Where, then, were the medieval and scholastic influences that could have generated an experimentalist attitude? A possible answer is that these resulted from gradual changes of mentality that came about during the fourteenth, fifteenth, and sixteenth centuries as the works of the Mertonians were studied and reevaluated in centers as diverse as Paris, Padua, and Salamanca.

II. DEVELOPMENTS ON THE CONTINENT

The Paris development was the first chronologically, taking place within a few decades of the Mertonian contribution. Here the work of John Buridan,[14] Albert of Saxony,[15] and Marsilius of Inghen[16] quickly led to an incorporation of Mertonian ideas within a more realistic framework. This development has been sufficiently studied by Duhem (1913) and Maier (1949, pp. 81–154; 1958, pp. 59–144) and need not be detailed here. The Paris terminists, for example, developed the theory of impetus precisely because of their concern over the reality of motion, which made more meaningful for them the question of its causes and effects. The basic problems of dynamics were certainly broached by Thomas Bradwardine, but they were not taken seriously before

John Buridan. The study of motion *quoad causam* and *quoad effectum* was probably mentioned by Richard Swineshead,[17] but it was Albert of Saxony and Marsilius of Inghen who took these terms seriously and effectively divided the study of motion into two areas, one *penes causam* and corresponding to dynamics, the other *penes effectum* and corresponding to kinematics.[18] This is not to say that either Albert or Marsilius abandoned nominalism in their attempts to treat motion realistically. No, they tried to be nominalist and realist at the same time, and they were not completely successful in resolving the latent contradictions that this implied.[19] But they were more consistently Aristotelian than were the Mertonians. They thought, for example, of applying "calculatory" techniques to the cases of falling bodies and the movement of the heavens. Such cases were used by them to illustrate the very types of motion that had been treated by the Mertonians only in kinematic fashion.[20] Thus they took the first step toward investigating the real world, the world of nature, with the new mathematical techniques.

The next step, as it appears, took place at the University of Padua in the mid-fifteenth century. Paul of Venice studied at Oxford in the latter part of the fourteenth century, and, on his return to Italy, propagated Mertonian ideas among his students.[21] Most important of these students for purposes of this study was Gaetano da Thiene, who wrote an extensive commentary on Heytesbury's *Regule.*[22] The difference between Gaetano's mentality and that of Heytesbury becomes clear on reading the following small part of Heytesbury's text, and then contrasting this with the corresponding portion of Gaetano's commentary. Heytesbury points out, at the beginning of his treatment on local motion, the distinction between uniform and nonuniform motion. He then explains how one goes about measuring a uniform velocity:

> In uniform motion . . . the velocity of a magnitude as a whole is in all cases measured by the linear path traversed by the point which is in most rapid motion, if there is such a point. And according as the position of this point is changed uniformly or not uniformly, the complete motion of the whole body is said to be uniform or difform. Thus, given a magnitude whose most rapidly moving point is moved uniformly, then, however much the remaining points may be moving non-uniformly, that magnitude as a whole is said to be in uniform movement. . . .[23]

The language, as is easily recognized, is that of kinematics. Heytesbury is talking of moving bodies and moving points, but these he conceives very abstractly, and one is hard put to see how they apply in any way to the order of nature.

Commenting on this section, however, Gaetano's imagination takes a realistic and practical turn. To exemplify Heytesbury's reasoning he proposes the case of a rotating wheel that expands and contracts during its rotation.[24] He talks also of a cutting edge placed against a wheel that continually strips off its outermost surface.[25] Another of his examples is a wheel whose inner parts are expanding while its outer surface is being cut off.[26] Gaetano speaks too of a disk made of ice rotating in a hot oven; here the outermost surface continually disappears and the velocity at the circumference becomes slower and slower, whereas the inner parts expand under the influence of heat and their linear velocity increases.[27] Yet another of his examples is a wheel that rotates and has material gradually added to its circumference, as clay is added by a potter to the piece he is working. Here the velocity of rotation would be uniform but the linear velocity of a point on the circumference would increase, unless the entire wheel could be made to contract in the process, in which case the linear velocity of the outermost point might remain constant.[28]

These examples, it should be noted, are Gaetano's and not Heytesbury's. Heytesbury's kinematic doctrine is, of course, important, for without it, Gaetano would have had no reason to seek its exemplification. But the examples furnished by Gaetano are important too, for they show that Gaetano was convinced that Heytesbury's doctrine could be applied to the real world, and in fact was thinking of cases that were realizable in materials close at hand after the fashion of the experimenter. Gaetano did not perform experiments or measurements (at least as far as is known), but he took another step closer to their realization. And he, like Paul of Venice, was a realist, perhaps more in the Averroist than in the Scotist sense, but nonetheless unwilling to accept fully the nominalist philosophy of nature.[29]

A third step in the evolution of an experimental component for modern mechanics may be said to have taken place at the University of Paris in the early sixteenth century. Here the school of John Major, as exemplified particularly in the writings of Jean Dullaert of Ghent, Alvaro Thomaz, and Juan de Celaya, focused attention once more on the controversy between the realists and the nominalists.[30] Seemingly eclectic in their philosophical views, these thinkers actually sought a *via media* that would be acceptable to partisans of the old controversy. And, in so doing, they attempted to incorporate the entire Mertonian tradition, as reworked particularly by Alvaro Thomaz, into the *Physics* of Aristotle.[31] Dullaert[32] and Celaya[33] thought that the proper place to do this was in their questions on the third book of the *Physics*, where Aristotle, significantly, treats the definition of motion. The tracts *De motu* they produced in this context are recognizably closer to modern mechanics than those of any of their predecessors. Certainly they included both dynamical and kinematical questions, and exemplified these with cases drawn both from nature and from artifacts along lines suggested by Gaetano da Thiene, whose work they knew and, in the case of Celaya, even cited.[34]

The writings of these Paris masters exerted a great influence in Spain within a few decades, and this led to what may be regarded as the final stage in the preparation for an experimental mentality.[35] The figure who best characterizes this development is Domingo de Soto, who himself had studied at Paris under Celaya, and further developed the doctrines while teaching at Alcalá and Salamanca.[36] Soto was not himself a physicist in the modern sense. He was primarily a theologian and a political theorist, but by avocation he happened to be also a teacher of physics. The circumstances under which he composed his physics course, it seems, required him to be both a simplifier and an exemplifier. Again, by philosophical heritage he was a moderate realist; he wished to steer a middle course between the *nominales* and the *realissimi*, as he called them, acknowledging elements of truth in these extremes.[37] Both his practical and his ideological bent, under such influences, put him yet another step closer to the mentality of the seventeenth-century scientist.

An illustration may serve to make the point. The "questionaries" on the *Physics* used at Paris while Soto was a student there all employed the Mertonian terminology with regard to uniform and difform motions. For some curious reason, when exemplifying these motions most writers used a system of classification that may be traced back to Albert of Saxony.[38] This included, among others, motions that are uniform with respect to time and difform with respect to the parts of the moving object, and motions that are difform with respect to time and uniform with respect to the parts of the moving object. The first was by then commonly exemplified by a wheel or by a heavenly sphere, which rotate uniformly with respect to time but whose parts move with greater velocity as they are located farther from the center of the pole and toward the outermost periphery. The second was similarly exemplified by the falling body, whose velocity of fall increases with time, but all of whose parts move with the same velocity at any instant. With very few exceptions, the authors before Soto who attempted to illustrate uniform or difform motions did so with examples that employed this two-variable schema. They always spoke of variations that take place both with respect to time and with respect to the parts of the object moved, and spoke of either being uniform or difform, in all the possible combinations.

Soto's advance here, it would seem, was one of simplification. He thought of discussing motion that is uniform merely with respect to time or uniform merely with respect to the parts of the object moved, and gave simple illustrations of these. He exemplified also motions that are difform with respect to time alone, and then went further to seek examples from nature illustrating how some motions are uniformly difform with respect to time, whereas others are difformly difform in the same respect. In other words, Soto substituted a one-variable schema for a two-variable schema, and restricted himself to one variable at a time when furnishing realistic (as opposed to imaginary) examples. This simple device, apparently, was what enabled him to adumbrate Galileo's work in the association of uniform acceleration with actual falling bodies, along the lines indicated in Essay 6 *infra*.

Did Soto ever measure the distance covered by a falling body to see if his exemplifications were correct? Certainly he did not. There are indications in his writings that he performed what later thinkers would call "thought experiments," particularly relating to the vacuum, but he seems not to have done any measuring or experimenting himself.[39] What is significant about his contribution is that he laid the groundwork, that he prepared the ideas, that he simplified the examples, so that someone else might see that here was a case that is experimentally tractable, and finally put a mathematical law of motion to empirical test.

III. THE SEQUEL IN NORTHERN ITALY

The final chapter in this development, of course, was written at the turn of the century in northern Italy – at Padua, fittingly perhaps, in view of the work done there earlier by Gaetano da Thiene. And other influences were undoubtedly present, apart from those deriving from the Mertonians and the schoolmen of the fifteenth and sixteenth centuries. These have been adequately discussed elsewhere (Randall, 1961; Drake & Drabkin, 1969; Rose & Drake, 1971). They include other traditions in mechanics, such as the study of dynamical problems deriving from the *Questions of Mechanics* of the Aristotelian school, the Archimedean study of statics and hydrostatics in the light of precise geometrical principles, the Alexandrian concentration on theoretical mechanics typified in the works of Hero and Pappus, the medieval science of weights associated with the name of Jordanus Nemorarius, and an on-going technological tradition with its roots in antiquity but developing rapidly in the sixteenth century through the efforts of civil and military engineers. To the last-named tradition may be assimilated the work of craftsmen and mathematicians done largely outside the universities – Niccolò Tartaglia and Giovanni Battista Benedetti come to mind in the latter category – who provided the proximate materials for the development of an experimentalist mentality.

All of these traditions somehow merged in the complex personality, activity, and literary productivity of Galileo. Analyzing this productivity in one of his celebrated theses, Alexandre Koyré discerned three stages in Galileo's intellectual development: the first stage, that of his early work at Pisa, was concerned mainly with the philosophy of motion of the later schoolmen under the influence of his Pisan professor Francesco Buonamici; the second stage, exemplified by his preliminary attempts at Pisa to write a definitive treatise *De motu*, was concerned with the anti-Aristotelian impetus mechanics of Benedetti; and the third stage, associated with his move to Padua, was one in which he became more markedly Archimedean than Benedetti and laid the foundations for the *Two New Sciences* of his declining years (1939, p. 10). In this analysis, the least that can be claimed for the movement we have been tracing throughout this essay is that it provided the point of origin, the springboard, for Galileo's distinctive, but later, contributions. My own further researches have unearthed additional evidence that can be used to argue for an even greater scholastic influence on the thought of Galileo, and this is detailed in the essays that constitute Part III of the present volume.

Even without taking such additional evidence into account, however, the line of argument pursued in this study suggests a modest conclusion. Bradwardine's *Tractatus de proportionibus* and its successors laid the mathematical foundations that made the seventeenth-century accomplishment in northern Italy a possibility. Less noticeably, perhaps, they introduced the problematic of how motions can be conceived and analyzed mathematically, and at the same time studied in nature or in artificially contrived situations. Scholasticism may have been in its death throes by the time the full solution to this problematic could be worked out, but withal the schoolmen were not completely sterile in the influences they brought to bear on its statement and eventual resolution.

NOTES

1 The *terminus ad quem* is the goal or end that terminates a movement or change. Scholastics generally held that every motion or change is "specified," i.e., given its species, from its *terminus ad quem*. In this they were merely following Aristotle, *Physics*, Bk. 5, ch. 1, 225b 6–10.

2 Thus the study of statics is omitted. This portion of the science of mechanics has been well treated by M. Clagett and E. A. Moody in *The Medieval Science of Weights* (Madison, 1952), where they show that the contributions of the medieval period to statics resulted from the interpenetration of two Greek traditions, the Aristotelian and the

Euclidean-Archimedean. The influences studied here lie predominantly within the Aristotelian tradition.

[3] William Heytesbury composed his *Regule solvendi sophismata* (Rules for Solving Sophisms) at Merton College, Oxford, *c.* 1335. Ch. 6 of this work is entitled *De tribus predicamentis* (On the Three Categories); here he discusses the three Aristotelian categories in which motion can be found, treating at length of the velocity of local motion in the section entitled *De motu locali*. Apart from Wilson's work on Heytesbury (1960), consult the extracts from the *Regule* in Clagett (1959).

[4] John Dumbleton wrote his *Summa logicae et philosophiae naturalis* (Sum of Logic and Natural Philosophy) *c.* 1349, likewise at Merton College, Oxford. Part 3 is concerned with *De motu*, and gives rules for calculating the velocity of local motion. The work exists only in manuscript, and apart from the brief selection in Clagett (1959), has never been edited or translated into English. Weisheipl analyzes portions of it in his unpublished dissertation, 'Early Fourteenth-Century Physics of the Merton 'School' with Special Reference to Dumbleton and Heytesbury' (Oxford, 1956).

[5] Richard Swineshead, not to be confused with John or Roger Swineshead. Richard was likewise a fellow of Merton College; his *Liber calculationum* (Book of Calculations) was composed *c.* 1350, and contains many rules for calculating the velocities of motions. There are early printed editions of the work but no critical edition or translation, apart from excerpts in Clagett (1959); see, however, the complete summary and analysis of this work by J. Murdoch and E. Sylla, *DSB* 13 (1976), pp. 184–213.

[6] The *via antiqua*, or "old way," was that of Aristotle and his commentators of the Middle Ages such as Averroës, Albertus Magnus, Thomas Aquinas, and John Duns Scotus. The *via moderna*, or "modern way," was that of William of Ockham; for a discussion of Ockham's views as these relate to motion, see H. Shapiro, *Motion, Time and Place According to William Ockham* (St. Bonaventure, N.Y., 1957).

[7] How this transition came about is not easy to explain, involving as it did a rejection of such basic principles as the Euclidean condition for any ratio, viz, that it must be between entities of a single kind. Some of the factors that perhaps account for the transition have been documented in ch. 5 of Weisheipl's unpublished Oxford dissertation; they relate to the various attempts to locate successive or continuous motion in one or other of the Aristotelian categories, e.g., *passio* or *ubi*. Associated with these attempts was the question whether motion should be viewed more properly as a *forma fluens* or as a *fluxus formae*; on this, see Anneliese Maier, 'Die scholastische Wesensbestimmung der Bewegung als forma fluens oder fluxus formae und ihre Beziehung zu Albertus Magnus,' *Angelicum*, **21** (1944), a study enlarged in Maier (1949) pp. 9–25. Such problems were being discussed by Avicenna, Averroës, and Albertus Magnus appreciably before the fourteenth-century development, and by John of Jandun, William of Alnwick, and John Canonicus early in the fourteenth century. Attention was thereby focused on the relative (as opposed to the absolute) character of motion, and the way prepared for viewing motion itself merely as an *ens rationis* in the sense of a relation or a negation. Possibly the association of motion with a ratio emerged from a thought-context in which motion was being implicitly subsumed under the category of relation, which is precisely the category in which ratio would have to be situated.

[8] John Dumbleton, for example, in his *Summa logicae et philosophiae naturalis*, Part 3, chs. 22–25, identifies the matter of local motion as the distance traversed (*spatio acquisita*) and implies that the formality of motion (*ut est forma realis vel imaginata*) is

the velocity with which that distance is traversed. In Dumbleton's view, increase of local motion is nothing more than increase of velocity, which itself means greater distances being traversed in equal times. Thus in Part 3, ch. 7, Dumbleton equates an increase or decrease of motion with an increase or decrease of ratio ("latitudo motus et proportionis inter se equaliter acquirunter et deperduntur" – MS Vat. Lat. 6750, fol. 40vb). More explicit is the statement in an anonymous fourteenth-century *Tractatus de motu locali difformi*, contained in Cambridge, Caius College MS 499/268, fol. 212ra–213rb, whose author (possibly Richard Swineshead) identifies the *causa materialis seu materia* of motion as the *ipsum acquisitum per motum* (i.e., in the case of local motion, the space acquired or traversed), the *causa formalis* as a *transmutatio quedam coniuncta cum tempore* (i.e., the time rate at which the space is acquired or traversed), and the efficient cause as the *proportio maioris inequalitatis potencie motive super potenciam resistivam* (i.e., the ratio by which the motive force exceeds the resistance). Here not only the formal cause but also the efficient cause seem to be identified with ratios. Weisheipl, who has generously allowed me to use his reading of this text, translates the entire passage as follows: "The material cause of motion is whatever is acquired through motion; the formal cause is a certain transmutation conjoined with time; the efficient cause is a proportion [= ratio] of greater inequality of the moving power over resistance; and the final cause is the goal intended." – *Development of Physical Theory* (1959), p. 76.

[9] Ockham, as is pointed out in the following essay in this volume, thought that the succession involved in local motion could be adequately accounted for by the negation of all parts of the motion not yet acquired. Since such a negation is not a *res*, but only an *ens rationis*, there is no reality to motion over and above the existing *res permanentes* (for him, quantified substance and qualities). Ockham conceded that those who use the term "motion" as an abstract noun imagine that it signifies a distinct reality, but he regarded this as an error, the *fictio nominum abstractorum*; for related texts, see Shapiro (1967), pp. 36–53. Wilson finds essentially the same teaching in William Heytesbury: "For Heytesbury, the real physical world consists only of objects; point, line, surface, instant, time, and motion are *conceptus mentis*. These affirmations (or perhaps they are better termed 'negations') are in accord with the nominalist or terminist position, developed at length in William of Ockham's work on the *Physica*." – *William Heytesbury* (1960), p. 24. Since *entia rationis* (or *conceptus mentis*) such as relations and negations were constantly invoked in these analyses of motion, it is not surprising that the reality of motion itself was denied and an ontological claim made only for the object moved. Yet this did not prevent highly imaginative mathematical analyses of various motions, particularly in terms of the ratios they involved. As Wilson observes, "It is of some interest, then, that the reductive tendency in nominalism – its tendency to deny real existence to what is not observable – does not operate as a prescription against speculation concerning the *imaginabilia*. Quite the reverse: in the discussion of hypothetical physical problems, Heytesbury and his contemporaries frequently multiply *formalitates* in the Scotian manner. The result is a kind of mathematical physics which at times runs strangely parallel to modern physics, but which neither seeks nor claims to have application to the physical world." – *ibid.*, p. 25. The impact of these nominalist analyses on Galileo's immediate precursors is discussed in the next essay in this volume.

[10] Crosby (1955), pp. 52–54, gives the evidence in support of this thesis, which he offers as a mild corrective to Maier's analysis. See also note 8 to Essay 4, *infra*.

[11] Thus Ockham rejected the motor causality principle, *omne quod movetur ab alio*

movetur, as applying to local motion, precisely on the grounds that "local motion is not a new effect" – Shapiro (1957), p. 53.

12 Even a cursory examination of the *Regule* and the *Liber calculationum* will show this. Wilson explains Heytesbury's frequent use of the phrase *secundum imaginationem* and his abstract, logical treatment in the work cited (1960), p. 25. Again, as M. A. Hoskin and A. G. Molland point out in 'Swineshead on Falling Bodies: An Example of Fourteenth-Century Physics,' *The British Journal for the History of Science,* **3** (1966), pp. 150–182, the author of the *Liber calculationum* uses impressive mathematical techniques to reach a null result that, for him, justifies an Aristotelian principle to which he already subscribes. They conclude: "The tractate therefore ends with the frustrating spectacle of an author using sophisticated techniques of applied mathematics in order to show that in the problem at issue mathematics is inapplicable" (p. 154). Yet, paradoxically, it was the very development of these "inapplicable techniques" that provided the mathematical apparatus earlier identified in this paper as a major contribution of late medieval writers to the developing science of mechanics. Cf. Wilson as cited toward the end of note 9 above.

13 Such a mentality lay behind Newton's 'Rules of Philosophizing,' and also the elaborations of scientific methodology by Francis Bacon, John F. W. Herschel, and William Whewell. See R. M. Blake, *et al.*, *Theories of Scientific Method: The Renaissance Through the Nineteenth Century* (Seattle, 1960) and my two volumes on *Causality and Scientific Explanation* (1972 and 1974a).

14 The more important of Buridan's works, for purposes of this study, are his *Subtilissime questiones super octo physicorum libros Aristotelis* (Paris, 1509; reprinted Frankfurt a. M., 1964) and his *Questiones super libros quattuor de caelo et mundo*, ed. E. A. Moody (Cambridge, Mass., 1942). Significant excerpts are given in Clagett (1959).

15 For Albert's teachings consult his *Tractatus proportionum* (Paris, *c.* 1510), his *Acutissime questiones super libros de physica auscultatione* (Venice, 1516) and his *Questiones subtilissime . . . in libros de caelo et mundo* (Venice, 1520). Excerpts are again to be found in Clagett (1959), while a summary of Albert's position on the reality of motion is given in the next essay in this volume.

16 Marsilius's teachings are contained in his *Questiones . . . super octo libros physicorum secundum nominalium viam* (Lyons, 1518; reprinted Frankfurt a. M., 1964) and his *Abbreviationes super octo libros physicorum* (Venice, 1521).

17 The distinction is contained in *De motu* commonly ascribed to Richard Swineshead; the text is in Clagett (1959), p. 245.

18 For Albert of Saxony, see his fourth question on the sixth book of the *Physics*, fol. 66va. Marsilius of Inghen gives a similar distinction in his fifth question on the sixth book of the *Physics*, fol. 68rb.

19 For Albert of Saxony's difficulties, see the next essay in this volume. Maier gives a similar analysis of Marsilius of Inghen's apparently contradictory position in her *Zwischen Philosophie und Mechanik* (1958), pp. 139–140, esp. fn. 100.

20 For a complete analysis and documentation of the various examples used by fourteenth- to sixteenth-century writers to illustrate the kinds of local motion discussed by the Mertonians, see Essay 6 *infra.*

21 A summary of Paul's teaching is contained in his *Summa philosophiae naturalis* (Venice, 1503), which was widely used as a textbook.

22 Gaetano's commentary is to be found in Heytesbury's *Tractatus de sensu composito et diviso, Regulae cum sophismatibus, Declaratio Gaetani supra easdem*, etc. (Venice, 1949).
23 The translation is from Clagett (1959), pp. 235–236.
24 "Notandum quod illa conclusio habet veritatem primo propter corruptionem punctorum extremorum, ut dicit magister [Hentisberus]. Secundo propter condensationem forme circularis ab intra et rarefactionem ab intra. Tertio per condensationem ab intra et additionem ab extra . . . " – *ed. cit.*, fol. 38rb.
25 "Que conclusio declaratur sic. Ponatur gladius supra rotam ut prius et dolet continue partem extremam rote . . . " – *ibid.*
26 "Deinde volo quod quelibet pars citra ultimam remotam et dolatam rarefiat, ita tamen quod non transeant magnitudinem dolatam . . . " – *ibid.*
27 "Adhuc posset considerari alius casus de aqua congelata et ponatur in furno calidissimo et continue volveretur . . . " – *ibid.*
28 "Que conclusio probatur sic. Ponatur quod una rota moveatur et continue in superiori parte addantur alie partes sicut sit in rota figuli cui additur glis circumquamque simul . . . " – *ibid.*, fol. 38va.
29 Commenting on Heytesbury's nominalism as implied in the statement: " . . . in rerum natura non est aliquid quod est instans ut instans, nec tempus ut tempus, aut motus ut motus. . . . " Gaetano writes: "Hoc dixit quia credidit quod motus non distingueretur realiter a mobili. . . . " *ed. cit.*, fol. 26ra, 28vb.
30 For details, see Hubert Élie, 'Quelques maîtres de l'université de Paris vers l'an 1500,' *Archives d'histoire doctrinale et littéraire du moyen âge*, 18 (1950–51), pp. 193–243.
31 Alvaro Thomaz, a Portuguese, was the "calculator" *par excellence* of the sixteenth-century Paris group, as can be seen from even a rapid perusal of his *Liber de triplici motu* . . . (Paris, 1509). Grant gives a brief appraisal of this treatise in his work on Oresme (1966), pp. 56n–58n, 70–72, 319–320.
32 *Questiones super octo libros physicorum Aristotelis necnon super libros de caelo et mundo* (Lyons, 1512); an earlier edition appeared at Paris in 1506.
33 *Expositio . . . in octo libros physicorum Aristotelis, cum questionibus . . . secundum triplicem viam beati Thomae, realium, et nominalium* (Paris, 1517).
34 Celaya refers to Gaetano in the edition cited, fol. 95ra. That Dullaert knew of Gaetano's work is evident from his furnishing the same examples and the same method of dividing the types of local motion; details are given in Essay 6 *infra*.
35 Here are omitted many details relating to the transmission of the Paris teaching to Spain; these are contained in Essay 5 of this volume.
36 For biographical details, see V. Beltrán de Heredia, *Domingo de Soto: Estudio biográfico documentado* (Salamanca, 1960). Soto wrote many logical, philosophical, and theological works; here interest is focused on his *Super octo libros physicorum Aristotelis questiones* (Salamanca, 1545?; first completed ed. 1551).
37 Soto spoke frequently of the *nominales* and the *reales*, meaning by the latter term Scotists. In question 2 on book 2, however, he refers to the *realissimi*, and likewise in question 1 on book 4. Soto's middle position is outlined in the following essay in this volume.
38 Details are given in Essay 6.
39 Discussion of this point will be found in an article by Charles B. Schmitt, 'Experimental Evidence for and against a Void: The Sixteenth-Century Arguments,' *Isis*, 58 (1967), pp. 352–366.

4. THE CONCEPT OF MOTION IN THE SIXTEENTH CENTURY

Few topics in intellectual history are as poorly understood as the concept of motion, and particularly the changes that this concept underwent during the scientific revolution of the seventeenth century. Part of the difficulty stems from the vaguely defined status of the concept in the sixteenth century, when the groundwork was being laid for the contributions of Galileo and his associates. At least three different speculative views of motion were being discussed during this period, and in academic circles as diverse as those in Britain, France, the Low Countries, Germany, Italy, Spain, and Portugal. Moreover, each speculative view had a distinctive practical import, and thus brought a different influence to bear on the science of motion, or mechanics, that was soon to undergo such extensive development. So complex was the resulting situation in the sixteenth century that any attempt to characterize it in a brief essay must run the risk of being a considerable over-simplification. This risk can perhaps be minimized by following a procedure similar to that adopted in the previous essay of this volume, viz, by restricting attention to the one problem of the entitative status of local motion. More precisely, we aim to examine the question as to how local motion may be said to differ both from the moving object and the terminus it attains, and to discuss the answers being given to it in the early part of the century at the University of Paris, or by thinkers who studied then at the university.

The three different speculative views of motion comprised two extremes, i.e., the nominalist and the realist views, and a third or intermediate view that attempted to reconcile the differences between the two. The extremist views had been current since the fourteenth century, when William of Ockham proposed his nominalist analysis of motion and was met by the replies of more traditional Aristotelians, whose analyses were thereupon labelled as realist. The intermediate view, on the other hand, was something quite new, and may be said to be distinctive of the early sixteenth-century development. How it came about makes a fascinating story wherein the University of Paris once again becomes the intellectual center of Europe, from which radiates a stream of scholars who reflect a type of unanimity, or of tolerant eclecticism, with regard to the basic philosophical issues of the day. In Hubert Elie's analysis,[1] the man who forged this unity was the Scottish scholar, philosopher, and theologian, John Major of Haddington, known in Latin as Joannes Maior, or, alternatively, in French, as Jean Mair. By accident or design, Major numbered among his students Scots, Belgians, French, Germans, Spaniards, and Portuguese; the only important nation not represented was Italy, and this because her universities already constituted another pole of attraction for scholars from all over Europe (Elie, 1950–51, p. 195). But if Italians were not themselves present at Paris, their works were, for the school of John Major quickly became a center for editing critically, and publishing anew, the writings of earlier scholastics. Augustinians such as Jean Dullaert of Ghent undertook to edit the works of Paul of Venice, another Augustinian; Peter Crokart of Brussels, a Dominican, prepared editions of Aquinas and Peter Paludanus, and commented on the *Summule* of Peter of Spain; George Lokart, a Scot, edited the physical works of Albert of Saxony and Jean Buridan; and Major himself, indefatigable worker that he was, produced editions of Duns Scotus's *Reportata* on the *Sentences* and Adam Wodham's commentary on the same (Elie, 1950–51; Villoslada, 1938). The old issues that divided the nominalists and the realists were thus revived, and the scene was set for another confrontation that would produce happier results than those of the fourteenth century. A scholastic revival, in fact, took place, and contributed in no small measure to the "second scholasticism" that was to emanate from the Iberian penninsula later in the sixteenth century.

Our concern here can only be with the concept of motion, and for this we must restrict ourselves to commentaries and "questionaries" on the *Physics* of Aristotle that were produced in fair numbers by Major, by his disciples, and by those he influenced – among whom we enumerate Dullaert, Luis Coronel, Juan de Celaya, and Domingo de Soto. All five of these writers dealt explicitly with local motion, arguing whether it differs from the object moving or from the terminus it attains. Unfortunately, for the purposes of a brief survey, they argued both compendiously and diffusely: compendiously, because each felt somehow compelled to list all the arguments that had ever been offered on any side of the argument; and diffusely, because no one refrained from elaborating his own views, or from digressing on related subjects, when this seemed able to advance the cause. Consequently, there was much repetition, and many of the arguments – even those proper to each writer – are hardly worth reporting. In the interests of economy, perhaps the best procedure is to list the common arguments of the nominalists and the realists that run through all the treatises, and then to reflect on the types of rapprochement that were offered, with their possible influence on the emerging science of mechanics.

I. NOMINALIST AND REALIST POSITIONS

The nominalist position in this matter placed reliance on two passages in Aristotle, the first in Book 1 of the *Physics* which implies that more entities are not to be posited than are absolutely needed to explain physical phenomena (189b 15–29), and the second in Book 3 to the effect that motion is nothing but the terminus attained (200b 33). Although Gregory of Rimini interpreted the latter text to mean that local motion is nothing but the space traversed by the moving object,[2] the more popular nominalist view was that of William of Ockham, who identified local motion with the object moved (Shapiro, 1959, pp. 36–44). Ockham's arguments were directed against those who held motion to be a kind of flux, or absolute entity, made up of a continuous flow of parts from being to non-being. Such parts, he held, were either simultaneously existent, and then motion itself would have actual quantitative dimensions (which no one would concede); or else they were nonexistent, and in this case motion would itself be a nonentity, since no real being can be composed merely of nonbeings (an equally unacceptable consequent). To explain local motion, he maintained, one need only have recourse to the moving body and to its successive states; the phenomenon "can be saved by the fact a body is in distinct places successively, and not at rest in any." (*ibid.*, p. 40) And, to complete his case, Ockham held that the motor causality principle, "whatever is moved, is moved by another," is not applicable to local motion, on the grounds that "local motion is not a new effect . . . since it is nothing but that a mobile coexist in different parts of space." (*ibid.*, p. 53)

Other nominalist arguments made use of theological premises. If local motion is really distinct from the moving object, wrote Marsilius of Inghen, God, by his absolute power, could separate the one from the other, and then an object would be moving without there being motion.[3] Many variations of this argument appeared, as did other variations relating to the paradoxes of the continuum and the indivisible, all of which led to impossible consequences if one attempted to hold that local motion, as a successive entity, enjoys any manner of real existence.

The realist counterattacks, on the other hand, came first of all from the established schools within scholasticism, Thomism and Scotism. In the later fifteenth and early sixteenth centuries, the Thomists assumed the ascendancy at Paris under the leadership of Peter of Brussels,[4] although the Scotists had their spokesman in France also in the person of Peter Tateret.[5] The more extreme realist position, however, was traceable to Paul of Venice, and seems to have been dominant in northern Italy, where it received encouragement from the allied views of Augustinians, Dominicans, and Franciscans, as well as from Italian Averroists. Paul argued that God could annihilate everything in the universe but the ultimate sphere of the heavens; then, if this continued moving, it would not traverse any new space but would still continually acquire a motion distinct from itself, which must therefore be more than a mere relationship.[6] If local motion is to be identified with the moving body, on the other hand, curvilinear motion would be rectilinear motion, and uniform motion would be difform motion, because the same identical body could be involved in each case. Paul likewise employed the continuity paradoxes to show that local motion cannot be an "indivisible motion," and that it cannot be a "fixed accident" in the object moved. His positive conclusion was that local motion must be "a successive and flowing (*fluxibile*) accident" that really inheres *in* the moving object: the "in" here cannot mean a relationship of predication only – it must designate a relationship of actual ontological inherence. And, since local motion is a real and novel effect, it must have its own proportionate cause, and the motor causality principle is still valid for this type of motion.[7]

Each of these extreme views, the nominalist and the realist, exerted its particular influence on the developing new science of mechanics. The nominalist view, equating motion with space or the quantified object that moves, and treating it as no more than a mathematical relationship, encouraged the growth of kinematics.[8] As seen in the previous essay, velocity in this view became essentially a ratio, and the way was prepared for fairly sophisticated treatments of the relationships that obtain between velocity, time of travel, and distance traversed. All motions were conceived simply as taking place in an imaginary space, and the complicating factors that arise from dynamical considerations were generally ignored. Some nominalists, it is true, spoke of local motion in terms of its cause and effect, equating the former with motive power (or force) and the latter with distance travelled, and they also discussed resistive media and other factors that might impede local motion. Yet they did this in a purely mathematical way, and as a consequence seemed completely disinterested in the physical factors that might bring about, or impede, movement – as, in all rigor, their speculative view allowed none.

The realists, on the other hand, were concerned first and foremost with the real world, the world of nature. They would indulge in imaginative (or thought) experiments, and were not completely adverse to the use of mathematics in their physics, although they *were* adverse to any simple equating of motion with a quantitative ratio. Paul of Venice was explicit on this point:

"Motion is not a ratio, because a ratio is only a relative accident, whereas motion is an absolute accident."[9] Realists used the complex terminology of the nominalists relating to the latitude of forms, but their concern was not with quantitative definitions alone; rather they sought cases in the order of nature that would exemplify such abstract definitions. It is this mentality that still dominated in Italy when Galileo did his work, and that partially explains his early concern with the causes of projectile and falling motion as well as his later preoccupation with experiment. Realists could not help but inquire about the dynamical factors that influenced motion, which they regarded as a real entity requiring its own causes, and producing its own effects. And seeing how complex were most of the motions observable in the physical universe, one should not be surprised that they "multiplied categories" and saw no simple way to subsume all of nature's variety under a single mathematical rule. Kinematics was interesting to them, but somehow irrelevant, unless it could be joined with dynamics to produce an integral account of actual physical motions.

II. JOHN MAJOR AND HIS SCHOOL

Coming now to the third speculative view, that intermediate between the nominalists and the realists – and one that possibly brought about the desirable blending of the mathematical and the physical so necessary for the new science of mechanics – we see little evidence of this before the sixteenth century. Walter Burley[10] and Jean Buridan[11] were clearly realist in their understanding of local motion, whereas Marsilius of Inghen and George of Brussels[12] were clearly nominalist. The only thinker to present an ambivalent attitude was Albert of Saxony,[13] and this in rather a strange way, although significant for our purposes, since Albertutius, as he was called, was consistently a favorite among sixteenth century writers. He devotes two questions on Book 3 of the *Physics* to the topic: in the first, Question 6, he inquires "whether anything that might be a certain flux distinct from both the moving object and [its] place is required for something to move locally," and answers this in the negative, i.e., as a nominalist.[14] Then, in the second, Question 7, he repeats the question with a qualification, viz, "Whether, admitting 'divine cases,' one would have to concede that local motion is a thing distinct from the object moved and from [its] place," and answers this in the affirmative, i.e., as a realist.[15] Thus Albert really subscribes to two positions: following logical and natural reasoning, he sides with the nominalists, but "according to truth and the faith," he sides with the realists.[16]

Possibly it is this precedent that determined John Major's way of handling the problem, and thus provoked similar treatments by his students and their associates. Like Albert of Saxony, Major has two questions on Book 3 of the *Physics* dealing with this topic, viz., Questions 2 and 3.[17] Both raise exactly the same query, "Whether local motion is a successive entity that is distinct from anything permanent?" Question 2 notes that "there are sides" on this, and the first is that "of the realists," who hold for the affirmative. Major thereupon explains the realist solution, and then raises 9 objections against it, some of which are theological; each objection he considers in turn, and explains how it can be answered "by this school" (*secundum hanc viam*), thus concluding the question. Question 3 then follows immediately, and treats the same problem, only now from the negative side, for those who regard the difficulties raised in the objections as not being adequately solved.[18] Major notes that there are various schools, too, on the negative. The first is that of Gregory of Rimini (although Gregory is not named); this is "the less popular school," and Major leaves it alone on that account. The second is the common teaching "of the nominalists" (actually William of Ockham's), which he explains, and then raises thirteen objections against it, including the "divine cases proposed by Albert of Saxony."[19] He concludes by answering all thirteen objections, and then goes immediately to the next question, without any comment about, or justification of, his seemingly eclectic procedure. That he is sincere in seeing some truth on both sides, however, and in regarding the difference as mainly terminological, seems indisputable. And Major is consistent in his later treatment of motion, for he considers in subsequent books of the *Physics* all the topics relating to the kinematics of motion that were customarily discussed by nominalists, as well as dynamical problems, such as "How the velocity of local motion is ascertained from its cause," that exhibit realist concerns.[20]

Major's Flemish disciple, Jean Dullaert of Ghent,[21] shows the same dualistic tendency as his master, although he goes into the problem in more detail, devoting some twenty pages to it. He poses the question, "Whether motion is something successive distinct from any permanent thing," and replies:

On this there are various opinions, and first, beginning with local motion, there are many opinions as to what it is. Some 'reifiers' say that local motion is one accident really inhering in the movable body. And these are further divided. Some say that it is a 'respective accident' – Burley follows this view; others say that it is an 'absolute accident,' and Paul of Venice takes this position. Still others, like the nominalists, deny that local motion is such a successive accident, and these too are further divided. Some, like Gregory of Rimini, hold that local motion is the space itself over which the movable object moves; others say that local motion is only the movable object.[22]

With this statement, Dullaert first defends the realist positions, citing Buridan and Paul of Venice, and, in one place, accusing "almost all the nominalists" of inconsistency and stupid argument.[23] Then, without explanation or apology, he briefly exposes and defends Gregory of Rimini's position, noting at the end that "few nominalists" follow this, and so he goes on to present a third opinion. This last exposition, likewise fairly brief, cites Albert of Saxony and George of Brussels, and concludes with the summary statement:

> Among these opinions, the first is more subtle and more consonant with the sayings of the Philosopher [Aristotle]; the second is less popular; and the third is regarded as true and is more common among the moderns.[24]

Thereupon follows an extensive analysis of ratios, required for studying 'the velocity of local motion,' and then a full exposition of the teachings of Richard Swineshead, William Heytesbury, and Nicole Oresme on the intensities of forms and the velocities of alteration and augmentation. All in all, a very considerable portion (62 of the 151 folios) of Dullaert's *Questiones* is thus devoted to the matter of Book 3 of Aristotle's *Physics*.

The ambivalence of Major and Dullaert is also discernible in the *Physice perscrutationes* of Luis Nuñez Coronel, one of Major's earlier Spanish disciples.[25] Coronel lists the three, by now classical, positions, viz, those of the realists, the less popular nominalists, and the more popular nominalists, and declares his intention first to defend all three, and then to give his judgment as to which is "more probable." In his exposition of the three positions he cites many authorities: Scotus, Buridan, Paul of Venice, James of Forlivio, Walter Burley, a 'Dominus Cameracensis,' Gregory of Rimini, and George of Brussels. Finally, he appends two sections to his exposition, the first devoted to how local motion produces heat (using arguments from St. Thomas Aquinas), and the second to a lengthy treatment of his personal views on impetus. Then he concludes:

> Having exposed and defended the varous views concerning local motion, with some omissions, there remains the task of selecting the 'more probable' view. But this I leave to the judgment of others. The first position is older and [more] subtle; the second is extraneous and uncommon; the third is easier and better appearing. The fourth (which we did not wish to enumerate at the outset) is intelligible and not completely improbable – it satisfies very well the three arguments we raised against the third position, and is no less able to explain the heat resulting from motion, the 'aptitude' left after motion, and the immovable impetus produced. And this suffices for the first part of this third book.[26]

The so-called "fourth position" is not completely clear, although it seems to propose a teaching intermediate between the realist and the more common nominalist position.[27] It is, moreover, the first explicit indication we have of a new view of the entitative status of local motion to emerge in the sixteenth century. Yet, despite this innovation, Coronel does not exhibit great interest in the problems associated with local motion. He discusses at length the intensification of qualities and the latitude of forms, in a manner reminiscent of Nicole Oresme, but when he comes to treating the velocity of motions, he is extremely brief: "We proceed very briefly and succinctly in this disquisition, because I do not think it worthwhile to dwell on such matters."[28]

III. CELAYA AND SOTO

The suspicion that some kind of rapprochement was originating from the ambivalent treatments of John Major and his disciples is possibly confirmed by the title of the work of another Parisian master, also a Spaniard, but not a direct disciple of Major. This was Juan de Celaya, who wrote his *Expositio . . . in octo libros phisicorum Aristotelis, cum questionibus . . . secundum triplicem viam beati Thome, realium, et nominalium* at Paris in 1517.[29] What is significant about Celaya is not only his mention, in the title, of a Thomistic position as a "third way" different from those of the realists and the nominalists, but also his numbering among his students a Spanish layman, Francisco de Soto, who was later to put on the Dominican habit and take the name of Domingo.[30] Like Coronel, however, in the final analysis Celaya is eclectic; he rests content with enumerating the different positions, without taking sides, and supplies a compendious treatment of all matters that would interest a nominalist, a realist, or anyone inclined to see elements of truth in either position.

Celeya's exposition is similar to Dullaert's for its bulk, as it occupies 74 of the 201 folios that comprise Celaya's *Physics*. First there is a treatment of St. Thomas's analysis of motion, then "the opinion of Scotus and of other realists," and finally "the opinion of the nominalists," with its various divisions.[31] No resolution of the difference is attempted; rather Celaya launches directly into a 'treatise on ratios,' and follows this by lengthy discussions (both *de motu penes causam* and *de motu . . . penes effectum*) of the topics discussed by Heytesbury, Swineshead, Albert of Saxony, Nicole Oresme, and the Italian commentators on Heytesbury whose works are contained in the Venice edition of 1494.[32] Thus Celaya, following in the footsteps of Major and Dullaert, became himself an encyclopedist who transmitted to his students the entire Mertonian and terminist tradition relating to the science of motion.

The culmination of the development we have been tracing in this study comes in a work of Domingo de Soto, *Super octo libros physicorum Aristotelis quaestiones*, composed at Salamanca *c.* 1545,[33] but undoubtedly based on lectures by Soto at the University of Alcalá in the early 1520's.[34] As we have noted in previous essays, Soto's importance in the history of science derived from his having been the first to formulate a correct description of what was later to be known as Galileo's "law of falling bodies." There may be reason to believe, however, that the two contributions are not unrelated, and this is at least suggested as a conclusion to be drawn from the present essay.

Soto raises the question, "Whether motion is something distinct from the thing moved and the form or terminus [attained]?" and exposes, in the Parisian manner, both the realist and the nominalist replies, with their better-known variations.[35] His own answer is that both replies contain elements of the truth, and that the difference between the realist and the nominalist positions is mainly one of terminology.[36] If one wishes to apply the notion of real distinction only to substances that are numerically different from each other, then local motion is not really different from the object moved or from the location reached. Yet, even though all of these exist "identically" in the same subject, they are not to be formally identified, since each has a different *ratio* or definition. At the least, they are different in the mind's way of considering them, even though they exist in one and the same body. Soto is even willing, so as to avoid further dispute, to call the distinction that St. Thomas and the older Aristotelians referred to as a real modal distinction, merely a "distinction of reason."[37] This, he thinks, is closer to the "connotations" that are spoken of by the nominalists, and a "distinction of reason," when properly understood, is sufficient to save not only the phenomena, but also the different ways of speaking about local motion, the object moved, and the space traversed. But Soto would avoid both the realist and the nominalist extremes: both "sin through excess," in his estimation.[38] He does not believe one should multiply entities, but neither should one dispense completely with the categories – without them, meaningful discourse becomes impossible. And he is explicit that motion itself, while only rationally distinct from the object moved, is not on this account to be regarded as a mere *ens rationis.*[39] Like a quality, it does require a cause, and it does produce distinctive effects, so the principle "whatever is moved, is moved by another" still applies to local motion.

This summary of Soto's view on local motion must suffice for our later purposes. Perhaps it should be emphasized, however, that Soto is not eclectic in his teaching; he does not merely *report*, but rather takes a consistent position with regard to the concept of motion.[40] This position, as already mentioned, recognizes elements of truth in both the nominalist and the realist extremes.[41] More important, it provides a workable basis for a consistent treatment of motion in both its kinematic and its dynamic aspects. Soto, having eliminated the logical quibbling of many of his predecessors, can still treat the quantitative aspects of local motion, and he does so in his "digression on ratios" and in his analysis of how the velocity of motion is to be ascertained "from its effect."[42] This is the standard kinematical treatise of the Mertonians, only with this difference, that Soto presents it, *not* as an abstract and imaginative mathematical exercise, but rather as an analysis that applies to motion in the physical universe. It is precisely for this reason that he wishes to exemplify all of the Mertonian distinctions with cases found in nature, a procedure, as we argue in Essay 6 *infra*, that leads him to associate motion that is *uniformiter difformis* with falling bodies, thereby adumbrating Galileo's great discovery by many decades. Moreover, this same concept of motion gives him a consistent reason, whether rightly or wrongly, for ascertaining the velocity of motion "from its cause,"[43] and thus for taking up also the dynamical problems that were to be hotly argued in northern Italy by Galileo's predecessors and contemporaries, and by the great Italian himself in both his Pisan and Paduan periods.

How influential Soto himself was on Galileo awaits treatment in the subsequent essays of this volume. Whatever this influence, however, it seems clear that Soto had already arrived, by the mid-sixteenth century, at a concept of motion that was capable of assimilating the experimental discoveries of Galileo's Paduan period to the earlier mathematical treatments of the Mertonians, and that could supply a consistent speculative basis for the new science of mechanics in both its kinematic and dynamic aspects.

NOTES

1 Hubert Elie, 'Quelques maitres de l'université de Paris vers l'an 1500,' *Archives d'histoire doctrinale et littéraire du Moyen Age*, **18** (1950–51), pp. 193–243, esp. 205–212.

2 Gregory of Rimini, *In II Sent.*, dist. 1, quest. 4 (Venice: 1522), fol. 15ra–21ra.

3 Marsilius of Inghen, *Questiones subtilissime Johannis Marcilii Inguen super octo libros phisycorum* [*sic*] *secundum nominalium viam* (Lyons: Harion, 1518), lib. 3, q. 7, 42ra–b.

4 For a description of Peter of Brussels' influence, see Elie, (1950–51), pp. 218–220.

Peter's views on the topic being discussed here are contained in his *Argutissime, subtiles et fecunde questiones phisicales* (Paris: 1521), lib. 3, q. 1, art. 3.

5 According to Elie (p. 202), Tateret's *Questiones super sex libros physicorum Aristotelis* were published at Rouen in 1496.

6 Paul of Venice, *Expositio Pauli Veneti super octo libros phisicorum Aristotelis necnon super comento Averois cum dubiis eiusdem*, Lib. 3, comm. 18, dub. 2 (Venice: 1499), no foliation.

7 *Ibid.*, Lib. 7, comm. 2.

8 The use of the term "kinematics" for this development within late medieval science is anachronistic, but it accurately corresponds to the work done. Marshall Clagett defends the usage in the following passage: "The mechanics of large bodies comprises two main divisions. The first is *kinematics*, which studies movements taken in themselves, i.e., the spatial and temporal aspects or dimensions of movement without any regard for the forces which engender changes of movement. The second part of mechanics considers movement from a different point of view. Called *dynamics*, it studies movements in relationship to their causes, or as is more commonly said, in relationship to the *forces* associated with movement. Although they did not have the Newtonian concept of force, the mechanicians of the high and late Middle Ages nevertheless evolved this fundamental division in the manner of treating movement." Marshall Clagett, *The Science of Mechanics in the Middle Ages* (Madison: 1959), p. 163. See also Essay 9 in this volume.

9 Paul of Venice, *Expositio*, Lib. 3, comm. 18, dub. 2, ad finem.

10 Walter Burley, *Expositio in libros octo de physico auditu* (Venice: 1491) Lib. 3, cap. 1: "Hec est quarta paticula [sic] secunde partis principalis huius Capituli . . . ," no foliation.

11 Jean Buridan, *Subtilissime questiones super octo phisicorum libros Aristotelis diligenter recognite et revise a magistro Johanne Dullaert de Gandavo* (Paris: 1509), Lib. 3, quest. 7, "Utrum motus localis est res distincta a loco et ab eo quod localiter movetur," fol. 50ra–51ra.

12 George of Brussels, *Expositio magistri Georgii super libris Physicorum Aristotelis necnon totius philosophie naturalis* (Paris: n.d.), Lib. 3, quest. 1, dub. 1, "Utrum motus sit res successiva a qualibet re permanente distincta," fol. 26va–27rb; quest. 3, dub. 4, "Utrum motus localis sit res distincta a mobili et a loco," fol. 30vb–31va.

13 Albert of Saxony, *Acutissime questiones super libros de physica auscultatione* (Venice: 1516), Lib. 3, quest. 6 et 7, fol. 36vb–38ra.

14 *Ibid.*, quest. 6: "Utrum secundum Aristotelem et eius Commentatorem ad hoc quod aliquid moveatur localiter requiratur aliqua res que sit quidam fluxus distinctus a mobili et loco," fol. 36vb.

15 *Ibid.*, quest. 7: "Utrum admittentes casus divinos oporteat concedere quod motus localis sit alia res a mobili et loco," fol. 37va. The reference to "casus divinos" is to the type of argument that admits the absolute power of God in one of the premises, either to create a vacuum or to annihilate parts of the universe. All of these cases arise at least indirectly from the famous "articulus Parisiensis" (cited by both Jean Buridan and George of Burssels in the places already mentioned, fn. 11 and 12), wherein the Bishop of Paris, Étienne Tempier, in 1277 had condemned a proposition relating to the localization of the eighth sphere because it seemingly placed God's omnipotence in doubt. Although the condemnation was revoked in 1325, it continued to exert influence through the sixteenth century.

[16] *Ibid.*, quest. 6, ad finem, fol. 37va: ". . . rationes que non sunt naturales, sed secundum veritatem et fidem probantes motum localem non esse mobile nec locum."
[17] John Major, *Octo libri physicorum cum naturali philosophia atque metaphysica Johannis Maioris Hadingtonani theologi Parisiensis* (Paris: 1526), Lib. 3, quest. [2], "Queritur an motus localis sit entitas successiva a qualibet re permanente distincta . . . ," quest. [3], "Queritur: idem titulus questionis cum questione precedenti," no foliation.
[18] *Ibid.*: "Responsio negativa patet per rationes contra opinionem precedentem factas, que si minime solvantur hanc conclusionem monstrant."
[19] *Ibid.*: "Ad octavum: aliqui dicunt, ut Albertus Saxo in hoc tertio, tales casus divini probant motum localem distingui a mobili, sed hoc nihil est dictu nullum verum necessario repugnat."
[20] *Ibid.*, Lib. 7, quest. [8]: "Penes quid debet attendi velocitas motus localis penes causam."
[21] Jean Dullaert of Ghent, *Questiones super octo libros phisicorum Aristotelis necnon super libros de celo et mundo* (Lyons: 1512), Lib. 3, quest. 1, "Utrum motus sit res successivus a qualibet re permanente distincta," fol. 54ra–63va; foliation, faulty in this edition, is here corrected.
[22] *Ibid.*, fol. 54ra: "Circa hoc varie sunt opiniones. Et primo inchoando a motu locali, quid videlicet sit motus localis, multiplex est opinio. Aliqui realisantes dicunt quod motus localis est unum accidens realiter inherens corpori mobili. Et isti sunt adhuc bipartiti. Aliqui dicunt ipsum esse accidens respectivum, et hanc opinionem insequitur Burleus; alii vero dicunt quod est est [sic] accidens absolutum, quam opinionem insequitur Paulus Venetus. Alii, sicut nominales, negant motus localem esse tale accidens successivum, et isti adhuc sunt bipartiti. Aliqui, sicut Gregorius de Arimeno, dicunt quod motus localis est ipsum spacium super quod ipsum mobile movetur; alii vero dicunt quod motus localis est ipsummet mobile."
[23] *Ibid.*, fol. 57rb: ". . . illud communiter omnes ferme nominales concedunt, et sic stulte impugnant reales de distinctione motus a mobili . . ."
[24] *Ibid.*, fol. 63rb–va: "Inter istas opiniones, prima est subtilior et magis consona dictis Philosophi [Aristotelis]; secunda minus usitata; tertia hoc tempore reputatur vera et communior." That Dullaert himself, however, regarded the dispute as basically one over terminology, is intimated by the statement that immediately precedes this: "Sed quia istud videtur stare in nomine, pertranseo." (But, because this [argument] seems to be a matter of words, I am disregarding it.)
[25] For information on Luis Coronel, see Élie, (1950–51), pp. 212–213. The edition of the *Physice percrutationes* used here was published in Lyons (n.d.); the first edition appeared in Paris in 1511. The topic being discussed is: "De entitate motus localis et an mobili localiter identificetur vel ab eodem distinguatur."
[26] *Ibid.*, fol. 60va: "Restabat probare de motu locali positionibus, et utcumque defensatis, aliis neglectis, probabiliorem amplecti. Sed hoc aliorum iudicio relinquitur. Prima positio antiquior est et subtilis; secunda extranea et non communis; tertia facilior et bene apparens. Quarta (quam in principio nolumus numerare) intelligibilis et non omnino improbabilis – secundum quam optime satisfit illis tribus argumentis que contra tertiam positionem formavimus, et non plus equo ponderavimus, de calefactione, videlicet, ex motu resultante, et de aptitudine post motum derelicta, et de impetu immobili producto. Et ista sufficant pro huius tertii libri parte prima."
[27] There are some indications that Luis Coronel even favored the Thomistic opinion in

this matter; in this connection, it is noteworthy that Luis's brother, Antonio, edited at Paris in 1512 the first part of the second part of Aquinas's *Summa theologiae*, and dedicated the edition to Peter of Brussels.

28 Coronel, *Physice*, fol. 84vb: "De triplici motu consideratione habita, penes quid cuiuscunque eorum attendatur velocitas disserendum occurit. Breviter admodum et succincte in huiusmodi perscrutatione procedimus, quia inutile in talibus immorari reputo." The "De triplici motu" with which this citation commences echoes the title of a work by the Portuguese, Alvarus Thomas, *De triplici motu* . . . , largely a commentary on Swinehead, which appeared at Paris in 1509.

29 Juan de Celaya, *Expositio* . . . Citations here are from this edition.

30 In fact, as Villoslada, (1938) p. 207, records, Celaya's *expositio* . . . *in libros posteriorum Aristotelis* . . . , the second edition of which appeared in Paris in 1521, contains laudatory letters by two of Celaya's students, Juan de Fonseca de Bobadilla and Francisco de Soto.

31 Celaya, *Expositio*, fol. 60rb–63vb: Lib. 3, quest. 2, "An motus distinguatur a mobili."

32 *Hentisberi de sensu composito et diviso*, etc. (Venice: 1494). This edition is described by Wilson, (1960), p. 4, fn., and *passim.*

33 The first edition of this work, which appeared *c.* 1545, is incomplete, lacking some portions of B. 7 and all of B. 8; the first complete edition to contain the parts of B. 7 discussed here appeared at Salamanca in 1551. Citations are given, however, from the more widely diffused edition of Salamanca, 1555.

34 On this point, see V. Beltrán de Heredia, O.P., *Domingo de Soto: Estudio biográfico documentado* (Salamanca: 1960), p. 23.

35 Soto, *Super octo libros*, fol. 49ra–52ra, Lib. 3, quest. 2, "Utrum motus sit res distincta et a mobili et a forma seu termino."

36 *Ibid.*, fol. 50ra–b: ". . . unde in re non differimus ab schola nominalium, sed solum in modo loquendi."

37 *Ibid.*, fol. 50rb: "Quod enim ipsi appellant connotationes terminorum, nos appellamus distinctionem rationis. . . . S. Thomas et antiqui philosophi in hoc sensu appellabat has distinctiones reales . . ."

38 *Ibid.*: "Utrique ergo (si ego non fallor) per extremum peccaverunt. Reales quidem, quia videntes illas propositiones esse falsas, crediderunt significare res distinctas; nominales vero, credentes non distingui, concesserunt res diversorum predicamentorum formaliter de seipsis predicari . . ."

39 *Ibid.*: "Unde non vocamus distinctionem rationis (ut supra dicebamus) eo quod motus vel actio sit ens rationis. Sunt enim identice entia realia, puta qualitas; sed quia per solam considerationem intellectus distinguimus inter calorem et motum, considerando illa tanquam duo."

40 Thus we disagree with the statement of Alexandre Koyré in his essay on science in the Renaissance in René Taton (ed.), *History of Science: The Beginnings of Modern Science from 1450 to 1800*, trans. A. J. Pomerans (New York: Basic Books, 1964), p. 94: "De Soto was not a great philosopher – his physics was traditionalist and eclectic. Hence it is surprising that he held that [freely falling bodies accelerate uniformly with respect to time]."

41 Soto is also basically Thomistic, although he differs from other Thomists in this particularly matter, viz, Capreolus, Hervaeus Natalis, and Diego de Deza (all Dominicans),

whom he regarded as excessively realist or nominalist in their views, although he himself was accused of being too sympathetic to nominalism by his Dominican student, Domingo Báñez.

42 The *disgressio de proportionibus* occurs in fol. 92rb–94rb, Lib. 7, at the end of quest. 2, "Utrum motus quicunque cuicunque alii motui sit comparabilis," fol. 90ra–92rb. It is followed by quest. 3, "Utrum velocitas motus ab effectu attendatur penes quantitatem spatii quod pertransitur."

43 Fol. 94rb–95vb, Lib. 7, quest. 4, "Utrum velocitas motus attendatur ex parte causae penes proportionem proportionum, quae sunt velocitatum ad suas ipsarum resistentias."

5. THE *CALCULATORES* IN THE SIXTEENTH CENTURY

The two essays immediately preceding have concentrated on the conceptual foundations of the science of motion in the sixteenth century. They have argued that these foundations contained seeds of development for both kinematics and dynamics, while at the same time they opened up the possibility of an experimental approach to the study of natural motions. Reference has also been made to the mathematical tools required for a new science of mechanics that were gradually being forged at this time within the "calculatory" tradition. It is the latter topic that now requires further elucidation. This will be undertaken in the present and the following essay, first in a general way for the early sixteenth century as a whole, and then with application to the details of Domingo de Soto's work on falling bodies and the sources from which it derived.

Soto's contribution was correctly regarded by Pierre Duhem (1913, pp. 263–583) as the culmination of a line of development that began in earnest with the *Doctores Parisienses* but which had earlier anticipations in the writings of the English *Calculatores*. As will become apparent from the essays in Part III of this volume, considerably more steps were involved in the transmission of this medieval knowledge to Galileo than Duhem realized, though such additions do not negate his general thesis. And Domingo de Soto did play a key role in the overall process. Without doubt this Spanish Dominican was in the center of two important sixteenth-century movements that involved the *Calculatores*, one at Paris and the other at the Spanish universities of Alcalá and Salamanca.[1] He also influenced thought in northern Italy in the latter part of the sixteenth century, as will become clear in Essays 8 through 13 *infra*. The latter influence, as it turned out, was more pronounced among the Jesuits and other religious orders than in the Italian universities, which generally did not further the "calculatory" movement to the same extent as French and Spanish universities in the early half of that century.[2] Preparatory to sketching this later development in subsequent essays, we propose in this study to examine how the "calculatory" tradition was transmitted to Soto *via* the scholastic revival at Paris and its reverberations in Spanish *studia* and universities in the early part of the sixteenth century.

As a young man, Soto studied first at the University at Alcalá de Henares, not far from what is now Madrid. He then journeyed to France and continued his education at the University of Paris in the College of Santa Barbara, where he was taught by the famous Spaniard, Juan de Celaya (Beltrán de Heredia, 1960, pp. 16–17). This latter contact was most important because, as we shall see shortly, Celaya was in the midst of the "calculatory" group at Paris. After this Soto returned to Alcalá, where he finished his studies under the mathematician Pedro Ciruelo and then stayed on to teach with Ciruelo and Fernando de Encinas, the nominalist from Valladolid who had taught for some time at the College of Beauvais in Paris (*ibid*., pp. 26, 34).[3]

Shortly thereafter, Soto entered the Dominican Order and, after a brief stay at the College of St. Paul in Burgos, was sent to Salamanca to teach. Here he again encountered Ciruelo and others sympathetic to the nominalist tradition who had moved from Alcalá to Salamanca (Muñoz, 1964, pp. 77–88). Those among the Salamancan professors who contributed most to the development of physics in this period include Juan Martínez Silíceo, Pedro Margallo (a Portuguese), Pedro de Espinosa, and Alonso de la Veracruz. Through his Dominican contacts Soto also came to know Diego de Astudillo, who taught physics at the College of St. Gregory in Valladolid, and, through Astudillo, with the work of Diego Diest, who earlier taught physics at the University of Saragossa. In what follows an attempt will be made to characterize the thought of these individuals as representative of the "calculatory" tradition in early sixteenth-century physics.

Before coming to the development in Spain, however, it will be well to sketch its background in terms of the contributions of Juan de Celaya and his associates at Paris, because most of those with whom this study is concerned had either studied or taught at Paris prior to their work in the Spanish universities.

I. THE UNIVERSITY OF PARIS

While himself a student in the College of Montague at Paris, Celaya had been taught by Gaspar Lax and Jean Dullaert of Ghent, both in turn pupils of the renowned Scottish nominalist, John Major.[4] Major himself had collaborated with the Spanish logician, Jeronimo Pardo, and had numbered among his disciples the two brothers from Segovia, Luis and Antonio Coronel, both of whom also taught at Paris. The presence of these Spanish professors in the colleges of the University of Paris undoubtedly explains the continued influx of Spanish students there during the early decades of the sixteenth century.

Of Celaya's teachers, Lax is important for his knowledge of mathematics;

he was at the center of a school that included Encinas, Ciruelo, Silíceo, and the Dominican Juan de Ortega, who collectively produced most of the mathematics textbooks used in the universities of the period.[5] (Another student of Lax was the Aragonian Juan Dolz del Castellar, who composed a work entitled *Cunabula omnium fere scientiarum et praecipue physicalium difficultatum in proportionibus et proportionalibus* [Montauban: 1518]; although a classmate of Celaya, Dolz was not regarded highly by him, and indeed felt prompted to preface his *Cunabula* with an extended reply to Celaya's attacks, which he entitled *Invectiva . . . in Iohannem de Celaya veritatem obnubilare volentem.*) After his studies at Montague, Celaya taught for several years at the College of Coqueret, where his associates were the Portuguese Alvaro Thomaz and the Scot Robert Caubraith; later he transferred to the College of Santa Barbara, where he taught Domingo de Soto and the Portuguese Juan de Ribeyro.[6] Silíceo himself records that he studied under Caubraith and Jean Dullaert of Ghent: he does not mention Celaya in this context, but it is known that Celaya employed Silíceo as a young student in order to assist him financially (Villoslada, 1938, p. 191).

This array of professors and students surrounding Juan de Celaya provided considerable expertise in the mathematical and logical arts that had been employed by the English *Calculatores*. In such an atmosphere, it is not surprising that an attempt was made to provide a mathematical background for understanding the *Calculationes* of Swineshead. This was precisely the task undertaken by Alvaro Thomaz in his *Liber de triplici motu.*[7] Earlier, in Italy, Bassanus Politus had made a similar attempt, but his results did not satisfy Alvaro. Accordingly, the Portuguese scholar composed a treatise on ratios and their application to the study of motion, which was similar in structure to the *Tractatus proportionum* of Albert of Saxony, but was explicitly intended to serve as a commentary on the text of Swineshead. In his treatment Alvaro shows himself well acquainted with the English sources, the elaboration by the Paris terminists, and the Italian commentators.[8] He is also aware of the disputes between the nominalists and the Scotist realists, and the somewhat intermediate position of St. Thomas Aquinas as explained by Capreolus. On points of detail Alvaro disagrees with Swineshead, Heytesbury, and Nicole Oresme.[9] His advance over these writers is mainly in the matter of systematization and in his treatment of mathematical problems relating to the convergence of series.[10]

Through this work, Alvaro Thomaz became the "Calculator" *par excellence* at Paris at the end of the first decade of the sixteenth century. It is a tribute to him that a significant portion of his treatise was excerpted and abbreviated for inclusion in two of the physics texts written at Paris in the second decade of the sixteenth century: the first of these was the *Perscrutationes physice* of Luis Coronel, published in 1511;[11] and the second, the *Expositio in libros physicorum* of Celaya, which appeared in 1517.[12] Both Coronel and Celaya used the same sources as Alvaro Thomaz, and passed substantially the same judgment on them. In at least one place, however, Coronel seems to insinuate that studying the rules for calculation provided by Swineshead would be a waste of time – at least as contrasted with Heyesbury's rules, "which are good and easy enough".[13] Celaya, on the other hand, has a fuller incorporation of Alvaro's treatise, and this in the context of an extended commentary on the third book of Aristotle's *Physics*. Celaya was clearly enthusiastic about the possibilities of a mathematical physics, and transmitted this enthusiasm to his disciples, Domingo de Soto included. Celaya thought, too, that mathematics could be applied to medicine, and even to sacred theology: all one would have to do would be replace such terms as "to move" by "to become feverish" or "to merit".[14] Like Coronel, Celaya was opposed to the logical subtleties and the quibbling that was evoked by many of the *sophismata* of the English school; yet both wished to incorporate the basic mathematical insights of the *Calculatores* into the physical science of their day (Duhem, 1913, pp. 548–549).

II. SPANISH UNIVERSITIES

Moving now to the universities of Spain in the early sixteenth century, one finds essentially the same spirit taking root there. In the first few decades of the sixteenth century, first at Alcalá and then at Salamanca, there was a great interest in nominalist logic (Muñoz, 1964). By 1530, however, this interest began to wane under the attacks of the more realist schools, who had pressing philosophical and theological problems to solve and were afraid of being lost in logical hair-splitting. On the other hand, a steady interest in the problems of mathematical physics continued until the end of the sixteenth century, and extended itself to most of the Spanish universities in the "Golden Age" of Spanish thought.[15]

Pedro Ciruelo and Juan Martínez Silíceo were the primary originators of this development of mathematical physics, and, as a consequence, are entitled to be called the first Spanish *Calculatores*. Ciruelo, who had a teaching career at the University of Paris prior to his going to Alcalá, and then to Salamanca, wrote a treatise on practical arithmetic that was published in Paris at the turn of the century.[16] He also composed four courses on mathematics in the liberal arts tradition that were extensively used in the three universities already mentioned. These included a paraphrase of Boethius's *Arithmetica*, "more clearly and carefully edited than that of Thomas Bradwardine"; a brief compendium of Bradwardine's geometry, "with some additions"; another brief compendium of John Pecham's *Perspectiva communis*, "to which also have been added a few glosses"; and two short treatises on squaring the circle.[17] Ciruelo composed also a commentary on the *Sphere* of Sacrobosco while at Paris, and produced a revised edition of this later at Alcalá.[18] While at Salamanca he composed a series of "paradoxical questions" dealing with a variety of logical, physical, and theological matters; these reflect a certain sense of frustration, which might indicate that Ciruelo's reputation as a philosopher and theologian never quite matched his fame as a mathematician.[19]

Silíceo, who was later to become Archbishop of Toledo, composed and edited while at Paris a series of commentaries of Aristotle's logic, as well as a course in arithmetic.[20] Later, at Salamanca, he first taught and wrote in logic, then passed to the chair of natural philosophy. At about the time of transition, he edited the first version of Swineshead's *Liber Calculationum* to appear in Spain.[21] This undoubtedly became the primary source for first-hand knowledge of the English *Calculatores* in the following three or four decades.

Another Spaniard who studied at Paris and who imported "calculatory" techniques into Spain was Diego Diest.[22] Diest taught at the Franciscan college in Saragossa, and published there, in 1511, his *Questiones phisicales*, "touching on all matters where difficulties arise in theology and other disciplines".[23] Diest attempted to integrate Scotistic and Thomistic thought with "more modern" currents arising within nominalism.[24] He pays considerable attention to Ockham and to Gregory of Rimini, but he also cites Walter Burley, Swineshead, and Heytesbury. Among the Parisians he is acquainted with the work of Buridan, Nicole Oresme, Albert of Saxony, Marsilius of Inghen, and (possibly) John Major.

Also worthy of mention is the Portuguese Pedro Margallo, who was an associate of Silíceo at Paris for part of his studies, then returned to Salamanca to complete them, and, after a short period at Valladolid, returned to the University of Salamanca to teach logic and physics, and subsequently to become rector of the College of Saint Bartholomew (Muñoz, 1964, pp. 122–126; Villoslada, 1938, p. 397). He later returned to Portugal to teach theology at the University of Coimbra. Margallo's *Physices compendium*, published at Salamanca in 1520, is very concise, but it reveals a knowledge of Burley and

Heytesbury, as well as of Swineshead.[25] His references to the "Calculator" are almost as numerous as those to St. Thomas Aquinas, Scotus, and Gregory of Rimini; Margallo mentions also Jeronimo Pardo, makes several references to John Major (including one to his disciples), and speaks of Alvaro [Thomaz] by name.[26]

Among the Spanish Dominicans there does not seem to have been appreciable interest in the *Calculatores* before Domingo de Soto.[27] A notable exception, however, is Diego de Astudillo, who taught with the more famous Dominican, Francisco de Vitoria, at the College of St. Gregory in Valladolid.[28] Astudillo composed questions on the *Physics* and on the *De generatione et corruptione* of Aristotle, wherein he showed a remarkable acquaintance not only with the Thomistic and the Scotistic traditions, but also with the "calculatory" tradition at Paris and in northern Italy.[29] He makes extensive use of Walter Burley and Albert of Saxony, and also of John Major and his school. Jean Dullaert of Ghent and Luis Coronel are quoted by him frequently; Juan de Celaya and Peter of Brussels (a Dominican) are also mentioned.[30] Astudillo's questions on the *De generatione et corruptione*, moreover, show a great preoccupation with the writings of Swineshead, as well as with the adaptations introduced by Alvaro Thomaz.[31] Finally, he makes fairly frequent reference to Amadeus Meygret, a Dominican who studied under Peter of Brussels at Saint-Jacques, and who had earlier composed a commentary on the *De coelo et mundo.*[32]

At least one statement of Astudillo relating to the *Calculatores* is worth mentioning. At the end of the seventh book of the *Physics*, having discussed the rules for the comparison of motions, and having given some arguments against them, with their solutions, he concludes:

> Other arguments are frequently brought against these rules, but these can be solved from the solution of the foregoing. I have set aside "calculatory" disputations, lest I should confound the judgments of beginners, who generally are ignorant of mathematics. I did this especially lest my writing should be cause for my readers spending time wastefully on such useless questions, especially if they are theologians, who must be zealous for the salvation of souls. I have written these physical questions to facilitate their task, so that those things that are usable for theological matters may be understood.[33]

This statement of Astudillo is revealing, and goes far to explain why the *Calculatores* enjoyed relatively brief popularity in the Spanish universities. Academic life in Spain centered, at this time, largely around the training of clerics, many of whom were being prepared for missionary activity in the New World and could afford little time for detailed work in the physical sciences.

Domingo de Soto, of course, knew of the work of Astudillo, and gives evidence of having perused it carefully, although he does not mentioned Astudillo by name.[34] Before discussing Soto's *Questions on the Physics of Aristotle*, however, it will be well to discuss the writing of another philosophy professor at Salamanca, Pedro de Espinosa, who was influenced by Soto's work in logic, but whose publications in physical science seem to antedate those of the Spanish Dominican.

Little has been known about Espinosa and his contributions, although a study has appeared in Spanish which analyzes his logical works.[35] These were published at Salamanca in the early 1530's and show that at this time Espinosa was familiar with the *Summulae* of Fernando de Encinas and Domingo de Soto. He cites also the logical writings of Margallo, Silíceo, Celaya, Coronel, Jean Dullaert of Ghent, and Jeronimo Pardo (Muñoz, 1967, p. 195). In addition to logical works, it is known that Espinosa composed a *Tractatus proportionum*, a commentary on the *Sphere* of Sacrobosco, and a *Philosophia naturalis*, the last of which was published at Salamanca in 1535. (Rey Pastor, 1926, p. 156; Solana, 1941, **3**: 612). The dedicatory letter of the latter work was addressed to Juan Martínez Silíceo, "prince of the Spanish masters", whom Espinosa acknowledges as his teacher. The work itself is divided into three parts: the first summarizes all of natural philosophy, following the text of Aristotle and adding "questions" where necessary to explain the thought; the second is devoted to "Calculatoria"; and the third is an alphabetical listing of questions and problems relating to physical science.[36]

Apart from this, there is in the Biblioteca Nacional at Madrid an undated edition of the *Tractatus proportionum* and the commentary on the *Sphere* of Sacrobosco, together with a fragment of a work entitled *Questiones phisice*, all of which were bound with a 1526 edition of Ciruelo's mathematics course. These works reveal that Espinosa was well acquainted with all of the treatises on ratios and on the *Sphere* of Sacrobosco that were circulating in early sixteenth-century Europe.[37] Espinosa mentions Bradwardine, Swineshead, and Heytesbury, among the English *Calculatores*, and knows of Nicole Oresme's treatise on the ratio of ratios; he cites also Gaspar Lax and Alvaro Thomaz, as well as Luis Coronel and other members of the school of John Major. Like many writers of the period, Espinosa seems to be in the tradition of Diego Diest and Juan de Celaya, for he conscientiously attempts to take account of the Thomistic, Scotistic, and nominalist positions on the major philosophical issues of the day.

Coming finally to Domingo de Soto, one finds the culmination of the Paris and the Salamancan traditions incorporated by him into a complete physics course designed for the arts faculty at Salamanca.[38] Soto published an incomplete version of this at Salamanca around 1545, and a complete edition in 1551. In it he attempted to supply a minimum of mathematical apparatus for calculations relating to what was to become the science of mechanics. And, as has been noted in our earlier essays, he provided the earliest known formulation of what was later to be known as Galileo's "law of falling bodies", applying the Mertonian "mean-speed theorem" to the case of free fall.

Soto rarely mentions his sources by name, and thus it is not a simple matter to identify them. He does cite the more classical authors, such as Walter Burley, Heytesbury, Paul of Venice, Albert of Saxony, and Gregory of Rimini, but he generally refrains from mentioning the names of living contemporaries. He has occasional references to the *Calculatores* and numerous disparaging references to the *sophismata*, which he regarded as fruitless logical quibbling. It is probable that Soto expurgated considerable portions of the treatises of Alvaro Thomaz and Juan de Celaya along lines already suggested by Diego de Astudillo, and yet presented the essentials of their course to the arts students at Salamanca, most of whom belonged to religious orders and were preparing to be theologians. Soto's text became quite popular, and went through many editions before it was finally supplanted by the more "manual type" textbook in natural philosophy, which was to be used in the training of scholastics for the next four centuries, but which contained little or no reference to mathematical physics.[39]

Soto never completed his projected commentaries and questions on Aristotle's *De coelo* and *De generatione et corruptione*. One of his Dominican successors at Salamanca, however, who was to gain renown as a theologian, viz, Domingo Bañez, wrote a treatise on *De generatione* that shows considerable acquaintance with the *Calculatores.*[40] Thus, for a brief period after Soto's death, the tradition he preserved continued, although it was later to disappear under the pressure of intensified theological and missionary activity.

This survey of the influence of the *Calculatores* in Spain will have to terminate with the mention of a final work that shows a heavy dependence on Soto. This is the *Physica speculatio* of Alonso de la Veracruz, a Spanish Augustinian who journeyed to Mexico as a missionary, and is referred to by some as "the father of Mexican philosophy".[41] Veracruz's text was intended for use in the University of Mexico, founded only in 1553, and is very brief compared to Soto's work. Veracruz mentions in the introduction that he knows of the *Physice compendium* of Franz Titelmans, and that he is patterning his textbook on this course.[42] In his prologue, Veracruz is severely critical of the waste of time spent by writers such as Swineshead in discussing natural maxima and minima and the ratios of motion, which are epitomized in the *De triplici motu* of Alvaro Thomaz. In Veracruz's opinion, one could aptly apply the biblical text, "We have laboured all the night and taken nothing," to Alvaro's work.[43] Yet, in his own treatment, Veracruz summarizes practically everything that is in Soto,[44] and makes frequent references to writers such as Espinosa, Ciruelo, Silíceo, and Luis Coronel. Veracruz's exposition is clear and logical, and in general is an excellent compendium. Unfortunately, it was probably wasted on Mexicans who found it difficult to absorb humanist culture generally, and were even less prepared to apply the new methods of mathematical physics to a study of nature than were their Spanish professors.

This summary account of the diffusion of the ideas of the English *Calculatores* into sixteenth-century Spain shows that there was a considerable literature that could have prepared for the evolving science of mechanics in the latter part of the sixteenth century. By the time Galileo was a student at Pisa, Soto's *Questions on the Physics* had been published in Venice and was available in northern Italy.[45] More important, perhaps, the works of two Spanish Jesuits, Franciscus Toletus[46] and Benedictus Pererius,[47] the first of whom surely studied under Soto, were already widely diffused throughout the Italian peninsula. Spain's pre-eminence as a world power during the sixteenth century assured the diffusion of her books, as well as the exportation of her scholars. Galileo, in his early notebooks, mentions the works of Soto and Pererius, so these must have been known to him at least in a general way. Thus the elements were at hand from which the youthful Italian genius could have formulated his new science of mechanics. How in fact he did utilize these materials will be investigated in the essays contained in the third part of this volume.

NOTES

1 For a chronology of the details of Soto's life, see Vicente Beltrán de Heredia, O.P., *Domingo de Soto*: Estudio biográfico documentado. Biblioteca de Teologos Españoles, vol. 20 (Salamanca, 1960); for a history of the Spanish universities, see Hastings Rashdall, *The Universities of Europe in the Middle Ages*, 3 vols., ed. F. M. Powicke and A. B. Emden (Oxford, 1936), **2**, 63–114.

2 One of the important sources for sixteenth-century studies in mechanics are the various expositions of Heytesbury written by Italian commentators and published at Venice in 1494; this work is described in Wilson (1960). Another source is a compilation of treatises on ratios (*proportiones*) published at Venice in 1505, including the *Tractatus proportionum introductorius ad calculationes Suisset* by Bassanus Politus and the commentary on Albert of Saxony's *Tractatus proportionum* by Benedictus Victorius

Faventinus, as well as the better-known treatises by Bradwardine and Nicole Oresme. Writers of Soto's own Order who discussed methods of calculating ratios include Isidorus de Isolanis, O.P., whose *De velocitate motuum* was printed at Pavia in 1522, and Chrysostomus Javellus, O.P., whose *Quaestiones in libros Physicorum*, completed before 1532, appears in printed editions of Venice, 1564 and Lyons, 1568. Other significant "calculatory" works by authors who were known to Galileo and his teachers are the opusculum *In quaestione de motuum proportionibus* of Alessandro Achillini, in the *Omnia opera* printed at Venice in 1545, and the *Opus novum de proportionibus numerorum, motuum, ponderum, sonorum, aliarumque rerum mensurandarum* of Girolamo Cardano, printed at Basel in 1570. Two studies that supply additional details are C. B. Schmitt, 'Hieronymus Picus, Renaissance Platonism, and the Calculator,' *International Studies in Philosophy* **8** (1976), 57–80, and C. J. T. Lewis, 'The Fortunes of Richard Swineshead in the Time of Galileo,' *Annals of Science* **33** (1976), 561–584.

3 On Encinas (Enzinas), see Vicente Muñoz Delgado, O. de M., *La Logica Nominalista en la Universidad de Salamanca (1510–1530)*. Publicaciones del Monasterio de Poyo, **11** (Madrid, 1964), 130–131.

4 The best account of Celaya is to be found in R. G. Villoslada, S.J., *La Universidad de Paris durante los estudios de Francisco de Vitoria, O.P. (1507–1522)*. Analecta Gregoriana (Rome, 1938), **14**, 180–215; on John Major, see Hubert Elie, 'Quelques Maitres de l'université de Paris vers l'an 1500', *Archives d'histoire doctrinale et littéraire du moyen âge* **18** (1950–1951), 193–243.

5 Many of these textbooks are described in D. E. Smith, *Rara Arithmetica*: A Catalogue of the Arithmetics written before the year MDCI with a description of those in the library of George Arthur Plimpton of New York (Boston/New York, 1908). Apart from a series of works on logic, Lax produced several mathematical treatises that were quite influential, including his *Arithmetica speculativa* (Paris, 1515) and *Proportiones* (Paris, 1515); he wrote also a series of *Questiones phisicales* (Saragossa, 1527). For details concerning Lax, see Marcial Solana, *Historia de la Filosofía Española*: Epoca del Rinacimiento (Siglo XVI). 3 vols. (Madrid, 1941), **3**, 19–33.

6 Ribeyro, in particular, was a devoted disciple of Celaya, as attested by Villoslada (1938) pp. 209–211. Soto wrote a laudatory preface to Celaya's exposition of the *Posterior Analytics*, where he identifies himself as a disciple. On Celaya's close association with Alvaro Thomaz, see Villoslada, p. 190.

7 The full title is *Liber de triplici motu proportionibus annexis magistri Alvari Thome Ulixbonensis philosophicas Suiseth calculationes ex parte declarans* (Paris, 1509).

8 Among others, Alvaro cites by name Burley, Bradwardine, Heytesbury, the "Calculator" (Swineshead), Albert of Saxony, Oresme, Paul of Venice, Gaetano da Thiene, James of Forli, John of Casali, Peter of Mantua – referring to his treatise *De primo et ultimo instanti* – and Bassanus Politus.

9 For particulars on Alvaro's disagreement with Oresme, see Edward Grant, *Nicole Oresme: De proportionibus proportionum* and *Ad pauca respicientes* (Madison, 1966), fn., 56–58.

10 Some details are given in J. Rey Pastor, *Los Matemáticos Españoles del Siglo XVI* (Teledo, 1926), 82–89.

11 This work is essentially a commentary on the eight books of Aristotle's *Physics*, interspersed with questions on the more controversial topics being discussed in the schools of the period.

[12] The fuller title is *Expositio . . . in octo libros Phisicorum Aristotelis cum questionibus ejusdem, secundum triplicem viam beati Thome, realium et nominalium*; the "three ways" are those of Thomism, Scotism, and Ockhamism respectively.
[13] Coronel states: ". . . Ferme omnia que dicta sunt de difformibus qualitatibus possunt applicari difformi motui, quapropter in istis non insisto. Videantur regulae Hentisberi in tractatu de motu locali, que sunt satis bone et faciles, et qui in vacuum vellet tempus terere videat regulas Suyset: quia ego inutile reputo peramplius in his insistere." – Lib. 3, pars 4, fol. 86r.
[14] In Celaya's words: ". . . Conclusiones non solum ad medicinam, verum ad sacram theologiam applicari valent mutando illum terminum 'moveri' vel 'motus' in aliquem istorum terminorum, scil., 'febris' vel 'meritum' vel 'mereri'." – Lib. 3, fol. 88rb.
[15] Soto falls approximately in the middle of this development. His predecessors and contemporaries are sketched in what follows, while his students and disciples, such as Francisco Toledo, Pedro de Oña, Domingo Bañez, and Diego Mas were still influential at the end of the century.
[16] *Tractatus arithmetice practice qui dicitur Algorismus* (Paris, 1495). Rey Pastor (1926) cites further editions of 1502, 1505, 1509, 1513, and 1514 – p. 155; see also pp. 54–61.
[17] The title of this work is *Cursus quattuor mathematicarum artium liberalium*. It appeared first at Alcalá in 1516, and subsequent editions followed in 1523, 1526, and 1528.
[18] The Paris edition of 1498–1499 is described by Lynn Thorndike, *The Sphere of Sacrobosco and Its Commentators* (Chicago, 1949), 39 and fn. 78; the title of the Alcalá edition of 1526 reads *Opusculum de Sphera mundi Joannis de Sacro Busto: cum additionibus et familiarissimo commentario Petri Cirueli Darocensis, nunc recenter correctis a suo autore, insertis etiam egregiis questionibus Petri de Aliaco.*
[19] See the prefatory letter to these *Paradoxae quaestiones* (Salamanca, 1538), addressed by Ciruelo to his students at the University of Salamanca, where he explains the title: "Quia fere omnia erunt preter communem doctorum opinionem, ea vocabulo greco 'paradoxa' censui noncupanda."
[20] The arithmetic is entitled: *Liber arithmetico practice astrologis, phisicis, calculatoribus admodum utilis* (Paris 1513); other editions appeared in 1514, 1519, and 1526 under slightly different titles. For the contacts between Silíceo and Soto, consult the index to Beltrán de Heredia (1960).
[21] The title reads in part: *Calculatoris Suiset anglici sublime et prope divinum opus . . . cura atque diligentia philosophi Silicei* (Salamanca, 1520). Villoslada (1938) notes another edition of 1524, p. 191, fn. 19.
[22] Diest is mentioned by Villoslada, but otherwise has been unnoticed by those working in this area. He is important for having transmitted some elements of Oresme's and Albert of Saxony's teaching to Spain in the early sixteenth century. See Essay 6 *infra*.
[23] The full title reads: *Magistri Didaci Diest questiones phisicales super Aristotelis textum sigillatim omnes materias tangentes in quibus difficultates que in theologia et aliis scientiis ex physica pendent discusse suis locis inseruntur* (Saragossa, 1511).
[24] As he notes in his introduction, Diest taught the arts course "in collegio fratrum minorum de observantia" at Saragossa; his intention is to cover "omnia quae difficultatem penes Aquinatem Thomam, Joannem Scotum, Guillermum Okam, Gregorium Ariminensem, ceterosque moderniores horum sequaces in artibus facere possunt" (fol. 1^{v}).

[25] The work contains a very brief summary of the Aristotelian corpus on natural philosophy, a synopsis of the *Sphere* of Sacrobosco, a treatise on ratios, and a somewhat disorganized discussion of selected topics relating to matter, form, privation and alterative qualities, including the intension and remission of forms.

[26] References to the "Calculator" occur on folios 24r, 28r, 29v, 30r, 30v, and 32r; the reference to "Neotericus Albarus" is on fol. 29r.

[27] Apart from Juan de Ortega, however, one should note the Dominican Tomas Durán, who edited Bradwardine's arithmetic and geometry at Valencia in 1503, and Diego Deza, a Dominican who later became Archbishop of Seville and is usually referred to in the literature of the day as "Hispalensis", who composed a commentary on the *Sentences* (Seville, 1517) that discussed topics of nominalist and "calculatory" interest.

[28] On Astudillo, see S. M. Ramírez, O.P., "Hacia una renovación de nuestros estudios filosóficos: Un indice de la producción filosófica de los Dominicos españoles", *Estudios Filosóficos,* 1 (1952), pp. 8–9.

[29] The title reads *Quaestiones super octo libros physicorum et super duos libros de generatione Aristotelis, una cum legitima textus expositione eorundem librorum* (Valladolid, 1532).

[30] Peter (Crokaert) of Brussels composed a series of *Questiones phisicales*, published at Paris in 1521; a convert to Thomism from nominalism, Peter seems responsible for the dialogue between the Dominicans and the disciples of John Major at Paris.

[31] References to "Sysset Calculator" are to be found on folios 14va, 22vb, 33va, 35vb, and 39va; those to "Alvarus Thomas" are on folios 14va, 22vb, 26rb, and 39rb.

[32] A copy of Meygret's *Questiones Fratris Amadei Meygreti Lugdunensis Ordinis Predicatorum in libros De celo et mundo Aristotelis* (Paris 1514) is in the university library at Salamanca; its presence there, and the use indicated by Astudillo's references, attests to the fact that the work of the Paris Dominicans was known in Spain during the early sixteenth century.

[33] The Latin reads: "Alia fieri solent argumenta contra alias regulas, sed ex istorum solutione solvi poterunt. Calculatorias autem disputationes reliqui, ne confunderem incipientium indicia, qui communiter mathematicam nesciunt. Potissimum autem ne scriptura mea causa sit legentibus ut in talibus inutilibus questionibus tempus vane consumant, specialiter theologi, qui saluti debent animarum consulere. Ad quod melius faciendum, illa phisicalia scripsi, dumtaxat que ad theologalia utilia esse videri possunt" – fol. 133rb. Later, in Question 11, article 2, on the first book of *De generatione*, he remarks similarly: "Quantum ad secundum articulum, principales secundi articuli sunt due opiniones que communiter deffenduntur a doctoribus calculantibus, quas breviter recitabo: tum quia mihi inutile videtur in calculationibus tempus consumere; tum quia false mihi apparent" – fol. 24vb.

[34] One reference of Soto, in Question 4 on Book 7 of the *Physics*, where he states, "Miror tamen quosdam schole nostre, qui aiunt regulam primam, contrariam huius conclusionis, veram esse in universum, posita constantia potentie motive", can only be directed at Astudillo and his students.

[35] Vicente Muñoz Delgado, 'La Logica en Salamanca durante la primera mitad del siglo XVI', *Salmanticensis* **14**, (1967), 171–207, especially pp. 192–195.

[36] The title page reads: "Philosophia naturalis Petri a Spinosa artium magistri: opus inquam tripartitum quod continent tres partes. Prima pars erit emporium refertissimum bone philosophie, currens per omnes textus Philosophi cum aptis questionibus ibidemque

propriis. Secunda pars erit Calculatoria: quam appello Roseam. Tertia pars erit Flos campi, Lilium agri, continens omnes naturales questiones ordine alphabetico, Nil optabis quod hec philosophia non clare tibi ostendat. Si textum ibidem habes expositionem lucidissimam. Si questiones ad idem. Si calculationes habes eas in secunda parte. Si denique problemata habes omnia ordine alphabetico: quo sit tibi minor labor inveniendi quod vellis." – Unfortunately only the first part of this work is preserved in the copy I have used at the Biblioteca Nacional in Madrid.

37 In the Biblioteca de Santa Cruz at Valladolid there is a collection of treatises on the *Sphere* of Sacrobosco entitled *Spherae tractatus* (Venice, 1531) which contains all of the works cited by Espinosa; he may have used this particular edition.

38 This appeared in two volumes: the first, *Super octo libros physicorum commentarii*, provided the exposition of Aristotle's doctrine; the second, *Super octo libros physicorum quaestiones*, took up special problems and applications in the context of sixteenth-century thought.

39 Apart from his own writings, Soto's teachings were promulgated by his Dominican students and supporters at Salamanca, among whom should be noted the following seventeenth-century writers: Diego Ortiz, Cosme de Lerma, Froilán Díaz, Jacinto de la Parra, and Domingo Lince. Practically all of these, unfortunately, paid no attention to the more mathematical portions of Soto's works.

40 The work is entitled *Commentaria et quaestiones in duos Aristotelis Stagyritae de generatione et corruptione libros* (Salamanca, 1585). In the preface, Bañez points out that the work was composed some thirty years previously and dictated to his students at that time; this would coincide roughly with the completed edition of Soto's *Physics*, and may have been prepared for use in conjunction with it.

41 The first edition appeared in Mexico in 1557; the second, at Salmanca in 1562. On Veracruz, see the article by K. F. Reinhardt in the *New Catholic Encyclopedia* (New York, 1967), 14, 607.

42 An edition of this work was published at Lyons in 1551; it is a very brief summary of all of Aristotle's natural philosophy, including treatises on minerals, plants, and animals.

43 Veracruz's prologue, in fact, is deorgatory of the "calculatory" tradition. He writes: "Quis enim non ex animo doleat, quanta iactura temporis (quo nihil pretiosius) adolescentumque olei, et operis amissio sit in tractandis quae de maximo et minimo naturali multiplicantur argumentis, in illis voluendis, quae a Calculatore diffuse valde tractantur, atque de motuum et mobilium proportione, et ad invicem comparatione sophystice proponuntur. Atque (ut unico verbo multa dicam) quae de triplici motu ab Alvaro Thoma sunt excogitata? Hoc unum vere tales asserere posse affirmo: 'Per totam noctam laborentes nihil cepimus'."

44 With the exception, that is, of the adumbration of the "law of falling bodies," which apparently did not interest Veracruz.

45 The Venice 1582 edition of Soto's *Quaestiones* appeared the year after Galileo began his studies at the University of Pisa.

46 *Commentaria una cum questionibus in octo libros Aristotelis de physica auscultatione* (Venice, 1580).

47 *De communibus omnium rerum naturalium principiis et affectionibus libri quindecim* (Rome, 1576).

6. THE ENIGMA OF DOMINGO DE SOTO

The aim of this study is to cast light on what Alexandre Koyré has referred to as "the enigma of Domingo de Soto."[1] As noted in previous essays, Soto was a Spanish Dominican who, in the early sixteenth century, studied at the University of Paris, returned to Spain, and at the University of Salamanca composed a commentary and questions on the *Physics* of Aristotle (*c.* 1545) along with an imposing series of works on political philosophy and theology.[2] In a much-quoted passage in his questions on the *Physics* Soto associates the concept of motion which is *uniformiter difformis* – an expression deriving from the English *Calculatores* – with falling bodies, and he indicates that the distance of fall can be calculated from the elapsed time by means of the so-called Mertonian mean-speed theorem.[3] The casual way in which Soto introduces this association has led some to speculate that this was generally known in his day and that he merely recorded what had become common teaching in the early sixteenth century. "But if this is the case," writes Koyré, "why was de Soto alone in putting [these views] down on paper? And why did no one else before Galileo . . . adopt them?"[4] These questions, neither of which is easily answered, pose the enigma of Domingo de Soto.

The answer to the riddle should be forthcoming from a study of the teachings of Soto's predecessors and contemporaries, and it is such a study that provides the background for the present essay. The results show that Soto is still probably the first to associate the expression *uniformiter difformis* (with respect to time) with the motion of falling bodies. The association itself, however, is not completely fortuitous: it appears to be the result of a progression of schemata and exemplifications used in the teaching of physics from the fourteenth to the sixteenth centuries. The purpose of the essay is to trace this development and to show the extent to which Soto's presentation of the material was novel. Soto's uniqueness, it appears, consists in having introduced as an intuitive example the simplification that Galileo and his successors were later to formulate as the law of falling bodies. How Soto came to his result is a good illustration of the devious route that scientific creativity frequently follows before it terminates in a new formulation that is capable of experimental test.

For the sake of convenience the presentation is divided on the basis of its approximate chronology, and treats successively schemata and exemplifications used in discussing local motion in the fourteenth, fifteenth, and sixteenth centuries. The background is, of course, Aristotelian, but in the foreground are to be found the various Mertonian and nominalist distinctions that figured prominently in the emerging science of mechanics.

I. FOURTEENTH CENTURY

In the first schema to be discussed the basic distinctions were foreshadowed in the works of Gerard of Brussels and of Thomas Bradwardine, but they came to be known by the mid-fourteenth century through the writings of Bradwardine's disciples William Heytesbury and Richard Swineshead (Clagett, 1959). These distinctions were applied generally to the intensification of changes or motions; as applied to local motion "intension" became synonymous with velocity or its change, and thus various qualifying adjectives such as "uniform" and "difform" came to have kinematical significance. A uniform motion U is one with constant velocity v, whereas a difform motion D has a changing velocity. Further, a motion may be uniform in either of two senses: with respect to the parts of the object moved, symbolized $U(x)$, in the sense that all parts of the object move with the same velocity; or with respect to time, symbolized $U(t)$, in the sense that the velocity of the object as a whole remains constant over a time interval. This distinction may also be applied to difform motions, yielding the two corresponding types, $D(x)$ and $D(t)$. With difformity, moreover, a further series of distinction may be introduced. Motion that is difform with respect to the parts of the object moved may be either uniformly difform, $UD(x)$, in the sense that there is a uniform (spatial) variation in the velocity of the various parts of the object, or difformly difform, $DD(x)$, in the sense that there is no such uniform (spatial) variation. Again, motion may be uniformly difform with respect to time, $UD(t)$, or difformly difform in the same sense, $DD(t)$. Both of these in turn may be subdivided on the basis of the direction of the change – that is, whether it is increasing or decreasing – to yield uniformly accelerated motion, $UD_{acc}(t)$, and uniformly decelerated motion, $UD_{dec}(t)$, or alternatively, difformly accelerated motion, $DD_{acc}(t)$, and difformly decelerated motion, $DD_{dec}(t)$. The resulting eight possibilities, all of which are capable of exemplification, are shown in Schema I; the symbols on the right will serve to number the particular types of motion in the schema and the examples that were proposed to concretize their definitions.

SCHEMA I

v	U	$U(x)$			I–1
		$U(t)$			I–2
	D	$D(x)$	$UD(x)$		I–3
			$DD(x)$		I–4
		$D(t)$	$UD(t)$	$UD_{\mathrm{acc}}(t)$	I–5
				$UD_{\mathrm{dec}}(t)$	I–6
			$DD(t)$	$DD_{\mathrm{acc}}(t)$	I–7
				$DD_{\mathrm{dec}}(t)$	I–8

The complete articulation of all the subdivisions of Schema I was not given so far as is known, by any author before Soto, although the main lines of the division were already implicit in Heytesbury (1335). All of the English *Calculatores*, however – and this designation includes Heytesbury as well as Swineshead (*c.* 1340), the English logician Robert Feribrigge (*c.* 1367), and the pseudo-Bradwardine (the author of the *Summulus de motu*, published in 1505) – were content to define the various kinds of motion in abstract and mathematical terms, without illustrations from the physical universe. However, on the Continent, at the University of Prague, John of Holland (*c.* 1369) repeated most of the divisions in Schema I and when explaining his definitions provided four examples: motion of type I–1 he illustrated with a falling stone (*motus lapis deorsum*), all of whose parts move at the same speed; type I–2, with an object (*mobile*) moving in uniform translation, or, alternatively, with a sphere (*spera*) in uniform rotation; type I–3, with the motion of the ninth sphere (*nona spera*), some of whose parts move more slowly than others even though the whole rotates uniformly; and type I–5, with the example of Socrates (*Sortes*) continually accelerating his walking speed. In passing, when defining types I–5 and I–6, John of Holland referred to the definitions given by the *Calculatores*, a reference that serves to align him with the Mertonian tradition. His work provides the fullest exemplification of Schema I of those written in the fourteenth century.[5]

Writing at about the same time as John of Holland but at the University of Paris, Albert of Saxony and (somewhat earlier) Nicole Oresme utilize another way of classifying types of local motion. Both refer to motions being uniform or difform according to parts, $U(x)$ and $D(x)$, and according to time, $U(t)$ and $D(t)$, although Albert prefers to speak of the latter motions as "regular" and "irregular."[6] Again, both are concerned to supply examples and in so doing group the types of motion in a way that influenced later writers. Rather than consider one independent variable at a time, they take two variables together and speak of motion being, for example, "uniform and regular," which may be symbolized as $U(x) \cdot U(t)$, or "uniform and irregular," symbolized as $U(x) \cdot D(t)$, or "difform and regular," $D(x) \cdot U(t)$. The four possibilities that result from this classification are given in Schema II.

SCHEMA II

v	$U(x) \cdot U(t)$	II–1
	$U(x) \cdot D(t)$	II–2
	$D(x) \cdot U(t)$	II–3
	$D(x) \cdot D(t)$	II–4

The examples provided by Albert and Oresme are particularly interesting in that a falling body is used as an illustration of local motion. Thus Albert's example of type II–2 is a heavy object (*grave*) or a falling stone (*lapis*): in the latter case the motion is uniform, $U(x)$, because all parts of the stone move with the same velocity at any instant, but irregular, $D(t)$, because the stone moves faster "at the end than at the beginning." His example of type II–3, on the other hand, is a wheel (*rota*) whose motion is difform, $D(x)$, because "the parts close to the axle do not move as far as those close to the circumference," but regular, $U(t)$, because the angular velocity of the whole remains constant. A third example, type II–1, is described by Albert as follows:

> Similarly note a third possibility, that there is no difficulty in a motion being both uniform and regular at the same time: for when a heavy object descends in a medium whose resistance is so regulated that the heavy object covers equal distances in equal parts of time, the motion of the heavy object would be both uniform and regular.

This example is peculiar in that it is more complicated than it need be: a stone in uniform translational motion would satisfy the case, as it did for John of Holland. Albert's example, however, is consistent with the discussions of the English *Calculatores* relating to motion through various resistive media and was perhaps suggested by them.

In the texts analyzed, Albert gives no example of type II–4 – difform and irregular motion. Nicole Oresme, however, does exemplify all four types for Schema II, although he is cryptic in doing so. In place of the wheel for type II–3 he mentions the movement of the heavens (*celum*) as being difform and regular. Then he goes on: "Conversely, the movement downward of a heavy body can be uniform and irregular [II–2], and it can be also uniform and regular [II–1], or even difform and irregular [II–4]." Here Oresme makes the falling body cover all three remaining possibilities, although he gives no indication as to what modalities must be superimposed on its motion in order to satisfy the various definitions.[7]

To provide background for an interpretation of authors to be considered later in this essay, it will be convenient to note the mathematical descriptions of falling motion that were proposed by Albert and Oresme. In their commentaries on Book II of Aristotle's *De caelo et mundo* both discuss a variety of possibilities but do not use the terms *uniformiter, difformiter*, or *uniformiter difformiter*. Oresme mentions that the velocity of fall either increases with time arithmetically toward infinity or else increases with time convergently (as do proportional parts) toward a fixed limit; he elects for the first possibility. Albert also mentions these two possibilities and elects similarly, although in his first mention he is ambiguous as to whether he regards the velocity as varying linearly with the time of fall or with the distance of fall. Later he is more explicit and mentions three additional possibilities: (a) the velocity receives equal increments in the proportional parts of time (thus going to infinity exponentially within a finite time); (2) the velocity receives equal increments in the proportional parts of the distance traversed (thus going to infinity exponentially within a finite distance); (3) the velocity receives equal increments in equal parts of the space traversed (thus going to infinity linearly as a function of distance). Albert here elects the third possibility, showing that his earlier ambiguity should be resolved in favor of a spatial rather than a temporal variation. Clagett has analyzed all of these texts to show that none explicitly identifies falling motion as uniformly difform in such a way as to allow the Mertonian "mean-speed theorem" to be applied to it in unequivocal fashion.

Schema II, it should be noted, did not enjoy the same popularity among later writers as did Schema I. However, types of motion that would fit into one or another of its categories were mentioned by the pseudo-Bradwardine, by Gaetano da Thiene, and by Jean Dullaert of Ghent. The principal importance of Schema II is that it introduced the two-variable concept and that it became the major vehicle for presenting falling bodies as exemplifications of the types of velocity variation in local motion being discussed in the late Middle Ages.

II. FIFTEENTH CENTURY

Moving now to the fifteen century, we come to a series of Italian writers associated with Paul of Venice (d. 1429), several of whose disciples wrote commentaries on the portions of Heytesbury's *Regule* concerned with local motion. These are preserved in a Venice edition of 1494,[8] which served to keep the "calculatory" tradition alive on the Continent long after it had ceased to be of interest in England. The most interesting of these treatises is the commentary of Gaetano da Thiene (d. 1465), who showed a knowledge of the terminology of Schemata I and II, but who proposed yet another alternative for classifying the various types of local motion.[9] This, like Schema I, influenced many authors and in fact dominated the tradition through the first half of the sixteenth century, up to the time of Soto's writing.

Gaetano differentiates uniform from difform motion both with respect to time and with respect to the parts of the object moved, as had earlier authors. He departs from them, however, in introducing a sixfold grouping that makes use of the two-variable concept of Albert and Oresme but allows for two more possibilities. For Gaetano a motion may be uniform either with respect to the parts of the object moved *alone*, or with respect to time *alone*, or with respect to the parts and to time *taken together.*[10] Similarly, a motion may be difform in the same three ways. The resulting six possibilities, written in an improvised notation (in which ~ stands for negation), are given in Schema III.

SCHEMA III

v	U	$U(x) \cdot \sim U(t)$	III–1
		$U(t) \cdot \sim U(x)$	III–2
		$U(x) \cdot U(t)$	III–3
	D	$D(x) \cdot \sim D(t)$	III–4
		$D(t) \cdot \sim D(x)$	III–5
		$D(x) \cdot D(t)$	III–6

This schema is redundant: III–2 is equivalent to III–4, and III–1 is equivalent to III–5. Such redundancy, however, was quite common in medieval systems of division.

Having enumerated these possibilities, Gaetano gives an example of each one, thereby setting a precedent for most of those who were to adopt this particular method of classification. He illustrates type III–1 with the heavy object (*grave*) falling and type III–2 with the wheel (*rota*), as had Albert of Saxony for the analogous cases in his schema (II–2 and II–3); the same two illustrations but in the reverse order he attaches to types III–4 (*rota*) and III–5 (*grave*). His example for type III–3, on the other hand, is ambiguous.

Gaetano speaks of a body descending "in a uniform space" (*mobile descendit in spacio uniformi*) and regards this as a motion that is uniform with respect both to time and to the parts of the falling object (Wallace, 1968a, p. 390, n. 20). He makes no mention of the resistance increasing with the interval of fall, as does Albert of Saxony for the analogous case (II–1), and this seems in fact to be ruled out by his expression *in spacio uniformi.*[11] An alternative possibility could be an object being lowered at constant speed – not falling freely – but there is no positive suggestion of this in the text. Gaetano's example for the remaining case, type III–6, on the other hand is new – a wheel (*rota*) whose angular velocity is being continually increased (*movetur velocius et velocius*). Finally he mentions the example of a ball (*pila*) that falls and rotates as it does so: various components of its motion then illustrate the different types. Falling, it exemplifies type III–1; if rotating uniformly it exemplifies type III–2; if turning slower and slower, it exemplifies type III–6 (*ibid.*, n. 22).

There are other examples in Gaetano's commentary that are of interest, and that have already been discussed in the third essay of this volume. He mentions an object moving rectilinearly and supposes that it is neither contracting nor expanding as it moves, for the expansion or contraction would obviously cause a nonuniform motion of some of its parts. Along the same line he proposes the case of a wheel that rotates, but he now imagines the wheel to expand and contract during its rotation – a phenomenon that would explain further difformities in the motion of its parts. Another example is the placing of a cutting edge against a wheel to continually cut off the outermost surface, thus producing a difformity of the motion of the circumference. A more imaginative possibility is to have the inner parts of the wheel expand while the outermost surface is being cut off; this produces a more complicated variation in the difformity of the movement of the parts. Yet another example is a disc made of ice that is rotated in a hot oven: here the outermost surface continually disappears, and the velocity at the circumference becomes slower and slower; whereas the inner parts expand under the influence of the heat, and their linear velocity increases. A final example is that of a wheel that rotates and continually has material added to its circumference, as a potter might add clay to the piece he is working. Although the velocity of rotation is uniform, the linear velocity of a point on the circumference would increase, unless the entire wheel could be made to contract in the process, in which case it would remain constant.[12]

The foregoing examples can all be viewed as variations of the types of motion sketched in Schema III. Gaetano mentions also some of the types of motion that occur in Schema I; he gives definitions or kinematical descriptions of uniformly difform (I–5 and I–6) and difformly difform (I–7 and I–8) motions, but in these cases he follows the English *Calculatores* and gives no examples whatsoever – not even those already provided by John of Holland.

The remaining Italian commentaries on Heytesbury's *Regule* show more affinity with the latter part of Gaetano's commentary than with its earlier sections in which examples abound: the commentators restrict themselves, for the most part, to kinematical descriptions. Thus Messinus divides local motion into uniform and difform motion and gives a definition of uniform motion that applies only to uniformity with respect to time; he does stipulate, however, that the moving object must retain its quantitative dimensions throughout the motion, thereby implying that there be no change of the parts with respect to each other, and he further stipulates that the "space" passed over be neither contracting nor expanding during the motion.[13] When speaking of difform local motion he makes explicit the distinction between difformity with respect to time and difformity with respect to magnitude and says that an infinite number of possibilities exist for both types of difformity. The relation between distance and time can have any ratio one might wish, and the variation of velocities between respective parts of the moving object can be anything imaginable (Wallace, 1968a, p. 391, nn. 24–25).

Angelo da Fossombrone,[14] on the other hand, reflects some of the concern for exemplification that is found in Gaetano da Thiene. Angelo's characterization of local motion, for example, stresses its priority in the order of nature and its essential division into upward and downward, the distinctions of uniform and difform being considered accidental. Like Messinus he is concerned with eliminating physical factors that would cause the moving body to expand or contract or would change the dimensions of the space through which it moves, and he wishes to define the types of local motion so as to rule out such possibilities.

Angelo follows Gaetano's Schema III, moreover, but supplies only three examples: for type III–3 he gives a moving object (*mobile*) that does not change its parts through condensation or rarefaction but continues in rectilinear motion with unchanging velocity; for type III–2 he cites a wheel (*rota*) that revolves and is moved by a constant force (*potentia*) exerted upon it; and for type III–1 in place of the customary falling body he provides the more abstract example of an object (*mobile*) all of whose parts move with the same velocity as the whole while this velocity is changing with respect to time (*ibid*., p. 392, n. 27).

Bernardo Torni of Florence,[15] alone of this group, reverts to Schema II and mentions three of its four possibilities – the identical ones discussed by Albert of Saxony. He also mentions a few cases that would fit into Schema I. However, he gives no examples – intentionally, since he believes the reader knows enough to furnish his own (*casus tuipse scis formare* – *ibid*., n. 29).

From this survey of fifteenth-century Italy we can see that the development there was somewhat ambivalent. As evidenced in the work of Gaetano, there is a concern for exemplification, with insistence on the case of the falling body, but this appears in the setting of Schema III with its two-variable classification similar to that used at the University of Paris, where *uniformiter difformis* with respect to time does not appear. As evidenced in the remaining commentators on Heytesbury, on the other hand, there are occasional references to types of motion that fit into Schemata I and II (including *uniformiter difformis*), but for these there is only kinematical description, no exemplification.

III. SIXTEENTH-CENTURY PARIS

A pronounced revival of interest in physical problems took place at Paris in the early part of the sixteenth century under the influence of the Scottish nominalist John Major, whose disciples wrote a considerable number of "questionaries" on the *Physics* of Aristotle. In these it was customary to incorporate treatments of local motion that borrowed heavily from such writers as Heytesbury, and thus there was once again a fusion of Mertonian and Parisian thought. Again, in this period a considerable number of scholars from the Iberian peninsula were studying at Paris, and as a result there was a diffusion of the new developments into Spain and Portugal within a few decades of their discussion at Paris. The various schemata already discussed figure in this new movement, and there is a growth of exemplification that prepares for the association of uniformly difform motion with the case of falling bodies.

The first writer to prepare for this association was the Augustinian Jean Dullaert of Ghent, who edited the works of Paul of Venice, another Augustinian, and also wrote questions on the *Physics*.[16] Dullaert was a disciple of Major, and he seems to have brought about a blending of Parisian nominalist interests deriving from Albert of Saxony with the realist concerns that characterized the school of Paul of Venice.

Dullaert's exposition follows Schema III and gives illustrations for all six possibilities. For type III–1 there is the usual example of the heavy body (*grave*) falling through a uniform medium; this case illustrates type III–5 also. For types III–2 and III–4 Dullaert uses Oresme's example of the motions of the heavenly sphere (*sphera celestis*) rather than the customary wheel. The illustration of type III–3 is Albert of Saxony's: a body falling through space whose resistance is so proportioned that it has uniform velocity with respect both to time and to all the parts of the subject. Type III–6, finally, is exemplified by the wheel (*rota*) that accelerates its rotation – identical with the case provided by Gaetano da Thiene (Wallace, 1968a, p. 393, nn. 32–33).

Dullaert, however, does not stop here; he mentions some of the categories of Schema I and significantly furnishes a few examples also. For type I–3 he gives the motion of a heavenly sphere (*sphera celestis*) the parts of which move with uniformly increasing velocity as one goes from the pole to the equator. For type I–5 he cites the case of Socrates (*Sortes*) uniformily increasing his walking speed from zero to eight degrees – the example of John of Holland – and for type I–6 he mentions the converse case of Socrates decelerating his motion uniformly from a given speed to zero. He then explains that types I–7 and I–8 would be defined "in the opposite manner" (*opposito modo*), without giving examples (*ibid*., p. 394, n. 34).

Apart from these divisions Dullaert mentions twice the velocity of descent of a falling body. In the first instance he gives it as an illustration of Albert of Saxony's terminology, as being a uniform but not a regular motion. The second mention comes when Dullaert refers to his exemplification of falling motion as being faster at the end than at the beginning, saying that some wonder how this can occur, since it would seem that the same ratio of force over resistance is maintained and thus the motion should be uniform. Dullaert postpones discussion of this case but says that the motion is actually faster at the end than at the beginning because of the accidental impetus that is built up in the fall. He does not state that the motion will be uniformly difform, but is content to illustrate uniformly difform motion in terms of Socrates' walking speed (*ibid*., n. 35).

Writing shortly after Dullaert, the Portuguese Alvaro Thomaz prepared a lengthy treatise *De triplici motu* patterned on the work of Swineshead but also incorporating materials from Gaetano da Thiene and others.[17] Like the English Mertonians, Alvaro is more concerned with kinematical descriptions of various types of motions than he is with examples drawn from the physical universe. He follows the initial classifications of Schema I, dividing motion into uniform and difform, and difform in turn into uniformly difform and difformly difform. Without defining these he immediately subdivides uniformly difform into a threefold classification similar to that used by Gaetano da Thiene: with respect to the subject *alone*, which may be symbolized $UD(x) \cdot \sim UD(t)$; with respect to time *alone*, symbolized $UD(t) \cdot \sim UD(x)$; and with respect to both subject and time *together*, symbolized $UD(x) \cdot UD(t)$. He then gives the same threefold classification for difformly difform motion and mentions that it can also be applied to uniform motion. The implied schema, a variant of Schema III, is sufficiently different from this to be included here as something new, which will be designated as Schema IV.

SCHEMA IV

v	U		$U(x) \cdot \sim U(t)$	IV–1
			$U(t) \cdot \sim U(x)$	IV–2
			$U(x) \cdot U(t)$	IV–3
	D	UD	$UD(x) \cdot \sim UD(t)$	IV–4
			$UD(t) \cdot \sim UD(x)$	IV–5
			$UD(x) \cdot UD(t)$	IV–6
		DD	$DD(x) \cdot \sim DD(t)$	IV–7
			$DD(t) \cdot \sim DD(x)$	IV–8
			$DD(x) \cdot DD(t)$	IV–9

Of the nine possibilities contained here, Alvaro discusses only three in detail, types IV–4 through IV–6, and he gives only one example, for type IV–4, the motion of the potter's wheel (*rota figuli*), although he does give a kinematical description of type IV–5 (which approximates I–5), an object (*aliquod mobile*) that moves from zero to a given velocity, uniformly increasing its speed (*ibid*., pp. 394–395, nn. 37–38).

Another student of John Major at Paris was the Spaniard Luis Coronel, a townsman of Domingo de Soto (both were from Segovia), who was possibly teaching at Paris when Soto came there as a young student. In his *Physice perscrutationes* (Paris, 1511) Coronel discusses many topics that were commonly dealt with in "questionaries" on the *Physics*, but he is sparing in his treatment of the types of local motion. He mentions, in passing, the division of local motion into rectilinear and curvilinear, and in discussing how the relative motions of two objects are to be compared, he states that either a uniformly difform or a difformly difform velocity would have to be reduced to an average value before a comparison could be effected. He gives no definitions of these types of motions, however, and provides no examples. Rather he refers the reader to the treatises of Heytesbury and Swineshead, with which he seems generally to agree, and otherwise does not think it worthwhile to waste his time over such matters.[18]

A similar treatment is to be found in the *Physica* of Juan de Celaya, another Spaniard who taught at Paris and who definitely numbered among his students Domingo (then Francisco) de Soto. Celaya is not only acquainted with the writings of Heytesbury and Swineshead, but makes explicit mention of the Italian commentators on Heytesbury. What is of particular interest in Celaya's exposition, however, in his departure from the two-variable type of schema (i.e., Schemata II, III, and IV) that dominated these treatments from Albert of Saxony to Alvaro Thomaz, and his return to the one-variable classification of Schema I. Although, like Coronel, he gives no examples, Celaya defines six of the eight types of motion in Schema I.[19] Possibly he was the writer who influenced Soto to adopt this classificatory schema in preference to the others, for he is the first among Soto's immediate predecessors, so far as is known, to make use of it; besides, as already noted, he did teach Soto. Yet strangely enough neither he nor any other writer at Paris in his time seems to have thought to associate motion that is uniformly difform with respect to time with the case of falling bodies.

IV. SIXTEENTH-CENTURY SPAIN

The Spaniards discussed thus far studied at Paris and wrote their treatises while there. Other Spaniards under Parisian influence wrote questions on the *Physics* of Aristotle at universities in Spain: principal among these are Diego Diest,[20] whose *Questiones phisicales* appeared at the University of Saragossa in 1511; Diego de Astudillo,[21] who wrote at Valladolid in 1532; and Domingo de Soto, whose physical works were first published at Salamanca *circa* 1545. Since nominalist treatises seem to have received a less enthusiastic reception in Spain than they had in Paris, these writers, all of whom undertook to incorporate "calculatory" concepts into their courses on Aristotle, were careful to show that such concepts relate in some way to the physical universe. Partially for this reason and partially for pedagogical reasons they utilized considerably more exemplification than did those who wrote in Paris. This, coupled with the diversity of schemata that now seemed to require exemplification, set the stage for a new look at the old examples, for the introduction of some new ones, and for the eventual association of motion that is *uniformiter difformis* with the case of falling bodies.

Diest treats of uniform and difform motions at length, first in the context of Schema III, and then in a less systematic way that mentions elements of Schemata I and IV while discussing uniformly difform motion. He treats all six possibilities in Schema III, giving an example of each: for types III–1 and III–5 he cites the motion of a heavy object (*grave*) downward; for types III–2 and III–4 he gives the example of the wheel (*rota*); for type III–3 he mentions a heavy ball (*sperula gravis*) falling downward in a medium that continually offers more and more resistance so that the velocity remains uniform; and for type III–6 he suggests a wheel (*rota*) rotating with varying angular velocity (Wallace, 1968a, pp. 396–397, nn. 45–48).

Having given and exemplified Gaetano's division of difform motion, Diest returns to this subject and provides an alternative division of difform motion into uniformly difform and difformly difform. He then embarks on a discussion of *uniformiter difformis* that is extremely interesting, for in it he states that the *uniformiter* part of this expression may be understood in various ways, meaning by this uniform variation either in a linear sense or in a logarithmic sense, which is clearly an innovation when compared to previous applications of this expression to the velocity of falling motion.

Diest first explains "uniformly" in the linear sense, as it had been commonly understood by his predecessors. He then proceeds to a second way of defining uniformly difform motion as follows:

[This type of] uniformly difform motion occurs when a change in intensification, velocity, or quantity corresponds immediately to an extensive change in proportionable parts; briefly, when there is the same excess of the first proportionable part over the second as the second over the third, and so on (*ibid.*, p. 397, n. 49).

This statement is cryptic, and is explained in what follows immediately:

This appears in local motion: it is commonly taught that a heavy object falling downward increases its speed uniformly difformly, so that it moves with a greater velocity in the second proportionable part than in the first, and with greater velocity in the third than in the second, and so on (*ibid.*, n. 50).

The expression "proportionable part" (*pars proportionabilis*) is used by Diest to designate a geometric (or, in modern terminology, a logarithmic) part. Thus Diest is saying that the velocity increase is the same in the first half of the body's fall as it is in the next quarter, as in the next eighth, and so on. In other words, the velocity of a falling object increases geometrically with the distance of fall, going to infinity over a finite range. This, it must be noted, Diest proposes as common teaching (*dicitur communiter*), a statement that offers difficulty when one considers that Albert of Saxony had already considered this possibility only to reject it and that there seems to have been little or no discussion of it by the intervening authors. A way out of the difficulty would be to read into Diest's statement the understanding that the velocity increases by proportional parts, corresponding to the proportional parts of the distance traversed, which would be equivalent to holding that it increases linearly with distance of fall. This seems to have been "common teaching" from Albert of Saxony onward, and may have been what Diest intended, although the textual exegesis does not favor this interpretation.

This example, again, might appear to be an illustration of motion that is of type I–5, that is, uniformly difform with respect to time; actually it is not, for the independent variable in Diest's presentation is spatial (*partes secundum extensionem*) and not temporal – and here he foreshadows a difficulty that was to plague Galileo (in his early writings) and others in their attempts to formulate a correct law of falling bodies.

Diego de Astudillo, like Soto, was a Dominican, and he was a close friend of one of Soto's first Dominican professors, the eminent jurist Francisco de Vitoria.[22] Astudillo's questions on the *Physics* (1532) cite Diest and most of the authors we have already mentioned, although his treatment is briefer than Diest's and not of as great significance. He works for the most part within Gaetano's schema, defining and exemplifying all six of its types. For type III–1 he provides the example of "all natural movements, for a stone [*lapis*] falling downward moves with equal velocity in all its parts, although with respect to time the velocity is greater toward the end than at the beginning, as is obvious from experience." The reference to "experience" is significant, although it clearly cannot be taken in any metrical sense. The same example, as was usual, he associates with type III–5. For type III–2 and III–4 he prefers the illustration of the heavens (*celum*), as did Dullaert, giving as his (erroneous) reason that "the parts closer to the poles traverse more space than do those that are more remote." For type III–3 he gives Albert of Saxony's example, observing that "this motion only seems capable of occurring per accidens, by reason of a resistance variation; e.g., if a stone [*lapis*] falls downward and encounters increasing resistance in the same proportion as its velocity of descent would be naturally increased." The latter statement might be taken to imply that the stone's natural fall is uniformly accelerated, but at best this is only an inference; there is no clear indication, moreover, how "uniformly" should then be taken, particularly considering Diest's difficulties with the alternative meanings of this term. Finally, for type III–6 Astudillo introduces a new example, that of a "violent circular motion [*motus circularis violentus*] resulting from a projecting," evidently meaning by this some type of impelled rotation that comes gradually to rest, in which case there would be a velocity variation with respect both to parts and to time (Wallace, 1968a, pp. 398–399, nn. 53–56).

Astudillo, like Diest, gives the further twofold division of difform motion into uniformly difform and difformly difform and divides the first into two types, "with respect to the subject" and "with respect to time," thus touching on the various members of Schema I, types I–3 through I–8. He illustrates type I–3 with the heavenly body (*corpus celeste*), but merely defines the other types. After his definition of type I–5, that is, uniformly difform with respect to time, he adds cryptically, "as is apparent from the above" (*et patet ex dictis*). This could mean that Astudillo thought he had already discussed this case in terms of an example or that the definition was so clear in light of the foregoing examples that it needed no illustration. Which meaning one takes depends on how he evaluates Astudillo's examples of the falling stone already discussed. I favor the latter alternative.

This brings us finally to Domingo de Soto, who had read Astudillo and most of the other authors already discussed, although he generally refrained from mentioning them by name. What is most remarkable about Soto is that he breaks completely with his immediate predecessors in rejecting all the two-variable schemata (II, III, and IV) and returns instead to Heytesbury's one-variable schema (I), which had been used by Juan de Celaya alone of all of the sixteenth-century writers. Soto gives a full explanation of this schema and then, in the fashion that had by then become customary, supplies examples for all its types. It is in this setting that he finally associates falling bodies with *uniformiter difformis* motion, taking *uniformiter difformis* in the precise sense of motion uniformly accelerated in time, to which he can, and does, apply the Mertonian "mean-speed theorem."

Soto gives the complete division of Schema I, plus definitions of all its types. Here we can only enumerate his examples, most of which had already been used by one or more of his predecessors.[23] For type I–1 he gives the case of a foot-length of stone being drawn over a plane surface (*si . . . pedalem lapidem trahas super planitiem*), while for type I–2 he mentions the invariant motion of the heavens (*in regulatissimo motu celorum perspectum est*), and for type I–3, the rotation of a millstone (*mola frumentaria*). Type I–4 offers more difficulty: Soto is unable to supply an example that involves local motion and so gives one that involves changes of quality (alteration), and this for the case of heating. His example is "a four-foot long object that is so altered in one hour that its first foot uniformly takes on a degree of heat of one, and its second uniformly the degree of two, and its third the degree of three, etc." – for which he supplies the diagram of a step-function. He goes on to observe that "the present treatise is not at all concerned with this kind of alterative motion" but that he has given this example for the simple reason that a local motion of this type is hardly possible: rectilinear motion must be uniform in this respect, and rotary motion can only be uniformly difform – by its nature it cannot be difformly difform (*ibid.*, p. 400, n. 59).

Types I–5 and I–6 Soto defines in conjunction with each other and then observes that they are "properly found in objects that move naturally and in projectiles." He goes on:

For when a heavy object falls through a homogeneous medium from a height, it moves with greater velocity at the end than at the beginning. The velocity of projectiles, on the other hand, is less at the end than at the beginning. And what is more the [motion of the] first increases uniformly difformly, whereas the [motion of the] second decreases uniformly difformly.[24]

Then later on in the text while discussing the same case, "uniformly difform motion with respect to time," he removes any possible ambiguity as to his meaning by proposing the difficulty "whether the velocity of an object that is moved uniformly difformly is to be judged from its maximum speed, as when a heavy object falls in one hour with a velocity increase from 0 to 8, should it be said to move with a velocity of 8?"[25] His answer to this is clearly in terms of the Mertonian "mean-speed theorem," for he decides in favor of the average velocity (*gradus medius*) as opposed to the maximum. He justifies this with the illustration: "For example, if the moving object *A* keeps increasing its velocity from 0 to 8, it covers just as much space as [another object] *B* moving with a uniform velocity of 4 in the same [period of] time."[26] Thus there can be no doubt about his understanding of *uniformiter difformis* and how this is to be applied to the space traversed by a freely falling object.

To exemplify types I–7 and I–8, finally, Soto resorts to the motion of animals and to other biological changes, stating:

An example would be if something were to move for an hour, and for some part [of the hour] were to move uniformly with a velocity of one, and for another [part] with a velocity of two, or three, etc., as is experienced in the progressive motion of animals. This kind of motion frequently occurs in the alteration of animals' bodies, and perhaps it can take place in the motion of augmentation and diminution.[27]

The illustrations apply mostly to type I–7, but his mention of diminution at the end (although not strictly a local motion) indicates that he is also aware of decreasing variations and thus implicitly includes type I–8 in his exemplification.

The further details of Soto's analysis of falling motion, together with its influence on later thinkers, must await treatment elsewhere. The materials presented here, however, should help clear up at least part of "the enigma of Domingo de Soto." The contribution of the Spanish Dominican was not epoch-making, but it was significant nonetheless. Of the nineteen authors considered in this essay he alone thought of systematically providing examples for the simplest of the four schemata used – that which considers only one independent variable at a time. The others who were interested in exemplification – and these were mostly late-fifteenth-century or sixteenth-century writers – worked in the context of two-variable schemata, and this generally precluded the possibility of their even considering the case of motions that are uniformly difform with respect to time.[28] All of Soto's examples, of course, like those of his predecessors, were proposed as intuitive, without empirical proof of any kind. Moreover, he and Diest, of all those considered, were the most venturesome in attempting to assign a precise quantitative modality to falling motion. Of the two, Soto was without doubt the better simplifier. He seems also to have been the better teacher, and, as has been seen in the fourth essay in this volume, he was philosophically more interested in unifying the abstract formulations of the nominalists with the physical concerns of the realists of his day. Again, he had the advantage of time and of being able to consider more proposals. The strange alchemy of the mind that produces scientific discoveries requires such materials on which to work. It goes without saying that Soto could not know all that was implied in the simplification he had the fortune to make. But then, neither could Galileo, in his more refined simplification, as the subsequent development of the science of mechanics has so abundantly proved.

NOTES

[1] In his essay on science in the Renaissance in René Taton (ed.), *History of Science*, Vol. II: *The Beginnings of Modern Science, from 1450 to 1800*, trans. A. J. Pomerans (New York: Basic Books, 1964), pp. 94–95.

[2] The best documentary study of Soto's life and works is Vicente Beltrán de Heredia, O.P., *Domingo de Soto: Estudio biográfico documentado* (Salamanca: Convento de San Esteban, 1960).

[3] As will be explained, the application of the expression *uniformiter difformis* to falling bodies is equivalent to stating that the motion of such bodies is uniformly accelerated. The method of calculating the distance traversed in such a motion was first worked out at Merton College, Oxford, by a group of scholars usually referred to as the Mertonians, or *Calculatores*. Their method consisted in replacing the uniformly varying speed by an average or mean value and then using this to compute the distance of travel; the equivalence on which this technique was based has come to be known as the "mean-speed

theorem." Actually the method was applied in a broader context to all types of change, whether quantitative or qualitative, and the theorem could be referred to more generally as the "mean-degree theorem." Likewise the expression *uniformiter difformis* was applied to all kinds of change, but in what follows we shall restrict ourselves exclusively to applications to local motion and thus omit any discussion of the extensive literature that developed relative to other applications.

4 Koyré, in Taton (1964), p. 95.

5 For texts, see Clagett (1959), pp. 247–250, 445–462, 630–631.

6 Albert of Saxony, *Questiones super quatuor libros Aristotelis de celo et mundo* ... (Venice, 1520), lib. 2, quest. 13. For the Latin of the texts cited here and elsewhere in this essay, consult the original version published in *Isis* **59** (1968), pp. 384–401.

7 For the texts, see Marshall Clagett, *Nicole Oresme and the Medieval Geometry of Qualities and Motions*, Madison: University of Wisconsin Press, 1968, pp. 272–273. "Uniform and irregular" is obviously the case of a body falling freely in a uniform medium. "Uniform and regular" could be the case of a body falling in a medium whose resistance continually increases so as to prevent the possibility of a velocity increase (as in Albert of Saxony's example); whereas "difform and irregular" could be the case of a body falling freely in a uniform medium and rotating as it falls.

8 *Hentisberi de sensu composito et diviso* ... (Venice: Bonetus Locatellus, 1949). This edition is described by Curtis Wilson, *William Heytesbury: Medieval Logic and the Rise of Mathematical Physics* (Madison: Univ. Wisconsin Press, 1960), p. 4, n., and *passim*.

9 On Gaetano see P. Silvestro da Valsanzibio, O.F.M.Cap., *Vita e dottrina di Gaetano di Thiene, filosofo dello studio di Padova, 1387–1465* (2nd ed., Padua: Studio Filosofico dei Fratrum Minorum Cappuccini, 1949).

10 *Expositio litteralis supra tractatum* [*Hentisberi*] *de tribus* [*predicamentis*], *De motu locali* (Venice: 1494), fol. 37rb.

11 The words *in spacio uniformi* are themselves puzzling, since fifthteen-century writers had no conception to match the modern notion of isotropic space. The words probably refer to a uniform *medium* and not to a uniform *space* understood in the strict sense.

12 Gaetano, *Expositio litteralis*, fols. 37ra–40rb. These examples may seem overly fanciful to the modern reader, and yet without their help it becomes almost impossible to visualize the complex kinematical cases being discussed by Gaetano and his contemporaries.

13 *Questio Messini de motu locali* (Venice: 1494), fol. 52va.

14 *Supra tractatu*[*m*] *de motu locali* (Venice: 1494), fols. 64ra–73rb.

15 *In capitulum de motu locali Hentisberi quedam annotata* (Vanice: 1494), fols. 73va–77va.

16 *Questiones super octo libros phisicorum Aristotelis necnon super libros de celo et mundo* (Lyons: 1512); Élie (1950–51) cites two previous editions, Paris 1506 and Paris 1511.

17 *Liber de triplici motu proportionibus annexis* ... *philosophicas Suiseth calculationes ex parte declarans* (Paris: 1509).

18 Coronel, *Physice*, lib. 3, pars 3, fol. 86rb.

19 *Expositio* ... *in octo libros phisicorum Aristotelis, cum questionibus* ... *secundum triplicem viam beati Thome, realium, et nominalium* (Paris: 1517), fols. 81rb–83vb. Although giving no examples, Celaya does refer the reader to his treatment of uniformly difform qualities for a further understanding of these definitions: "Et qualiter iste

diffinitiones habent intelligi ex declaratione qualitatis uniformiter difformis inferius apparebit" (fol. 81va).

[20] Diest, a native of Bolea in Spain, studied at Paris in the latter part of the fifteenth century. The full title of his work is *Questiones phisicales super Aristotelis textum, sigillatim omnes materias tangentes in quibus difficultates que in theologia et aliis scientiis ex phisica pendent discusse suis locis inseruntur* (Saragossa: 1511). For some further details on Diest, see Villoslada (1938), pp. 401–402.

[21] Astudillo taught at the College of Saint Gregory in Valladolid, which was staffed by Dominicans; here he composed his *Questiones super octo libros phisicorum et super duos libros de generatione Aristotelis* (Valladolid: 1532).

[22] Vitoria also taught at Saint Gregory's in Valladolid. Of Astudillo, Vitoria graciously remarked that he knew far more than himself but was not as good at marketing his ideas: "Fray Diego de Astudillo más sabe que yo, pero no vende tan bien sus cosas." See Villoslada (1938), pp. 304–305.

[23] The following citations are from Soto's *Super octo libros physicorum questiones* (Salamanca: 1555), lib. 7, quest. 3; the earliest complete edition of this work was published at Salamanca in 1551 although an earlier printing, lacking parts of Bk. 7 and all of Bk. 8, appeared there *c.* 1545. There are no known manuscripts of the text.

[24] Soto, *ed. cit.*, fol. 92vb: Hec motus species proprie accidit naturaliter motis et proiectis. Ubi enim moles ab alto cadit per medium uniforme, velocius movetur in fine quam in principio. Proiectorum vero motus remissior est in fine quam in principio; atque adeo primus uniformiter difformiter intenditur, secundus vero uniformiter difformiter remittitur.

[25] *Ibid.*, fol. 93vb: Utrum velocitas mobilis uniformiter difformiter moti sit denominanda a gradu velocissimo, ut si grave decidat in una hora velocitate a non gradu usque ad 8, dicendus sit moveri ut 8?

[26] *Ibid.*, fol. 94ra: Exempli gratia, si *A* mobile una hora moveatur intendendo semper motum a non gradu usque ad 8 tantumdem spatii transmittet quantum *B*, quod per simile spatium eodem tempore uniformiter moveretur ut 4.

[27] *Ibid.*, fol. 92vb: . . . ut si ita res aliqua moveretur per horam, ut per aliquam partem uniformiter moveretur ut 1, et per aliam ut 2 vel 3, etc. Ut est experiri in motibus progressivis animalium. Que quidem species motus crebro accidit in alteratione corporum animalium, et potest forsan contingere in motu augmenti et decrementi.

[28] The exception is Alvaro Thomaz, who did mention motion that is uniformly difform with respect to time in the context of his two-variable schema. His division, however, was so complex as to discourage any attempts at simple exemplification with natural examples.

7. CAUSES AND FORCES AT THE COLLEGIO ROMANO

Thus far our studies of the sixteenth-century achievement in mechanics have focused on conceptual changes that encouraged an experimentalist attitude toward the study of motion and on the development of "calculatory" techniques that permitted a mathematical analysis of the observed results. With regard to the latter, the emphasis to this point has been on kinematics, i.e., on the quantification of spatio-temporal aspects of motion, without reference to the dynamical factors that produce and influence motions of various types. In this essay we propose to complement the studies already presented by considering how sixteenth-century natural philosophers, especially those associated with the Aristotelian tradition, viewed the causes and forces that initiate and otherwise determine the local motions that occur in the cosmos.

The linking of "causes and forces" in the title is suggested by a chapter in Annaliese Maier's *Die Vorläufer Galileis im 14. Jahrhundert* which she captioned "Ursachen und Kräfte."[1] In this she concentrated on the thirteenth and fourteenth centuries, and attempted to explain how the Latin terms *causa* and *vis* then had meanings that were partly the same as, and partly different from, their modern equivalents. At first sight it would appear that the concept of cause is characteristically medieval, whereas that of force is characteristically modern, but from Miss Maier's researches one can see that no such clearcut dichotomy obtains. Indeed, *vis* was quite commonly used by medievals to designate violent or external causes that force a motion from without; it was also used by them, though less commonly, to designate causes or forces that are operative in nature and that originate motions from within.[2] Significantly, for our purposes, this latter use reappeared and was reinforced in the sixteenth century, possibly for the reason that John Philoponus's commentary on Aristotle's *Physics*, which was published in Latin translation in 1539 and again in 1558, employs a definition of nature that explicitly incorporates the force concept.[3] In Aristotle, nature is defined as a principle and cause of motion and of rest in that in which it is primary and immediate, and not merely incidental.[4] In Philoponus's version, the classical definition is prefaced by the statement that nature is a kind of force that is diffused through bodies, that is formative of them, and that governs them; it is a principle of motion and of rest, and so on.[5] Instead of nature being only a *principium et causa*, as it was for Aristotle, for Philoponus it became *quaedam vis* that is in bodies and is the source of their natural motion and rest. With Philoponus, then, the attention of the Latin West was again directed to the possibility that nature might be a type of internal force, and we should not be surprised that writers of the later sixteenth century began to question how such a force could be operative in various natural motions.

Although the time frame of this essay, like that of the others in Part II of the volume, is the sixteenth century, to keep it within manageable limits we shall concentrate on the teaching on this subject that was current at the Collegio Romano during the latter part of the century. Such a concentration has a twofold advantage. In view of the synthesizing function exercised by the Jesuit professors at the Collegio, it summarizes the wide range of scholastic and Aristotelian thought on this subject in northern Italy to the end of the sixteenth century. Again, in view of the influence exerted by the Collegio on Galileo's early thought, as will be detailed in the studies in Part III *infra*, it provides an overview that should prove helpful when evaluating Galileo's statements on the causes of natural and projectile motion. Nature is an important concept in Galileo's writings, and we know from his early memoranda on motion that he was aware of it as an internal cause of local motion.[6] Precisely how it exercised this causal activity is never thoroughly explained by him. In a series of notes very similar to Galileo's, however, and written around the same time by a Jesuit professor at the Collegio Romano, by name Mutius Vitelleschi, this problem is discussed in some detail.[7] In what follows we shall summarize Vitelleschi's explanations, not only as typical of what was being taught in Italy in 1590, but as probably giving a good insight into the way the young Galileo conceived of nature as both a cause and a force behind its various operations.

I. NATURE AS A CAUSE OR FORCE

In defining nature Vitelleschi indicates that he is aware of Philoponus's emendation to Aristotle's definition and is concerned over its pantheistic overtones, for one might construe God to be the *quaedam vita sive vis* that is diffused through all bodies.[8] Yet he himself gives numerous indications that the force concept has become part of his own way of understanding nature's activity. Nature for him is a *principium internum* of motion, and in the sense defined by Aristotle it signifies an internal propensity (*internam propensionem*) for a particular type of motion. As internal it excludes art and other extrinsic principles of operation, for these do not properly effect motions, though they can modify them in one way or another. Natural things have their motions from a principle that inheres in them, and more specifically, they are moved by a natural motive power (*virtute motiva naturali*).[9] *Virtus* is here translated as power, but *virtus, potentia,* and even *qualitas* in these contexts have much the same connotation as *vis*. Vitelleschi uses these terms interchangeably, noting, for example, that natural things have within them a certain force (*vim quandam*) by which they effect motion within themselves, by which they bring themselves to their proper perfection, and at the same time cause motion in other things. Does this mean that such a natural force is to be considered an efficient cause? For Vitelleschi this presents a problem, for he would prefer to reserve the term efficient cause for agents that produce effects in objects other than themselves; but there is a sense, as we shall see, in which even the natural movements of the non-living come from nature with the connotation of an efficient agent.[10]

For the moment, therefore, a motion is natural if it proceeds from some inclination and propensity within the moving object (*secundum aliquam inclinationem et propensionem ipsius rei*). In the case of the natural motion of a heavy body downward, this propensity arises from a motive quality (*qualitas motiva*) the body has from nature. This quality, moreover, need not be an active principle of the body's movement; it suffices that it be simply a passive principle whereby the body receives its propensity for a particular type of motion.[11]

A motion is violent, as opposed to natural, if it is imposed from without and the thing acted upon contributes no force at all (*nullam vim conferenti passo*). In addition to this, however, for violence in the strict sense the action must be opposed to the natural propensity of the moving thing. But if the motion is imposed on a body from without in such a way as not to oppose the body's natural inclination, then the motion is intermediate between the natural and the violent, and so may be regarded as neutral, i.e., as beyond nature but neither according to it nor contrary to it. The statement that no unnatural motion can be perpetual is then to be understood of the violent only in the strict sense, for this type of violence takes away the force of nature (*afferat vim naturae*), with the result that it depletes the body's source of motion.[12]

The precise way in which Vitelleschi conceives nature to be a motive force can be seen in his discussion of what it is that moves the elements. We have already noted that gravity, as a motive quality, need only be a passive principle of the element's motion. In addition to this, must there also be an active principle within the body from which its motion proceeds, or is it sufficient that the active principle be an external mover, namely, the generator *per se* or the remover of impediments *per accidens*? Now Vitelleschi does not deny that external movers are necessary for the motion of the elements; he does note, however, that neither the generator nor the *removens prohibens* are in contact with the elements, and so must impress some quality that inheres within them (*imprimit qualitatem aliquam quae in ipsis inhaereat*). Thus external movers move the elements by this type of quality, just as the magnet attracts iron by impressing some quality on it. Gravity, then, is but an instrumental principle of the body's fall. Whatever is done instrumentally, moreover, can be said to be done by the principal agent behind the instrument, and in the case of an elemental body, this principal agent is the body's substantial form. Therefore an active internal principle is also required for this type of natural motion, and this principle is nothing more than the substantial form of the element.[13]

Such being the case, the question arises again whether such a substantial form may be looked upon as the efficient cause of the body's falling motion. To answer this Vitelleschi borrows a distinction from Jacopo Zabarella and says that there are really two kinds of efficient cause: one that acts on something different from itself and produces an effect in it, and this is said to be *proprie efficiens*; another that produces an effect in itself by simple emanation, and so is described only as *efficiens per emanationem*. The first type of agent requires that the body acted upon be really distinct from it, whereas the second type does not. Applying this distinction to the case at hand, as a true and proper efficient cause the substantial form of an element can produce effects in something different from itself, whereas by the improper agency of emanation it can produce motion in itself alone. In both cases, however, it is truly the efficient cause of the motions that result.[14]

To clarify in more detail the various agents that are involved, say, in falling motion, Vitelleschi would distinguish three different potencies that have to be actuated for such motions to occur, and then assign to each an appropriate efficient agent. The first potency is a potency to the substantial form itself, and this is actuated by the generator that produces the elemental body *per se*. The second potency is to the motion that follows on the form, and this is actuated *per accidens* by the remover of impediments. Finally, in the very process of its motion the element is further in potency to its particular terminus or natural place, and this last potency is activated by its motive force or quality as an instrument, but by the substantial form of the element as the active principle from which the motive quality flows. Aristotle, says Vitelleschi, when discussing the principle *omne quod movetur ab alio movetur* in the eighth book of the *Physics* spoke only of the first two agents, i.e., the generator and the *removens prohibens*, because he was interested only in how the elements are moved from without. Also, it is clearly the case that these first two agents are efficient causes of the body's falling motion in the strictest sense of the term agent. But apart from these the substantial form is also required, and this is not only the formal cause of the element but it is also the efficient cause of its motion, understanding efficient here in the weaker sense of causing, through a process of emanation, properties and accidents that are proper to a substance. In considering, therefore, the motion of a heavy body downward, the following additional causal sequence must be recognized: (1) when impediments are removed, the element is moved by its substantial form as by an agent, and its motion emanates from its motive force or quality as an instrument of that form; (2) not only the form but the entire element moves the surrounding medium, and in this action the element is truly and properly an efficient cause, the same as when it alters something through its active qualities; and (3) as the medium moves, the motion of the element is assisted by this motion because it effectively removes further impediments to the element's motion, as will be explained later when we treat of resistive forces.[15]

II. MOTIVE FORCES

To come now to a more detailed examination of motive forces or qualities, these are of two general types, natural forces such as *gravitas* and *levitas*, and the impressed force that is invoked by some to account for projectile motion, usually called *impetus* or *virtus impressa*. With regard to the natural forces, Vitelleschi is aware of current arguments that would reject *levitas* as a natural motive quality on the ground that light bodies are in reality only less heavy, and so do not move upward naturally but are propelled there by heavier bodies. He does not regard such arguments as convincing, however, and so treats of both *gravitas* and *levitas*, while giving greater attention to the former.[16]

One problem that interests him is whether these motive qualities are primary qualities in bodies, and if not, whence they arise. The active or alterative qualities, namely, the contrary pairs of hot-cold and wet-dry, were commonly regarded by Aristotelians as primary qualities and thus as most basic to the elements. Vitelleschi does not believe, however, that a simple answer can be given to such a question of priority, and suggests that the comparison be made in three ways: (1) in the order of perfection; (2) in the order of generation;

and (3) in the order of the ends served by the elements. In the order of perfection, he says, the alterative qualities are prior because they are instrumental in producing the substantial forms of the elements, they are the agents that bring about the formation of compounds, and through them the elements produce real qualities in others, whereas through the motive qualities the elements produce only motion in themselves. Yet one can still say that the motive qualities are more perfect *secundum quid*, because when the forms of the elements are considered precisely as natures, then the motive qualities are more properly their instruments. Again, in the order of generation the alterative qualities are prior, because they are instrumental in producing the substantial forms of the elements, on which motive qualities and other accidents depend. But in relation to the ends intended by nature, it is the motive qualities that are prior to the alterative, because they preserve the distinctive perfection of the elements, while also contributing to the integrity of the universe; the alterative qualities, on the other hand, are ordered mainly to the formation of compounds.[17]

Additional light is cast on motive qualities by contrasting the *gravitas* and *levitas* of the intermediate elements, air and water, with the motive qualities of the extreme elements, fire and earth. Some thinkers, such as Agostino Nifo, hold that the motive qualities of the intermediate elements are not simple but are composed of the qualities of the extreme elements. Vitelleschi disagrees with this, maintaining that the motive qualities of all four elements are simple and specifically different from the others. To justify this he notes that two things should be considered in any natural motion: (1) its terminus, and (2) its velocity. Specification is usually taken from the terminus, for acts are specified by their objects and powers by their acts; applying this principle, motions should be specified by their termini, and motive powers by the motions that proceed from them. Since the motions of the intermediate elements have distinctive termini, their motive qualities should also be specifically distinct. With regard to velocity, different gravities produce different velocities in motion, and so this may be a way of distinguishing motive powers also. Vitelleschi would note, however, that one gravity can be greater than another in two ways: (1) intensively, because it has more degrees and produces a greater effect than another; (2) extensively, because the object having this gravity has more parts of the same kind and heavy to the same degree. Some think that greater or less velocity arises from greater or less gravity understood intensively alone, and that extensive differences have no effect; this, it would appear, is an over-simplification. Yet Vitelleschi is convinced that the body descending with greater velocity is the heavier intensively, noting that a body that is heavier extensively will encounter greater resistance from the medium because it has more parts. In any event, motive forces can be quantified in at least two ways: (1) within the same species, because a body possesses more or less degrees of gravity of the same type; and (2) with different types, because a body possesses different degrees of gravities that are specifically distinct. The second kind of quantification is proper to compounds, whereas the first is proper to elements. And the motive qualities of intermediate elements are specifically distinct from those of the extreme elements, since they move the intermediate elements to distinctive termini, even though they need not always do this with the same velocity.[18]

A related problem is whether the elements gravitate or levitate within their own spheres.[19] There can be no doubt, says Vitelleschi, that when the elements are at rest in their proper places they still retain their motive qualities, and so they can be said to be heavy or light in first act. Thus, when elements are about to be moved, they do not immediately acquire a motive power (*virtutem motivam*), but the power they already possess then goes into second act.[20] The problem is whether, when elements are heavy and light in their proper places, they also have the secondary effects of those qualities, i.e., whether they gravitate or levitate. To answer this Vitelleschi notes that natural motive powers have three different effects: (1) they move to a proper place when nothing impedes them; (2) they provide a certain tendency (*quidam conatus*) to motion if something does impede them; and (3) they keep the element in its proper place, resisting any attempt to remove it therefrom. Vitelleschi records the experiments adduced by Girolamo Borri to prove that all the elements except fire gravitate within their own spheres, and reviews the arguments of others, such as Francisco Valles, who deny this.[21] Vitelleschi's position is that if by gravitation is understood the element's remaining in its proper place and resisting any effort to be removed from it, then one can say that the elements gravitate within their own spheres. But if gravitation is understood properly for a motion downward or for a *conatus* to such motion, then the elements do not gravitate – a fact that explains why we do not feel the weight of the air with which we are surrounded.[22]

So much for the natural motion of the elements. With regard to their violent motion, usually treated under the question of what moves projectiles, Vitelleschi has no consideration *ex professo*–probably not having reached this in his lectures. In discussing the principle *omne quod movetur ab alio movetur*, he does note a parallel between the proximate mover in the case of the projectile and the proximate mover of the falling body, but he is content merely to repeat Aristotle's teaching that the projectile is moved remotely by the thrower and proximately by the medium.[23] Perhaps noteworthy is the fact that earlier in the 16th century Domingo de Soto employed this same parallel to develop his teaching that *impetus* is a motive power completely analogous to *gravitas* in the falling body.[24] Soto's teaching was generally advanced by Dominicans, but the Jesuits at the Collegio Romano, following Benedictus Pererius, tended to reject it.[25] There are indications, however, that some Jesuits in the late 16th century were beginning to favor the *impetus* explanation. Thus Paulus Valla, who taught the *Physics* course at the Collegio the year before Vitelleschi, strongly defended the teaching on the *virtus impressa*, and his influence is detectable in the notes of Ludovicus Rugerius, who taught the same course the year following Vitelleschi.[26] While finally settling for the medium as the projectile's mover, Rugerius holds that it is not improbable that there be some type of impressed force in the projectile that accounts for its continued motion. Most of the arguments against this position, he says, can be solved by maintaining that this force is different from a natural motive force, that there is only one species, that it moves up and down and in other directions as well, that it can be intensified and diminished, and that it corrupts either from the contrary action of a natural motive force or from the projectile's being brought to rest. Rugerius admits finally that it might even be possible to see this as not really distinct from a natural motive force, but actually as a modification of the element's gravity that affects the direction and velocity of its motion.[27]

III. RESISTIVE FORCES

The resistance encountered by moving bodies is a recurrent theme in the late 16th century, and so we should not be surprised that Vitelleschi discusses also the causality exercised by resistive forces. His first mention of this is in the context of analyzing the regularity and the composition to be found in the motions of the non-living. The movements of the elements are not completely regular, he observes, for projectiles slow down and falling bodies speed up as they move. Vitelleschi enumerates five different causes that can produce such irregularities in the velocities of bodies. These are: (1) the mover, as this has greater or less force (*maiorem vel minorem vim*) to effect motion; (2) the thing moved, as this more or less resists the mover (*magis vel minus resistit moventi*); (3) the thing moved again, as this is endowed with more or less motive power (*plus vel minus virtutis motivae*); (4) the medium, as this is thicker or thinner and so variously impedes the motion; and (5) other factors, particularly the shape of the moving body, which makes it more or less suited to cut through the medium.[28] From this programmatic statement one can see that resistive force is placed by Vitelleschi on almost the same plane as motive force, and he develops this equivalence in some detail, as we shall explain presently when discussing acceleration in free fall.

The composition of motions is another context where resistance enters incidentally, though with a slightly different nuance. Vitelleschi notes, following Aristotle, that some motions are simple whereas others are composite. To understand, therefore, this simplicity or composition of a motion, he proposes to consider three different factors related to it: (1) the motion itself; (2) the power that causes it (*virtutem a qua fit*); and (3) the distance over which or around which it takes place. With regard to the third point, the distance traversed, Vitelleschi takes the position that straight and circular distances are simple whereas others made up of these, i.e., those partly straight and partly circular, are composite. Similarly, with regard to the first point, the motion itself, the downward motion of an element is simple whereas the progressive motion of an animal is composite. There remains then the second point of comparison, the power causing the motion, and from this viewpoint a motion is simple if it comes from a simple power, such as gravity or levity, composite if it comes from several powers, say, from both gravity and levity at the same time. The first kind of motion obviously characterizes the elements, for these have only one motive power (*una virtus motiva*). Compounds, on the other hand, can have two powers, for example, different degrees of gravity and levity. In their case, however, a particular element always predominates, with the result that the motive power of the predominant element exceeds any power opposed to it, and an effect is produced that is really the result of interacting forces.[29] This opposition of motive powers within compounds may therefore be seen as generating a type of internal resistance (although Vitelleschi does not use this expression) that affects the resulting motion and renders it composite. In the case of a compound, he notes elsewhere, its substantial form is the primary internal cause of its motion, though the motive quality of the predominant element serves as its instrument. So he maintains that the form of the compound, much like the elemental substantial form, has the force of effecting its motion (*vim efficienti hunc motum*) through a motive quality it derives from the form of its predominant element.[30]

But it is in the context of discussing what makes falling bodies move faster as they fall that Vitelleschi has his fullest discussion of resistive forces. He takes it as a fact of experience that bodies moved by nature accelerate as they move, whereas those that are moved from without by force gradually decelerate. With regard to the natural motion, the following seem to him to be the possible explanations: since two things are involved, the moving body and the medium through which it moves, either (1) the cause of the velocity increase is extrinsic to the falling body and is to be sought in the medium, or (2) it is in the body itself, and then it arises either (a) from the fact that proximity to its proper place increases the body's motive power (*virtutem motivam*), or (b) because some power opposing its fall from within is gradually diminished. Vitelleschi then eliminates various alternatives, and first he discards the explanation favored by Galileo in his early writings: contrary to Hipparchus and others, the velocity increase is not caused by a residual opposing force that is gradually overcome. Similarly, the greater velocity does not arise from the motive power's being increased or strengthened as the body gets closer to its natural place. Again, the velocity increase is not traceable to a decrease of resistance on the part of the medium, at least not in the sense that the medium becomes more easily separable or because the air closer to the earth is not as light and so offers less resistance to the descending body. Yet Vitelleschi admits that the medium is a big factor in explaining the acceleration, provided it is taken to be in interaction with the falling body itself. The explanation that he finally prefers, following Zabarella, is that the earlier part of a body's fall causes a greater velocity in the later part of its fall, because it then causes the medium to resist less. The basic mechanism is easily understood on the analogy of a body moving against a flow of water; obviously the body will move quickly if the water is at rest, and more quickly still if it moves in the same direction as the flow. Similarly, a falling body propels the first part of the medium, and then when it reaches the second part it propels that more quickly in the same direction, and the third part more quickly still; as a result the medium comes gradually to impede the movement less and less. And since the velocity of any motion results from an excess of the motive force over the resistance encountered, the velocity of the motion increases as the resistance grows less.[31] Stated otherwise, since the resistance of the medium continually decreases while the motive force remains constant, the difference between them increases, and with it the velocity of fall. In the case of projectile motion, on the other hand, the motive force (*virtus movens*) is always being weakened and lessened, and so it cannot effect any decrease in the resistance of the medium; that is why its motion is more rapid not at the end but at the beginning, for then its motive power is the stronger.[32]

Vitelleschi admits a difficulty with this explanation in that Aristotle seems to claim that the greater velocity at the end of falling motion results from an increase of gravity. His reply to this is that gravity can be taken in two ways: (1) in first act, and then it refers to the motive power that produces the motion; and (2) in second act, and then it refers to the motion that results. The first he would prefer to call *gravitas*, the second, *gravitatio*. And *gravitatio*, as he sees it, is nothing more than the excess of the motive force over the resistance of the medium. Thus, even though the *gravitas* remains the same, the *gravitatio* increases, and this is what Aristotle means. In other words, the velocity increase comes from the excess of the motive power over the resistance, and this excess derives not from an increase of gravity in first act, but rather from a decrease in the resistance encountered as the body falls.[33]

Like most commentators in the late sixteenth century, Vitelleschi considers also the phenomenon of action and reaction, and treats how reaction is related to resistance and how resistance itself should be defined. In his view, action and reaction are involved in all cases of alteration and in most cases of local motion, specifically those in which sublunary bodies are involved. Reaction is different from resistance, he observes, although the two are related. Resistance is formally a privative notion, in the sense that it signifies the non-acceptance of an action by the body acted upon, whereas reaction signifies the positive production of a quality in the agent by the body that reacts. The cause of resistance is not the body's matter, but rather its form, and this is also the source of its reactivity. In fact, nature seems to have endowed every form that can be acted on by contraries with a twofold force (*duplicem vim*): one whereby it conserves itself and wards off the action of the contrary, another whereby it reacts actively against the agent. Both resistance and reaction, as Vitelleschi conceives them, come from the form or nature of the resisting or reacting body as from a proper efficient cause; this substantial form is helped by the body's quantity, density, etc., in the sense that these augment the body's resistive forces (*vires*) and so enable it to resist agents that act against it.[34]

IV. OCCULT FORCES

Finally, a word remains to be said about occult forces as these may affect the motions of bodies. Vitelleschi's main treatment of these is when considering how the heavenly bodies influence the sublunary region, but he also mentions them when discussing the motion of the elements. In general he takes the position that the elements are not moved by the heavens, in the sense of being pushed down from above, nor are they drawn downward by their natural place. In explaining the causality of place, moreover, he sees no difficulty in a particular place attracting the falling body after the fashion of a final cause, but he explicitly denies that bodies are attracted to their natural places efficiently by some occult force (*per vim quandam occultam*).[35]

Although occasionally using the expression occult force, as just seen, Vitelleschi more commonly refers to powers of this type as *influenciae*, which he regards as emanating from the heavenly bodies. Apparently the Jesuits of his day were quite divided on this question, and so he is quick to recognize that astronomers are prone to multiply such influences beyond all reasonable claims. He takes the position, however, that the light coming from the heavens, and even the motions they produce in sublunary bodies, are not sufficient to explain the tides of the sea and other natural phenomena. He also sees some occult influences behind the massive floods that took place in 1589, inundating not only Rome but other cities of Italy and Spain as well.[36] Elsewhere, when discussing the art of alchemy and the exceptional powers (*virtutes*) found in metals and other compounds, he allows the possibility that alchemists might be assisted by the force of an influence (*vis influenciae*). As he sees it, influences and other natural forces (*vi causarum naturalium*) can account for many marvelous effects, but he admits that it is extremely difficult for men to apply them in the right proportion to transmute metals and obtain other desirable effects. Yet in principle he sees nothing against such influences, and indeed there are so many natural wonders that he is persuaded they must exist, even though little is actually known about them.[37]

This concludes what can be said here about causes and forces in the sixteenth century. Admittedly the treatment is truncated, as nothing has been said about those who were more decidedly under Platonic influences, deriving not only from Philoponus but also from the *anima mundi* tradition, such as Girolamo Fracastoro, Antonius Ludovicus, Bernardino Telesio, and others down to William Gilbert and Johannes Kepler. All of what has been presented is based on manuscript sources, and it represents a rather conservative Aristotelian development within Renaissance scholasticism, but also in line with the thought of Pietro Pomponazzi, Jacopo Zabarella, and others of the Paduan school. Mutius Vitelleschi, who has been the focus of this account, epitomizes in many ways the diversified tradition of the Collegio Romano, originating with Franciscus Toletus and Benedictus Pererius, and soon to be codified at the turn of the century in the famous *Cursus philosophicus* of the Jesuits of Coimbra.[38] Those who are acquainted with the fourteenth-century development in mechanics at Merton College, Oxford, and by Jean

Buridan and his disciples at the University of Paris may perhaps be surprised at how closely this sixteenth-century development reincarnates, as it were, the medieval mechanics of the Mertonians and the *Parisienses*. For, apart from the new concern with nature as a force, when one compares what Annaliese Maier wrote about "Ursachen and Kräfte" in the fourteenth century with what has been here presented, there turns out to be little difference between the two accounts. And if these strains of thought connecting causes with forces were being discussed in Jesuit colleges at the turn of the seventeenth century, it may not be surprising that they soon entered into other contexts and there prepared the way for the mechanical philosophy.

The immediate importance of all this, of course, derives from the fact that the *reportationes* of Vitelleschi's lectures that have been described parallel very closely the early treatises of Galileo. Galileo himself gives several indications that the notebooks surviving from his Pisan period are but a portion of more extensive notes he made, or planned to make, on the whole of the Aristotelian *libri naturales.*[39] It could well be, therefore, that what has been sketched is the type of material Galileo studied after leaving the University of Pisa in 1585, and even planned to teach in the late 1580's or early 1590's. This is not to say that the more mature development of the concept of force in Galileo's writings, which have been discussed in detail by Maurice Clavelin[40] and Richard Westfall,[41] or the more explicit development of Johannes Kepler, which Max Jammer treats in his *Concepts of Force,*[42] are the same as the ideas here presented. What these researches suggest, however, is that there are subtle connections between concepts of cause and concepts of force, and that the late sixteenth century was the period during which these sets of concepts, which had been used more or less interchangeably for centuries within the Aristotelian tradition, began to get sorted out and assume the form they now have in scientific discourse.

NOTES

1 Rome: Edizioni di Storia e Letteratura, 1949, pp. 53–78.

2 See, for example, the recently produced *Index Thomisticus*, a computer-made index of all the words in Thomas Aquinas's vast literary output, which reveals that Aquinas used the term *vis* and its inflected forms 2540 times – *Index Thomisticus*, Sancti Thomae Aquinatis Operum Omnium Indices et Concordantiae, ed. Robert Busa, S.J., Stuttgart: Friedrich Frommann Verlag, 1975, Sectio II, Concordantia prima, Vol. 23, #88534/vis, pp. 348–374. Frequently Aquinas associates the term with the notion of violence, with which *vis* is obviously connected etymologically, to designate an external efficient cause that forces an action contrary to the natural inclination of the agent, and

so is opposed to nature. At other times, however, Aquinas uses *vis* to refer to forces that are not contrary to nature but rather are part of nature's operation; these are best seen in living things, where the various forces or powers of the soul, such as the *vis cogitativa* and the *vis aestimativa*, initiate natural activities. It is noteworthy, moreover, that in his commentary on Aristotle's *Physics* Aquinas explicitly rejects an emendation of Aristotle's definition of nature that would make it a *vis insita rebus*. But, as Crisostomo Javelli notes, Aquinas here is only following the commentary of Avicenna, who reproves an unnamed predecessor for embellishing Aristotle's definition with the addition, *natura est virtus diffusa per corpora*, etc. This predecessor could well have been John Philoponus, as cited in note 5 below. See Chrysostomus Javellus, *Totius rationalis, naturalis, divinae ac moralis philosophiae compendium* . . . , 2 vols., Lyons: Apud haeredes J. Junctae, 1568, Vol. I, p. 529.

[3] Strangely enough, Philoponus's commentary on the first four books of the *Physics* was not translated into Latin until these dates. That of 1539 was made by Gulielmus Dorotheus, and that of 1558 by Ioannes Baptista Rosarius. The latter is used in what follows, viz, Aristoteles, *Physicorum libri quatuor*, cum Ioannis Grammatici cognomento Philoponi commentariis, quos . . . restituit Ioannes Baptista Rosarius, Venice: Hieronymus Scotus, 1558.

[4] Natura est principium et causa motus et quietis in eo in quo est primo et per se et non secundum accidens – *Physica*, Lib. 2, cap. 1.

[5] Natura est quaedam vita sive vis quae per corpora diffunditur, eorum formatrix et gubernatrix, principium motus et quietis in eo cui inest per se primo et non secundum accidens – *ed. cit.*, p. 67, col. b.

[6] See *Le Opere di Galileo Galilei*, ed. Antonio Favaro, 20 vols. in 21, Florence: G. Barbèra Editore, 1890–1909, reprinted 1968, Vol. I, p. 416: "Aristoteles, 7 Phys. t. 10, inquit, ad naturalitatem motus requiri causam internam, non externam, motus." Even as late as the *Two New Sciences* (1638), Galileo held that nature is the determining principle within bodies that makes their velocity increase uniformly with the time of fall; see *Le Opere* . . . , Vol. VIII, p. 197.

[7] For details of the similarities between Vitelleschi's and Galileo's notes, see my *Galileo's Early Notebooks: The Physical Questions*. A Translation from the Latin, with Historical and Paleographical Commentary. Notre Dame: Notre Dame University Press, 1977. Muzio Vitelleschi taught at the Collegio Romano from 1588 to 1591, and later served as general of the Jesuit Order; see R. G. Villoslada, *Storia del Collegio Romano dal suo inizio (1551) alla soppressione della Compagnia di Gesù (1773)*, Analecta Gregoriana, Vol. **LXVI**, Rome;Gregorian University Press, 1954. Other parallels between Vitelleschi's notes and Galileo's early treatises are discussed in Essays 10 through 15, *infra.*

[8] Lectiones R. P. Mutii Vitelleschi in octo libros *Physicorum* et quatuor *De caelo*, Romae, Annis 1589 et 1590, in Collegio Romano Societatis Jesu, Staatsbibliothek Bamberg, Cod. 70 (H.J.VI.21), In secundum *Physicorum*, Disputatio secunda, An in definitione naturae contineatur causa aliqua universalis vel accidentalis vel efficiens, fol. 122r. A portion of this *disputatio* has been edited as an Appendix to Essay 13, *infra.*

[9] *Ibid.*, Disputatio prima, De definitione naturae, fols. 109r–112r.

[10] *Ibid.*, Disputatio secunda, An in definitione naturae contineatur causa aliqua universalis vel accidentalis vel efficiens, fols. 112v–113r.

[11] *Ibid.*, Disputatio tertia, An secundum Aristotelem natura sit solum principium passivum, an solum activum, an utrumque, fols. 114v–115v.

[12] *Ibid.*, In quintum *Physicorum*, Disputatio tertia, De motu regulari et irregulari, veloci et tardo, naturali et violento, simplici et mixto, fols. 258r–259r. Implicit in this concession is the possibility that the circular motion of an element in its own sphere could go on forever, an adumbration of the concept of circular inertia.
[13] *Ibid.*, In libros *De caelo*, Tractatio tertia, De elementis, Disputatio quinta, A quo moveantur elementa, fols. 373r–374r.
[14] *Ibid.*, fol. 374v.
[15] *Ibid.*, fols. 374v–376v.
[16] *Ibid.*, Disputatio prima, An sint gravitas et levitas, et quomodo definiuntur, fols. 359r–360r. As the title indicates, Vitelleschi considers here only natural motive forces and has no formal treatment of *impetus*; but see below, note 32.
[17] *Ibid.*, Disputatio secunda, Unde oriantur gravitas et levitas, an potius sint qualitates primae, fols. 361r–363v.
[18] *Ibid.*, Disputatio tertia, De qualitatibus motivis mediorum elementorum, fols. 363v–367r. There is no mention of specific gravity in this discussion, but the notion seems to be implied; Galileo has an explicit treatment in his early treatises *De motu*, for example, *Le Opere* . . . , Vol. I, pp. 262–273.
[19] *Ibid.*, Disputatio quarta, An elementa gravitent et levitent in propriis sphaeris, fol. 369r.
[20] *Ibid.*, fol. 369r. The expressions "first act" and "second act" were used by scholastics to distinguish stages of actuation of operative powers. Thus a person who had been prepared to teach but was not yet actually teaching could be referred to as a teacher "in first act," in the sense that he or she had already actualized the power to teach; when the same person was in the classroom actually teaching, he or she would then be a teacher "in second act," for the power earlier acquired would now be actualized, i.e., it would be actuated beyond its initial acquisition, and so put into "second" actuation. The same terminology could be applied to causes: a cause in first act is one able to produce an effect, whereas a cause in second act is one actually causing. The latter is denominated as such only when the effect is being produced, and for this reason causes in second act are said to be simultaneous with their effects, whereas causes in first act are not.
[21] *Ibid.*, fols. 369r–270r; earlier, on fol. 365r–v, he expresses his uneasiness with such experiments because (1) it is difficult to make the resistance a body encounters and other disturbing factors exceedingly small, and (2) it is hard to discern differences in velocity unless a very large distance is traversed, and over a large distance many things can happen that affect the validity of the experiment.
[22] *Ibid.*, fol. 370r–v.
[23] *Ibid.*, Disputatio quinta, A quo moveantur elementa, fol. 375v.
[24] Dominicus Sotus, *Super octo libros physicorum Aristotelis quaestiones*, Salamanca: Andrea a Portonariis, 1555, In octavum librum, Quaestio tertia, Utrum omne quod movetur moveatur ab alio, fols. 99v–102r.
[25] Diego Mas, a Dominican writing at the end of the century, surveys the state of the discussion and reiterates Soto's solution to the impetus problem; see Didacus Masius, *Commentaria in universam philosophiam Aristotelis, una cum questionibus quae a gravissimis philosophis agitantur*, Valencia: Apud Petrum Patricium, 1599, 2 vols., Vol. II, pp. 1473–1477.
[26] Paulus Valla, Commentaria in libros Meteororum Aristotelis, Tractatus quintus de

elementis, Archivum Pontificiae Universitatis Gregorianae, Fondo Curia, Cod. 1710, no foliation, Quaestio sexta. A quo moveantur proiecta. Ludovicus Rugerius, Quaestiones in quatuor libros Aristotelis *De caelo et mundo*, Romae, in Collegio Societatis Jesu, 1591, Staatsbibliothek Bamberg. Cod. 62 (H.J.VI.10), fols. 200–203: An proiecta moveantur ab aliqua virtute seu qualitate impressa, an vero a medio. For more details, see Essay 15 and the Appendix to Essay 10, *infra.*

27 Rugerius, Quaestiones . . . , fol. 203.

28 Cod. Bamberg. 70, In quintum *Physicorum*, Disputatio tertia, De motu regulari et irregulari . . . , fols. 257r–258r.

29 *Ibid.*, fol. 259r–v.

30 *Ibid.*, In secundum *Physicorum*, Disputatio secunda, An in definitione naturae contineatur causa aliqua . . . , fol. 131r–v.

31 *Ibid.*, In libros *De caelo*, Tractatio tertia, De elementis, Disputatio sexta, An et cur gravia et levia moveantur velocius in fine quam in principio, fols. 380v–383v. This statement seems equivalent to the dynamic formula, $V = P - M$, attributed to Avempace by Ernest Moody in his much-cited article, 'Galileo and Avempace: The Dynamics of the Leaning Tower Experiment,' *Journal of the History of Ideas,* **12** (1951), pp. 163–193 and 375–422. Vitelleschi is aware of Avempace's teaching, and discusses it in his commentary on the fourth book of the *Physics*, Summa tertia, De vacuo, Disputatio tertia, An in vacuo si daretur fieri possit motus, fols. 215r–220r. He rejects Avempace's teaching, however, not because this dynamic formula is incorrect, but rather because he sees the role of the medium as essential to local motion, in the sense that without the medium the formula itself would not apply; see his fols. 217r and 219r–v. For a related critique of Moody's interpretation of Avempace, see Essay 16, *infra.*

32 *Ibid.*, fol. 383v; Vitelleschi's use of *virtus movens* in this context suggests that he is thinking of a *virtus impressa* in the projectile that is the source of its continued motion.

33 *Ibid.*, fols. 384v–385r.

34 This treatment occurs in Vitelleschi's lectures on the *De generatione*, which are preserved in the Archivum Pontificiae Universitatis Gregorianae, Fondo Curia, Cod. 392 (no foliation): Disputationes in libros *De generatione*, Tractatio de actione et passione, Disputatio sexta, An inter res naturales sit mutua actio et passio seu an detur reactio. For a survey of sixteenth-century teachings on action and reaction, see John L. Russell, 'Action and Reaction Before Newton,' *The British Journal for the History of Science*, **9** (1976), pp. 25–38.

35 In libros *De caelo*, Tractatio secunda, De caelo, Disputatio duodecima, An caelum alio modo agat in haec inferiora, Cod. Bamberg. 70 (H.J.VI.21), fol. 353v; Tractatio tertia, De elementis, Disputatio quinta, A quo moveantur elementa, fol. 376r–v.

36 *Ibid.*, Tractatio secunda, Disputatio duodecima, fols. 353v–356v.

37 *Ibid.*, In secundum *Physicorum*, Tractatio secunda, Disputatio secunda, De subiecto in quo est ars et de forma artificiosa et de quibusdam artis operibus, fol. 126r–v; cf. also fol. 355r.

38 For an account of the content and origins of this *Cursus*, see Friedrich Stegmüller, *Filosofia e Teologia nas Universidades de Coimbra e Evora no Seculo XVI*, Coimbra: Universidade de Coimbra, Instituto de Estudos Filosoficos, 1959, pp. 95–99.

39 See, for example, *Le Opere* . . . , Vol. I, pp. 77, 113, 122, 125, 127, 128, 129, 137, 138, and 150.

[40] *The Natural Philosophy of Galileo*, translated by A. J. Pomerans, Cambridge, Mass.: The MIT Press, 1974.

[41] *Force in Newton's Physics*. The science of Dynamics in the Seventeenth Century. New York: American Elsevier, 1971, pp. 1–55.

[42] Cambridge, Mass.: Harvard University Press, 1957.

PART THREE

GALILEO IN THE SIXTEENTH-CENTURY CONTEXT

8. GALILEO AND REASONING *EX SUPPOSITIONE*

Galileo has been seen, from the philosophical point of view, alternately as a Platonist whose rationalist insights enabled him to read the book of nature because it was written in 'the language of mathematics,' and as an experimentalist who used the hypothetico-deductive methods of modern science to establish his new results empirically (McTighe, 1967; Settle, 1967; Drake, 1970; Shapere, 1974). Both of these views present difficulties. In this essay I shall make use of recent historical research to argue that neither is correct, that the method utilized by Galileo was neither Platonist nor hypothetico-deductivist, but was basically Aristotelian and Archimedean in character. This method, moreover, was not merely that of classical antiquity, but it had been emended and rejuvenated in the sixteenth century, and then not by Greek humanist Aristotelians or by Latin Averroists but rather by scholastic authors of the Collegio Romano whose own inspiration derived mainly from Thomas Aquinas. Other influences, of course, were present, and these came from other medieval and Renaissance writers, but these need not concern us in what follows.

I. CURRENT ALTERNATIVES

Before coming to my thesis I must first explain my dissatisfaction with the two alternatives that have occupied the attention of historians and philosophers of science up to now. Of the two, the Platonist thesis is the more readily disposed of. This enjoyed considerable popularity owing to writings of Alexandre Koyré (1939, 1968), who criticized the experimental evidence adduced by Galileo for his more important results and argued that the experiments were either not performed at all, being mere "thought experiments," or, if they were performed, that they did not yield the results claimed for them but merely provided the occasion for Galileo's idealizing their results. Koyré's analyses had great appeal for many philosophers, and they were not seriously contested until Thomas Settle (1961) explained how he had duplicated Galileo's inclined-plane apparatus, actually performed the experiment himself, and shown that the results were not as poor as Koyré had alleged. Since then, Stillman Drake (1973a) has examined anew Galileo's unpublished manuscripts and discovered evidence of hitherto unknown experiments, James MacLachlan (1973) has actually performed one of Galileo's experiments described by Koyré as only imaginary, and Drake himself (1973b), followed by R. H. Naylor (1974) and others (Shea *et al.*, 1975) have variously analyzed and verified the measurements and calculations reported in the re-discovered manuscripts. The results of all this research show that Galileo was far from being a Platonist or a Pythagorean in his practice of scientific method. He was a prolific experimenter and, within the limits of the apparatus and facilities available to him, tried to place his 'two new sciences' on the strongest empirical footing he could find.

The method he used to do so is not easy to discover, despite the mass of materials now available for analysis. The simplest expedient would be to attribute to Galileo the hypothetico-deductive method generally accepted among philosophers of science as typical of modern scientific reasoning. In this view, Galileo would begin with certain hypotheses, such as the principles of inertia and of uniform acceleration in free fall, and from these deduce the type of motion one might expect from heavy bodies projected or falling under given circumstances. The experimental program would then be designed to verify the calculated characteristics, and if these proved in agreement or near agreement, Galileo would be justified in accepting his hypotheses as exactly or very nearly true.

This account, it may be noted, cannot be rejected out of hand as anachronistic, for Galileo was certainly aware of the possibilities of hypothetical reasoning and indeed explicitly made use of a postulate, or hypothesis, in the *Two New Sciences* (Wisan, 1974, pp. 121–122), just as Sir Isaac Newton was to acknowledge an important hypothesis in his *Mathematical Principles of Natural Philosophy*.[1] The problem with the hypothetico-deductive method as it was used in Galileo's time, however, is that this could never lead to true and certain knowledge. It could be productive only of *dialectica*, or opinion, since any attempt to verify the hypothesis – and here we use 'verify' in the strict sense of certifying its truth – must inevitably expose itself to the *fallacia consequentis*, the fallacy of affirming the consequent. Viewed logically, hypothetical reasoning was at the opposite pole from demonstrative or scientific reasoning. This being so, it is extremely difficult to reconcile the use of hypothetico-deductive argument with Galileo's repeated insistence in the *Two New Sciences* that he had actually discovered a 'new science,' one providing demonstrations that apply to natural motions. Galileo, as we know, was prone to speak of *scientia* and *demonstratio*, using these Latin terms or their Italian equivalents with great frequency, and to my knowledge never conferring on them a sense different from that of his peripatetic adversaries.

The difficulty with the hypothetico-deductivist interpretation of Galileo's experiments, then, lies not so much in its modern flavor as in the fact that such a method could never achieve the results claimed by Galileo for the techniques he actually employed. Thus we are left with the prospect of rejecting not only the Platonic interpretation but the hypothetico-deductivist as well, and searching for yet another alternative to describe Galileo's basic methodological stance.

I would like to propose such an alternative based on Galileo's repeated use of the Latin expression *ex suppositione* to describe the line of reasoning whereby he arrived at strict demonstrations on which a *nuova scienza* dealing with local motion could be erected. Now a peculiar thing about the Latin *ex suppositione* is that it translates exactly the Greek *ex hupotheseōs*, but at the end of the sixteenth century it could carry a meaning different from the transliterated *ex hypothesi*, which also enjoyed vogue at that time in Latin writings. Reasoning *ex hypothesi* was hypothetical reasoning in the modern understanding; arguments utilizing such reasoning could not be productive of science in the medieval and Renaissance sense *(scientia)* but were merely dialectical attempts to save the appearances – a typical instance would be the Ptolemaic theories of eccentrics and epicycles. Reasoning *ex suppositione*, on the other hand, while sometimes used to designate dialectical argument, had a more basic understanding in terms of which it could be productive of demonstration in both the natural and the physico-mathematical sciences. My alternative, then, is that Galileo was not basing his *nuova scienza* on an hypothesis, on a mere computational device that would 'save the appearances' of local motion, as a modern-day positivist might interpret it, but was actually making a stronger claim for demonstration *ex suppositione* and thus for achieving a strict science in the classical sense.

II. THE METHODOLOGY OF DEMONSTRATION *Ex Suppositione*

The expression *ex suppositione*, therefore, provides a clue to Galileo's actual methods, and that clue, if pursued historically, leads back to the medieval Latin commentators on Aristotle who first used the term, e.g., Robert Grosseteste and Albert the Great, but more particularly to Albert's disciple, Thomas Aquinas, who gave classical expression to the teaching on demonstration that it entails.[2] Aquinas did this mainly in his commentaries on Aristotle's *Physics* and *Posterior Analytics*, although he also used the expression *ex suppositione* frequently in his other writings.[3] The need for such a methodology arises from the fact that the physical sciences deal with a subject matter that is in the process of continual change, and that always might be otherwise than it is. Now science, for Aquinas as for Aristotle, has to be necessary knowledge through causes. Causes for them are indeed operative in nature, but the necessity of their operation presents a problem, for sometimes they prove defective and do not produce the effect intended. Is it possible, then, to have scientific or necessary knowledge of contingent natural phenomena? Aquinas devoted much thought to this question and finally answered it in the affirmative. Demonstrations in the physical sciences can circumvent the defective operation of efficient causes, he maintained, but they can do this only when they are made *ex suppositione*.

This technique, to schematize Aquinas's account, begins by studying natural processes and noting how they terminate in the majority of cases. For example, in biological generation it can be readily observed that men are normally born with two hands, or that olive plants are usually produced from olive seeds when these are properly nurtured. From this type of generalization, and the examples are Aquinas's, one can never be certain in advance that any particular child will be born with two hands, or that each individual olive seed will produce an olive plant. The reason is that the processes whereby perfect organisms are produced are radically contingent, or, stated otherwise, that natural causes are sometimes impeded from attaining their effects. But if one starts with an effect that is normally attained, he can formulate this as an ideal *suppositio*, and from this reason back to the causes that are able to produce it, whether or not it will ever actually be attained. In other words, one can use his experience with nature to reason *ex suppositione*, i.e., on the supposition of an effect's attainment, to the various antecedent causes that will be required for its production. It is this possibility, and the technique devised to realize it, in Aquinas's view, that permit the physical sciences to be listed among sciences in the strict sense. They can investigate the causes behind natural phenomena, they can know how and why effects have been produced in the past, and they can reason quite apodictically to the requirements for the production of similar effects in the future, even despite the fact that nature and its processes sometimes fail in their *de facto* attainment.

To illustrate this technique in more detail the favored example of medieval commentators is the causal analysis of the lunar eclipse. Such eclipses are not constantly occurring, but when they do occur they are caused by the earth interposing itself between the sun and the moon. So, on the supposition that a lunar eclipse is to occur, the occurrence will require a certain spatial configuration between sun, moon, and the observer on earth. Thus one can have necessary knowledge of such eclipses even though they happen only now and then and are not a strictly necessary or universal phenomena.

A similar contingent occurrence is the production of the rainbow in the atmospheric region of the heavens. The rainbow is more difficult to explain than the lunar eclipse, and this especially for the medieval thinker, since for him the regular movements of the celestial spheres do not guarantee its periodic appearance as they do that of the eclipse. In fact rainbows are only rarely formed in the heavens, and sometimes they are only partially formed; when they are formed, moreover, they come about quite haphazardly – it is usually raining, the sun is shining, and if the observer just happens to glance in a particular direction, lo! he sees a rainbow. These factors notwithstanding, the rainbow can still be the subject of investigation within a science *propter quid*, if one knows how to go about formulating a demonstration in the proper way.

The correct details of this process were worked out by Theodoric of Freiberg, who studied at Paris shortly after Aquinas's death and while there apparently used the latter's lecture notes on the portions of the *Meteorology* that treat of the rainbow (Wallace, 1959, 1974c). Rainbows do not always occur, but they do occur regularly under certain conditions. An observer noting such regularity can rightly expect that it has a cause, and so will be encouraged to discover what that cause might be. If he moves scientifically, according to the then accepted method, he will take as his starting point the more perfect form that nature attains regularly and 'for the most part', and using this as the 'end' or final result, will try to discover the antecedent causes that are required for its realization. The necessity of his reasoning is therefore *ex suppositione*, namely, based on the supposition that a particular result is to be attained by a natural process. *If* rainbows are to occur, they will be formed by rays of light being reflected and refracted in distinctive ways through spherical raindrops successively occupying predetermined positions in the earth's atmosphere with respect to a particular observer. The reasoning, though phrased hypothetically, is nonetheless certain and apodictic; there is no question of probability or verisimilitude in an argument of this type. Such reasoning, of course, does not entail the conclusions that rainbows will always be formed, or that they will necessarily appear as complete arcs across the heavens, or even that a single rainbow need ever again be seen in the future. But if rainbows *are* formed, they will be formed by light rays passing through spherical droplets to the eye of an observer in a predetermined way, and there will be no escaping the causal necessity of the operation by which they are so produced. This process, then, yields scientific knowledge of the rainbow, and indeed it is paradigmatic for the way in which the physical sciences attain truth and certitude in the contingent matters that are the proper subjects of their investigations.

The eclipse and the rainbow are obviously natural phenomena, but their understanding requires a knowledge of geometry in addition to observation of nature, and on this account demonstrations of their properties are sometimes referred to as physico-mathematical so as to distinguish them from those that are merely physical, or natural. A similar type of physico-mathematical demonstration was employed by Archimedes when demonstrating the properties of the balance. We shall have occasion to return to this later when discussing Galileo's use of Archimedes, but for the present it will suffice to note that with the balance, since it is an artifact, the question of the regularity of its occurrence in nature does not arise in the same way as with the eclipse and the rainbow. Thus the suppositional aspect of the regularity of 'weighing phenomena,' to coin a phrase, does not enter into the process of reasoning *ex suppositione* as we have thus far described it. Other suppositions are involved in its case, however, and these have more the character of a mathematical definition, such as the supposition that the cords by which the weights are suspended from the ends of the balance hang parallel to each other. The demonstration, in this case, obtains its validity on the strength of such a mathematical supposition and how it can be reconciled with the physical fact that the cords, if prolonged, must ultimately meet in a common center of gravity. This type of supposition relates to the closeness of fit between a physical case and its mathematical idealization, whereas the former relates to the actual occurrence of phenomena in the order of nature when these have a contingent aspect to them or can be impeded by physical factors. In both types, however, scientific knowledge of properties can be obtained through demonstration *ex suppositione*, and the reasoning is not merely dialectical, or hypothetical, as it would be if reasoning *ex hypothesi* alone were employed.

III. GALILEO'S EARLY NOTEBOOKS

Having made these methodological observations, and promising to return to them later to clarify the formal difference between *ex suppositione* and *ex hypothesi* reasoning, let me now come to the man who is commonly regarded as 'the father of modern science,' Galileo Galilei. My interest in Galileo began some years ago when studying a sixteenth-century figure, Domingo de Soto, who had been singled out by Pierre Duhem (1913, pp. 263–583) as the last of the scholastic precursors of Galileo. Duhem did this on the basis that Soto had anticipated the law of falling bodies in a work published in 1545, some ninety years before Galileo proposed the law in his *Two New Sciences*. I therefore set about the task of studying Soto's mechanics to find out where and how he formulated the so-called "law," and to trace possible lines of communication to Galileo (Wallace, 1968, 1969, 1971). This search led to a mass of early unpublished writings by Galileo, actually hundreds of folios composed in his own hand. Goodly portions of these materials have been edited, some under the title of *Juvenilia*, or youthful writings, and others under a general heading, *De motu*. (The latter notes are usually referred to as the *De motu antiquiora* to distinguish them from the treatise on motion that is discussed at length in Galileo's last work, the *Two New Sciences*; they have been analyzed recently by Fredette, 1972.) These early writings of Galileo are scholastic in style, they are very much concerned with the works of Aristotle, especially the *Posterior Analytics*, the *Physics*, the *De caelo*, and the *De generatione*, and they cite authors extensively – about 150 authors alone being mentioned in the *Juvenilia*. Among these we find the name of Domingo de Soto, and indeed of several authors whom Soto influenced, including a Jesuit who was teaching at the Collegio Romano in Galileo's youth, Benedictus Pererius. Also mentioned is one of Pererius's Jesuit colleagues at the Collegio, Christopher Clavius, and his erudite commentary on the *Sphere* of Sacrobosco. Not mentioned, but apparently studied by Galileo anyway (one of his textbooks survives in Galileo's personal library), was yet another Jesuit who taught at the Collegio at that time, Franciscus Toletus, who had been Soto's favored disciple at Salamanca before going to Rome.

My more recent studies have been concerned with tracking down the sources cited in these early writings, to ascertain the extent to which such sources were actually used by Galileo, and, if possible, to determine the dates of composition of the various tracts that make up the notes. The work is tedious, but thus far it has led to some interesting results. For example, Galileo refers eleven times to a "Caietanus," whose opinions he discusses at considerable length. In the portions of the notebooks that have been edited by Favaro this author is identified as the Paduan *calculator*, Caietanus Thienensis, or Gaetano da Thiene (*Opere* 1, p. 422). This identification turns out to be inaccurate; the author to whom Galileo was referring in most of these citations is the celebrated Thomist and commentator on Aquinas, Tomasso de Vio Caietanus, generally known simply as Cajetan. And not only are Soto and Cajetan mentioned in the notes, but other *Thomistae*, as Galileo calls them, are there as well: Hervaeus Natalis, Capreolus, Soncinas, Nardo, Javelli, and Ferrariensis. In fact, after Aristotle, whose name is cited more than 200 times, Galileo's next favored group is St. Thomas and the Thomists, with a total of 90 citations; then comes Averroës with 65, Simplicius with 31, Philoponus with 29, Plato with only 23, and so on down to Scotus with 11 and Ockham and assorted nominalists with six or fewer citations apiece (Wallace, 1974b; Crombie, 1975).

This, of course, is a most interesting discovery, for if Galileo truly composed these notes, and if he understood the material contained in them, his intellectual formation would be located squarely in an Aristotelian context with decided Thomistic overtones. But previous scholars who have looked over this material, more often than not in cursory fashion, have been unprepared to accept any such result. Favaro, for example, while admitting that the notes are clearly written in Galileo's own hand, refused to accept Galileo's authorship, maintaining that these were all trite scholastic exercises, copied from another source, probably a professor's notes transcribed by Galileo in 1584 while still a student at the University of Pisa (*Opere* 1, pp. 9–13; 9, pp. 273–282). Thus they are his "youthful writings," or *Juvenilia*, not his own work, material for which he had no real interest and indeed failed to comprehend, and so could have exerted no influence on his subsequent writings.

At the outset of my researches I was prepared to go along with this view, but now I suspect that it is quite mistaken. One piece of evidence that counts heavily against it was the discovery in June of 1971 by Crombie (1975, p. 164) that some ten of the hundred folios that make up the so-called *Juvenilia* were actually copied by Galileo, very skilfully, from Clavius's commentary on the *Sphere*. A detailed comparison of the various editions of this commentary with Galileo's manuscripts indicates strongly that these notes, and other compositions with the same stylistic features and dating from the same period, were composed and organized by Galileo himself. They were not copied from a professor's notes; in fact, in all probability they were not even done while Galileo was a student at Pisa, but while he was a young professor there between 1589 and 1591. It is known that in 1591 Galileo taught at Pisa the "hypotheses [*hipotheses*] of the celestial motions" (Schmitt, 1972, p. 262), and then at Padua a course entitled the *Sphere* in 1593, 1599, and 1603. His lecture notes for the Padua course have survived in five Italian versions, all similar but none of them autographs, and showing that the course was little more than a popular summary of the main points in Clavius's commentary on Sacrobosco. In one of the versions of these notes, moreover, there is reproduced a "Table of Climes According to the Moderns," which is taken verbatim from Clavius's commentary. Indeed, when these Paduan lecture notes are compared with the summary of Clavius contained in the so-called *Juvenilia* written at Pisa, the latter are found to be far more sophisticated and rich in technical detail. It is probably the case, therefore, that the notes labelled *Juvenilia* by Favaro represent Galileo's first attempt at class preparation, and that the course based on them subsequently degenerated with repeated teaching–a phenomenon not unprecedented in the lives of university lecturers.

This dependence on Clavius is not without further interest for possible influences on Galileo's methodology. Clavius was firmly convinced, and repeatedly makes the point in his commentary on the *Sphere*, that astronomy is a true sciénce in the Aristotelian sense, that it is not concerned mainly with "saving the appearances" but rather with determining the motions that actually take place in the heavens, and that it does this by reasoning from effects to their proper causes (Blake *et al.*, 1960, pp. 32–35; Duhem, 1969, pp. 92–96; Harré, 1972, pp. 84–86; Crombie, 1975, p. 166). There is little doubt that the young Galileo heartily subscribed to this methodological conviction of the famous astronomer of the Collegio Romano, which was consistent with the Aristotelian-Thomistic teachings of his fellow Jesuit philosophers at the Collegio. In fact, there is good reason to believe that this strong realist mind-set on the part of the young Galileo was what encouraged him to apply the canons of demonstrative proof to his discoveries and to claim that he had reached true *scientia* in both his middle period, when his over-riding concern was to demonstrate apodictically the truth of the Copernican system, and in his final period, when he made the claims we have seen for the *Two New Sciences*.

IV. GALILEO'S MIDDLE AND LATE PERIODS

With regard to Galileo's middle period, from 1610 to 1632, we must be brief. The period has already been studied in detail by William R. Shea (1972), who has shown abundantly the extent of Galileo's commitment to science as strict demonstration. Even though his revered colleague, Jacopo Mazzoni, had given an instrumentalist interpretation of eccentrics and epicycles, arguing that astronomy was not a strict science in the Aristotelian sense but merely a system of calculation for "saving the appearances," Galileo was firmly convinced of the opposite (Shea, 1972, p. 68; Purnell, 1972). In his letter on sunspots, significantly entitled *History and Demonstrations Concerning Sunspots and Their Phenomena*, Galileo admitted that the Ptolemaic eccentrics, deferents, equants, and epicycles are "assumed by pure astronomers [*posti da i puri astronomi*] in order to facilitate their calculations." But he went on,

> They are not retained as such by philosophical astronomers [*astronomi philosophi*] who, going beyond the requirement that appearances be saved, seek to investigate the true constitution of the universe – the most important and admirable problem that there is.

For such a constitution exists; and it is unique, true, real, and cannot be otherwise, and should on account of its greatness and dignity be considered foremost among the questions of speculative interest. (*Opere* 5, p. 102)

As this text shows, Galileo had no doubts that the structure of the universe is real and knowable, and that knowledge of it is a legitimate goal of scientific endeavor. This is not to claim, of course, that he was successful in attaining such knowledge. My point is essentially methodological: Galileo was in no sense a logical positivist or an instrumentalist; he was a realist, more Aristotelian than the peripatetics of his day, whom he regarded, to use Shea's phrase, as advocating "nothing more than a thinly disguised nominalism" in their own explanations of nature (1972, p. 72).

At the beginning of his middle period Galileo had written in 1610 to Belisario Vinta outlining what his plans would be should he leave Padua and get the appointment as chief philosopher and mathematician to the Grand Duke of Tuscany:

The works which I must bring to conclusion are these. Two books on the system and constitution of the universe – an immense conception full of philosophy, astronomy, and geometry. Three books on local motion – an entirely new science in which no one else, ancient or modern, has discovered any of the most remarkable properties [*sintomi*] that I demonstrate [*che io dimostro*] to exist in both natural and violent movement; hence I may call this a new science and one discovered by me from its first principles. Three books on mechanics, two relating to demonstrations of its principles and foundations and one concerning its problems... (*Opere* 10, pp. 351–352)

After the disastrous failure of the first of these projects, which culminated in his trial and condemnation in 1633, Galileo turned in his final period to the completion of the second project here mentioned, which he brought out in 1638 under the title of *Two New Sciences*. What is most remarkable is that, despite the rebuffs he had received and the rejection of the demonstrations he had offered in the *Two Chief World Systems*, in his final work he still firmly held to the ideal of *scientia* and strict demonstrative proof. Rather than abandon the ideal he was more intent than ever on preserving it, only now he would be more careful than previously to assure that his demonstrations would gain universal acceptance.

In what follows we shall focus attention on only one aspect of Galileo's final attempt to justify his claims, that, namely, of utilizing the technique of demonstration *ex suppositione*. With regard to the expression *ex suppositione* itself, it is noteworthy that Galileo recognized early in his middle period that it could carry two senses, one that is merely hypothetical and equivalent to an argument *ex hypothesi* that merely "saves the appearances," and the other standing for a supposition that is true and actually verified in the order of nature. He made the distinction, in fact, in reply to Cardinal Bellarmine's letter of April 12, 1615, addressed to the Carmelite Foscarini (*Opere* 12, pp. 171–172), in which Bellarmine had commended Foscarini and Galileo for being prudent "in contenting yourselves to speak *ex suppositione* and not absolutely" when presenting the Copernican system, and thus entertaining this as only a mathematical hypothesis, as he believed Copernicus himself had done. In his *Considerazioni circa l'opinione Copernicana* (*Opere* 5, pp. 349–370), written shortly thereafter, Galileo disavowed that this was either his own or Copernicus's intent, although one might gain such an impression on reading the preface to the *De revolutionibus*, which he noted was unsigned and clearly not the work of Copernicus himself (*ibid.*, p. 360). Galileo did not disclaim the *ex suppositione* character of his own arguments, however, but rather distinguished two different meanings of supposition:

> Two kinds of suppositions have been made here by astronomers: some are primary and with regard to the absolute truth in nature; others are secondary, and these are posited imaginatively to render account of the appearances in the movements of the stars, which appearances they show are somehow not in agreement with the primary and true suppositions. (*ibid.*, p. 357)

He went on to characterize the first kind as "natural suppositions" that are "established" and "primary and necessary in nature," (*ibid.*, p. 357) and the second kind as "chimerical and fictive,... false in nature, and introduced only for the sake of astronomical computation." (*ibid.*, pp. 358–359). The whole point of the *Considerazioni*, of course, was to advise Bellarmine that Copernicus's (and Galileo's) *suppositiones* are of the first kind and not of the second.

Coming now to the reasoning advanced in the *Two New Sciences*, we find that the Latin expression *ex suppositione* occurs in at least four crucial places where Galileo is explaining his thought, twice in the text itself, and twice in letters wherein he is elaborating, in fuller detail, the methodology behind the discoveries he records in that work. Of the two uses in the text of the *Two New Sciences*, the first occurs in the Latin treatise being explained and discussed on the Third Day, and leads to the definition of naturally accelerated motion and to the demonstration of the property that the distances traversed in free fall will be as the squares of the times of fall. The second use occurs in the Italian dialogue on the Fourth Day, following Galileo's enunciation of the theorem that the path followed by a heavy object which has been projected horizontally will be compounded of a uniform horizontal motion and a natural falling motion, and will therefore be a semiparabola. In the latter context, with which it is more convenient to begin our analysis, Sagredo states:

It cannot be denied that the reasoning is novel, ingenious, and conclusive, being argued *ex suppositione*; that is, by assuming [*supponendo*] that the transverse motion is kept always equable, and that the natural downward motion likewise maintains its tenor of always accelerating according to the squared ratio of the times; also that such motions, or their velocities, in mixing together do not alter, disturb, or impede one another... (*Opere* 8, p. 273)

Having conceded this, Sagredo then goes on to raise various objections to this demonstration based on the actual physical geometry of the universe, and concludes with the telling observation:

All these difficulties make it highly improbable that the results demonstrated [*le cosi dimostrate*] from such an unreliable supposition [*con tali supposizione inconstanti*] can ever be verified in actual experiments. (*ibid.*, p. 274)

At this point Salviati comes quickly to the rescue. Rising in Galileo's defense and speaking in his name, he admits "that the conclusions demonstrated in the abstract are altered in the concrete," and in that sense can be falsified, but that such an objection can even be raised against Archimedes' demonstration of the law of the lever, for this is based on the supposition "that the arm or a balance... lies in a straight line equidistant at all points from the common center of heavy things, and that the cords to which the weights are attached hang parallel to one another." (*ibid.*). One should recall, however, Salviati goes on, that Archimedes based his demonstrations on the supposition that the balance could be regarded as "at infinite distance" from the center of the earth (*ibid.*, p. 275); granted this supposition his results are not falsified but rather drawn with absolute proof [*con assoluta dimostrazione*]. When great distances are involved, moreover, abstraction can even be made from the small errors introduced by this simplifying supposition and the results are still found to apply in practice. Similarly, he continues, when treating of the dynamic cases taken up in Galileo's new science, as opposed to the old Archimedean statics,

it is not possible to have a firm science [*ferma scienza*] that deals with such properties as heaviness, velocity, and shape, which are variable in infinitely many ways. Hence to deal with such matters scientifically [*scientificamente*] it is necessary to abstract from these. We must find and demonstrate conclusions abstracted from the impediments [*impedimenti*], in order to make use of them in practice under those limitations that experience [*esperienza*] will teach us. (*ibid.*, p. 276)

From these texts of the *Two New Sciences* it can be seen that Galileo's "new science" of local motion was Archimedean in inspiration, but that it aimed to satisfy essentially the same classical requirements for demonstrative rigor and for application to the world of dynamic experience. On the latter point Galileo was well aware that he had to abstract from many more "impediments" than Archimedes had to, particularly the resistance of the medium traversed by falling and projected bodies, but he felt that he had sufficient experimental evidence to be able to do so. What that evidence was has long eluded historians of science, but Stillman Drake's recent re-discovery of folio 116v in BNF MS Galileiana 72 now supplies the missing link. The cases treated *ex suppositione* by Galileo in the Third and Fourth Days of the *Two New Sciences* were investigated by him in experiments he never reported, and found to be very nearly in agreement with what actually occurs in nature. This is why he could write, in the Latin treatise on naturally accelerated motion, at his first mention of the demonstration *ex suppositione* based on the definition of such motion:

Since nature does employ a certain kind of acceleration for descending heavy things, we decided to look into their properties [*passiones*] so that we might be sure that the definition of accelerated motion which we are about to adduce agrees with the essence [*essentia*] of naturally accelerated motions. And at length, after continual agitation of mind, we are confident that this has been found, chiefly for the very powerful reason that the properties [*symptomatis*] successively demonstrated by us [*a nobis demonstratis*] correspond to, and are seen to be in agreement with, that which physical experiments [*naturalia experimenta*] show forth to the senses. (*Opere* 8, p. 197)

Further clarification of the method Galileo used in this discovery is given by him in two letters, one written to Pierre Calcavy (or, de Carcavi) in Paris on June 5, 1637, while the *Two New Sciences* was still in press, and the other to Giovanni Battista Baliani in Genoa on January 7, 1639, after its publication. The letter to Calcavy is of particular interest because it is an answer to a query from Pierre Fermat, forwarded by Calcavy to Galileo, concerning a passage in the *Two Chief World Systems* wherein Salviati mentions the treatise on motion that was later to appear in the

Two New Sciences, but which he has already seen in manuscript form. Salviati there had explained Galileo's initial reasoning concerning the path that would be described by a heavy body falling from a tower if the earth were rotating in a direction away from the body's path of fall, and had described that path as compounded of two motions, one straight and the other circular, on the analogy of Archimedes' treatment of spiral motion. The path recounted at that time by Salviati, however, was not a semiparabola but a semicircle, and the composition of motions had obviously been made incorrectly, as Fermat was quick to notice. In his reply to Calcavy Galileo retracted the error – tried, in fact, to cover it up as a mere jest and not as a serious account (Shea, 1972, p. 135) – and then went on to explain how he had derived his new parabolic curve, again noting the analogy with Archimedes' method. In his works *On Weights* and *On the Quadrature of the Parabola*, writes Galileo, Archimedes is supposing [*supponendo*], as do all engineers and architects, "that heavy bodies descend along parallel lines," thereby leading us to wonder if he was unaware "that such lines are not equidistant from each other but come together at the common center of gravity." (*Opere* 17, p. 90). Galileo goes on:

From such an obviously false supposition [*falsa supposizione*], if I am not in error, the objections made against me by your friend [Fermat] take their origin, viz, that in getting closer to the center of the earth heavy bodies acquire such force and energy, and vary so much from what we suppose to take place on the surface, admittedly with some slight error, that what we call a horizontal plane finally becomes perpendicular at the center, and lines that in no way depart from the perpendicular degenerate into lines that depart from it completely. I add further, as you and your friend can soon see from my book which is already in the press, that I argue *ex suppositione*, imagining for myself a motion towards a point that departs from rest and goes on accelerating, increasing its velocity with the same ratio as the time increases, and from such a motion I demonstrate conclusively [*io dimostro concludentemente*] many properties [*accidenti*]. I add further that if experience should show that such properties were found to be verified in the motion of heavy bodies descending naturally, we could without error affirm that this is the same motion I defined and supposed; and even if not, my demonstrations, founded on my supposition, lose nothing of their force and conclusiveness; just as nothing prejudices the conclusions demonstrated by Archimedes concerning the spiral that no moving body is found in nature that moves spirally in this way. But in the case of the motion supposed by me [*figurato da me*] it has happened [*e accaduto*] that all the properties [*tutte le passioni*] that I demonstrate are verified in the motion of heavy bodies falling naturally. They are verified, I say, in this way, that howsoever we perform experiments on earth, and at a height and distance that is practical for us, we do not encounter a single observable difference; even though such an observable difference would be great and immense if we could get closer and come much nearer the center. (*ibid.*, pp. 90–91)

Galileo then describes an experiment by which he is able to verify, by sense observation and not by reasoning alone, the conclusion he has just stated.[4]

Turning now to Galileo's letter to Baliani after the appearance of the *Two New Sciences*, we find him repeating there in summary form what he had already written to Calcavy, and in so doing describing his steps in more accurate detail. Galileo states:

I assume nothing but the definition of the motion of which I wish to treat and whose properties I demonstrate, imitating in this Archimedes in the *Spiral Lines*, where he, having stated what he means by motion in a spiral, that it is composed of two uniform motions, one straight and the other circular, passes immediately to demonstrating its properties. I state that I wish to examine the characteristics associated with the motion of a body that, leaving from the state of rest, goes with a velocity that increases always in the same manner, i.e., the increments of that velocity do not increase by jumps, but uniformly with the increase of time. (*Opere* 18, pp. 11–12)

Proceeding on this basis Galileo notes that he comes "to the first demonstration in which I prove the spaces passed over by such a body to be in the squared ratio of the times..." (*ibid.*, p. 12) After a brief digression, he then resumes his main theme:

But, returning to my treatise on motion, I argue *ex suppositione* about motion defined in that manner, and hence even though the consequences might not correspond to the properties of the natural motion of falling heavy bodies, it would little matter to me, just as the inability to find in nature any body that moves along a spiral line would take nothing away from Archimedes' demonstration. But in this, I may say, I have been lucky [*io stato... avventurato*]; for the motion of heavy bodies, and the properties thereof, correspond point by point [*puntualmente*] to the properties demonstrated by me of the motion as I defined it. (*ibid.*, pp. 12–13)

Note here Galileo's explicit affirmation of the methodology of *ex suppositione* argumentation, and his further admission that he had actually discovered, though by a stroke of luck (it was *avventurato* – recall that in the letter to Calcavy it was by accident, *e accaduto*), that both his definition of motion and the properties resulting therefrom correspond point by point to what actually occurs in nature.

V. *Ex Hypothesi* VS. *Ex Suppositione* ARGUMENTATION

Let us now return to the problem of hypothetico-deductive methodology that was presented at the outset, to clarify how this differs from demonstration *ex suppositione*, and how the latter could achieve the results claimed by Galileo whereas the former could not. Both types of reasoning, it should be obvious, can be expressed in conditional form. Modern hypothetico-deductive reasoning takes the form "if p then q," where p formulates an hypothesis that does not pertain to the order of appearances, whereas q states a consequent that pertains to this order and so is empirically verifiable. The sixteenth-century parallel would be reasoning *ex hypothesi*, "if there are eccentrics and epicycles, then the observed planetary motions result." Here, as in the modern theory, the antecedent cannot be verified directly; one must work through the consequent, either by showing that it is not verified in experience and that the antecedent is therefore false, or that it is so verified, in which case the antecedent enjoys some degree of probability or verisimilitude. The latter alternative gives rise to the problems of contemporary confirmation theory, whereas the former is the basis for Karl Popper's insistence on techniques of falsification. Neither alternative, as is universally admitted, is productive of positive scientific knowledge that could not be otherwise, and so neither can produce *scientia* in the classical sense.

Demonstration *ex suppositione* employs conditional reasoning of a different type. It too can be expressed in the form, "if p then q," but here p stands for a result that is attained in nature regularly or for the most part, whereas q states an antecedent cause or condition necessary to produce that result. Unlike *ex hypothesi* reasoning and its modern hypothetico-deductive equivalent, p usually pertains to the order of appearances, for this is what can be observed to take place in nature regularly or for the most part. Again, with regard to p's content, no claim is made for the absolute necessity or universality of such an observational regularity, since there are always impediments in nature that can prevent the realization of any ideal result. The logical consequent, q, on the other hand, standing as it does for antecedent causes or conditions that produce the appearances, need not itself pertain to the order of appearances, at least not initially, although it may subsequently be found to do so, as in Theodoric of Freiberg's explanation of the rainbow and, as we now know, in Galileo's supposition of uniform acceleration. Unlike purely hypothetical reasoning, finally, this mode of argumentation can lead to certain knowledge and to *scientia* in the strict sense. The form of argumentation, "if p, then, if p then q, then q," will be recognized as one of the valid forms of the *modus ponendo ponens* of the conditional syllogism.

It now remains to show that Galileo's reasoning conforms to the latter pattern, and so could justify his claims for strict demonstration that is productive of a "new science" of local motion. The analysis of the rediscovered folio 116v shows that in his unreported experiments with free fall, as opposed to the inclined-plane experiment described in the *Two New Sciences*, Galileo was able to verify in a surprising way various properties of falling motion compounded of a rectilinear inertial component and an accelerated downward component in accordance with the times-squared ratios. The experiments consisted in dropping a ball from different heights to a deflector located at the edge of a table, at which point the ball was given different horizontal velocities depending on the distance of its fall. Apparently Galileo had computed the horizontal distances the ball should travel depending on the velocity imparted to it, and then had actually measured points of impact to verify his calculations. The accuracy of his results is truly remarkable considering the crude apparatus Galileo had to work with, but he was not able to verify them consistently, and particularly could not reconcile the exceptionally good results of the free-fall experiments with those made on the inclined plane. Galileo rightly discerned that the cause of the discrepancy arose from friction and air resistance, the first of which was particularly serious with the inclined plane, and so he took the various properties he had calculated as something that should be verified in the ideal case, but, as the medieval would have it, need be found true only generally and for the most part. On such a supposition it was a simple matter for him to demonstrate mathematically that the only kind of naturally accelerated motion that could produce the result he had observed, more or less, would be one whose velocity increases uniformly with time. Impediments and defects could, of course, prevent the ideal result from being attained, as Galileo realized, but this is true generally in the physical world, and it is in fact the reason why demonstration *ex suppositione* has to be employed when studying natural processes in the first place.

Still there is a difference in the nuances of demonstration *ex suppositione* as this is employed in an Aristotelian natural science such as biology and in the Archimedean type of science such as those that treat of the balance and falling bodies, and this must now be pointed out. In the former case the force of the demonstration is usually carried through efficient causality, and impediments are seen to arise through imperfections in the matter involved or in the deficiencies of agent causes. In the latter case the force of the demonstration is usually carried through formal causality, understanding this in the sense of mathematical form, where the relationships involved are those between quantifiable aspects of the subject under consideration. When these quantitative relationships are realized in the physical world, however, as opposed to the world of pure mathematical forms, they too can be found defective, or be "impeded," either by material conditions or by defects of agent causes. It is noteworthy that Galileo concentrates only on material defects arising from friction and air resistance and that he is not particularly concerned with the deficiency of agent causes in his analysis. The reason for this is that, along with the Aristotelians of his day, he regards nature as the basic internal cause of falling motion, which is why he always refers to the phenomenon of free fall as "naturally accelerated motion." So he acknowledges:

> ...we have been led by the hand to the investigation of naturally accelerated motion by consideration of the custom and procedure of nature herself in all her other works, in the performance of which she habitually employs the first, simplest, and easiest means. (*Opere* 8, p. 197)[5]

With the internal cause thus taken care of, Galileo's main burden of proof can become that of showing that this "simplest means" is to have velocity increase uniformly with the time of fall, and not with the distance of fall, and this he is able to demonstrate mathematically once he is assured experimentally that the distances traversed are really as he had calculated them to be. But he must still be able to account for the fact that even these results will probably never be realized perfectly in the concrete, and so, to take care of the "impediments," as he calls them, he must resort to the technique of demonstration *ex suppositione*.

To make more explicit the methodology here attributed to Galileo, let us note that it combines elements of Archimedean *ex suppositione* reasoning and of Aristotelian *ex suppositione* reasoning in the following way. At the outset, before the experimental confirmation was available to him, Galileo's demonstration could be expressed in the following logical form:

> If *p* (definition of motion laterally uniform and downwardly accelerated with time), then, if *p* then *q* (by mathematical reasoning), then *q* (properties of semiparabolic path, e.g., distance of fall, of horizontal travel, etc.).

Note that this is a scientific demonstration, not a merely hypothetical argument, even though it is expressed in conditional form; however, it pertains to the "old science," the Archimedian type of mathematics that is ideal even though it is applicable, in some way, to the physical universe. Now Galileo thought that he had advanced beyond this "old science" to a "new science," for he had been lucky enough to obtain experimental confirmation of the properties he had calculated, but not sufficiently complete confirmation to remove all possibility of error. The error, however, he had by this time come to see could be attributed to the *impedimenti*, understanding these not merely in the Aristotelian sense of physical defects that prevent perfect regularities from being observed in nature, but also in the Archimedian sense of the physical characteristics of the universe (such as its spherical geometry) that prevent simplified mathematical ideals from being applied there perfectly.[6] So, modifying the Aristotelian type of argument *ex suppositione*, Galileo went on to the second stage of his new type of demonstration, which may be expressed logically in the same form simply by interchanging the *p*'s and the *q*'s:

> If *q* (more or less, physico-mathematically), then, if *q* then *p* (by mathematical reasoning), then *p* (physically verified).[7]

Let us note parenthetically that the scholastic Aristotelian of Galileo's day did not customarily argue in the physico-mathematical mode suggested by the above formulation, but rather employed the following type of argument:

> If *q* (regularly and for the most part, physically), then, if *q* then *p* (by reasoning "philosophically" to a physical cause or necessary condition), then *p* (physically required, but able to be impeded).

Here the "regularly and for the most part" can be said to be approximate in the qualitative sense, but not in the quantitative sense suggested by the words "more or less" in the reconstruction of Galileo's argument. A primary aspect of Galileo's contribution, it would appear, is that he pointed out how one could legitimately make the transit from the qualitative "regularly" understood in a physical way to the quantitative "more or less" understood in a physico-mathematical way. If this be admitted, then other aspects of Galileo's contribution follow. For example, his use of limit concepts (which we have not been able to go into here, but which are discussed by Koertge, 1977) plus his use of precise experimentation and measurement are what made the above transition scientifically acceptable, if not to the conservative Aristotelians of his day at least to those who were willing to follow in his path. Again, apart from the empirical aspect, Galileo showed how mathematical functionality could serve as a valid surrogate for physical causality in manifesting the necessary connection between antecedent and consequent in a physical situation. This explains why he could proclaim, in the celebrated passage in the *Two New Sciences* (*Opere* 8, p. 202), his indifference to the precise physical cause of the acceleration observed in falling motion (Finnochiaro, 1972; Drake, 1974, pp. xxvii–xxix, 158–159; Wallace, 1974d, pp. 229–230, 239–240). This also clarifies the sense in which he was an Archimedean and could rightfully proclaim the power of mathematics as he had employed it in his *nuova scienza*. Yet again, and this may be the most ingenious aspect of his contribution, Galileo was able to show how, through the use of his physico-mathematical techniques, some "unobservables," i.e., the actual mode of velocity increase (e.g., whether uniformly with distance of fall or uniformly with time of fall, *p*), could actually be certified empirically, through the use of other mathematical relationships (e.g., the distances of horizontal and vertical travel, *q*) that were "observables" in the sense that they could be verified approximately in the experiments he had contrived. When all of these aspects of the *Two New Sciences* are taken into account, we see why Galileo can still rightfully be hailed as "the father of modern science." Even if he was not a complete innovator (and who is?), he knew at least in a general way the strengths and limitations of the Aristotelian and Archimedean traditions that had preceded him, and he had the genius to wrest from those traditions the combination of ideas that was to prove seminal for the founding of a new era.

To conclude, then, by resuming the theme stated at the outset of this study, the logic of *ex suppositione* reasoning was already at hand for Galileo, it was part of the intellectual tradition in which he had been formed, and it was capable of producing the scientific results he claimed to have achieved. Since the same cannot be said for hypothetico-deductive method in the modern mode, there is no reason to impose that methodology on "the father of modern science." Rather we should take

Galileo at his word and see him neither as the Platonist nor as the hypothetico-deductivist he has so frequently been labelled, but as one who made his justly famous contribution in the Aristotelian-Archimedean context of demonstration *ex suppositione.*

APPENDIX

Since the original publication of this essay, a number of references have been made to it in the literature (Machamer, 1978, pp. 161–180; McMullin, 1978, pp. 234–237; Wisan, 1978, pp. 47, 53–54). Since some of the reactions suggest that Galileo's reasoning does not have the cogency attributed to it by the author, it may prove helpful to restate the demonstration in more traditional Aristotelian terms, and then relate it directly to Galileo's statement in the Latin treatise being read at the beginning of the Third Day of the *Two New Sciences*. This treatise, entitled "On Naturally Accelerated Motion" (*De motu naturaliter accelerato*), opens with the following two paragraphs, not cited in their entirety in the original essay:

> Those things that happen which relate to equable motion have been considered in the preceding book; next, accelerated motion is to be treated of.
>
> At first, it is appropriate to seek out and explain the definition that best agrees with that [accelerated motion] which nature employs. For anyone may choose to make up any kind of motion and consider the properties [*passiones*] that follow from it; so, for example, some have constructed for themselves spiral and conchoidal lines arising from certain motions that nature does not employ and have commendably demonstrated *ex suppositione* the properties [*symptomata*] these curves possess. But since nature does employ a certain kind of acceleration for descending heavy things, we decided to look into their properties [*passiones*] so that we might be sure that the definition of accelerated motion we are about to adduce agrees with the essence [*essentia*] of naturally accelerated motions. And at length, after continual agitation of mind, we are confident that this has been found, chiefly for the very powerful reason that the properties [*symptomatis*] successively demonstrated by us [*a nobis demonstratis*] correspond to, and are seen to be in agreement with, the evidences physical experiments [*naturalia experimenta*] present to the senses. Further, it is as though we have been led by the hand to the investigation of naturally accelerated motion by consideration of the custom and procedure of nature herself in all her other works, in the performance of which she habitually employs the first, simplest, and easiest means. And indeed, no one of judgment believes that swimming or flying can be accomplished in a simpler or easier way than that which fish and birds employ by natural instinct. [*Opere* 8: 197; cf. Drake, 1974, p. 153]

In the second paragraph cited above, the words "spiral and conchoidal lines" are an implicit reference to Archimedes, and the demonstration *ex suppositione* that Galileo has in mind typifies what medieval and Renaissance Aristotelians would regard as that of an intermediate or mixed science (*scientia media seu mixta*). Since such a science is *epistēmē* in the unconditional sense, its reasoning can be expressed in the form of a demonstrative syllogism. This employs two premises, one of which is a *thesis* and the other (placed under it) a *hupothesis*; in the strict sense neither is arbitrary or conjectural. Both are either evident or demonstrable; usually, however, the one considering the *hupothesis* need not have seen its actual demonstration – it suffices that he be convinced of its likehood. In a mixed science this will usually be a definition, and rarely does one need to "prove" a definition. (Whether one can indeed "prove" a definition is debatable and is the subject of practically the entire second book of Aristotle's *Posterior Analytics*.) Moreover, when the *Posterior Analytics* was translated into Latin *thesis* was commonly rendered as *positio* and *hupothesis* correspondingly became *suppositio*.

Since Galileo, on this accounting, was working within an Archimedian context, his demonstration *ex suppositione* as first presented is strictly scientific in the same way as Archimedes' *On Spiral Lines* is scientific, i.e., in the sense of a mixed science. To apply a mixed science to the world of nature, however, implies a further step, and here is where difficulties come in. Perhaps such an application also leads to an ambiguity in the expression *ex suppositione*, for clearly a mathematical definition might be acceptable in itself and yet not be acceptable when said to apply to a natural phenomenon. And in the case of natural phenomena there is again a twofold problem: (1) whether there ever can be an exact fit between mathematics and nature, and (2) whether, because of the contingency of nature's operation and the impediments encountered there, one can ever have a strict demonstration or even a definition that is universally valid.

Galileo was certainly aware of the problem of the fit between mathematics and nature, and also of the problem of impediments that are encountered in nature's operations. Moreover, in some of his uses of *ex suppositione*, as noted in the essay above, Galileo argues that one can demonstrate results that are true in nature by making suppositions that are approximately verified in nature, provided these are within the degree of approximation of the desired result. This is what authenticates the application of the law of the lever to the operation of the balance, as explained in the essay. In this case, on closer examination, we find that a twofold *suppositio* is involved: (a) that of the definition of the lever or balance; and (b) that of the conditions under which such a definition will be approximately verified in nature.

Let us return now to the original Aristotelian sense of *suppositio* as that which is evident or demonstrable (note that we say "demonstr*able*" and not "demonstrat*ed*," as only the truth of the *suppositio* is here involved – not whether its truth is actually seen at the moment.) If a premise or a *suppositio* of this type can be demonstrated, it will be done in two ways: either *a priori*, from the principles of a superior science (e.g., mathematics with respect to mixed sciences), or *a posteriori*, from effects that are more known to us and that serve to reveal the truth of the premise (which itself, of course, is a principle of the properties that are subsequently demonstrated). Note now, and this is the crucial point, if Galileo is arguing for the truth of his *suppositio* about the definition of naturally accelerated motion, he cannot prove this *a priori* from mathematics. (He does suggest, in the last two sentences of the second paragraph cited above, a persuasive *a priori* argument based on the simplicity of nature's operation, but this he proposes as confirmatory rather than as independently convincing.) If he is to prove the *suppositio*, therefore, his basic argument must be *a posteriori*, arguing from what is actually found in physical experiments. But this again involves him in a difficulty because of the approximate character of his measurements and the impediments encountered in nature, either of which could falsity his results.

It is at this point that we have recourse to the medieval development of *ex suppositione* argument to show a possible way out of the difficulty. The basic technique is already in Aristotle's *Physics* and is touched on in the *Posterior Analytics*, but its fuller articulation is found in Aquinas, and was later to be taken up by Buridan (see Essay 16, *infra*). To circumvent the defective operation of nature one may demonstrate on the basis of a *suppositio* that abstracts from impediments, or, in Buridan's application, from God's suspension of the laws of nature so as to perform a miracle. Since this last type of *suppositio* is slightly different from the two already mentioned, to be more precise we should now list three possible meanings of *suppositio*, from which three different meanings of *ex suppositione* can be drawn:

(1) *suppositio* of a definition merely posited;

(2) *suppositio* of a condition under which a mathematical definition will be verified in nature to a determinate degree of approximation; and

(3) *suppositio* of the removal of impediments or of extraneous efficient causes that permits a definition to be verified as it ideally might be within the order of nature.

Illustrations of the three are as follows: (1) definition of a spiral motion; (2) that of a balance located at a distance sufficiently remote from the earth's center to permit simple mathematical calculations; and (3) that of a parabolic motion achieved in abstraction from impediments that otherwise might cause departures from the ideal path. All three are hinted at by Galileo in the *Two New Sciences* or in correspondence wherein he is discussing the methodology that lies at the base of that work. Moreover, techniques of demonstrating *ex suppositione* in all three understandings were known in the sixteenth century and were not foreign to the thought context in which Galileo operated. All three, indeed, are touched on by Galileo in his logical questions, as will become clear when these are transcribed and edited for publication. The first two, moreover, are already to be found in Pererius's lectures on the *De caelo* (Wallace, 1978a, pp. 127–128), whereas the third is explicitly discussed in its alternate forms by two authors known to Galileo: Ludovicus Buccaferreus in his exposition of the *De generatione* (Venice, 1571, fols. 2v–3v) for the cases involving the contingency of nature's operation, and Ludovicus Rugerius in *reportationes* of his lectures at the Collegio Romano on the *De caelo* (Cod. Bamberg.) 62.4, pp. 40, 69) for cases involving the divine power (on Rugerius, see Wallace, 1977a, pp. 13, 19–20).

Returning now to the text with which this appendix begins, we see that Galileo's initial use of demonstration *ex suppositione* is based on the first meaning of *suppositio* listed above; suffice it to note that Galileo excludes this because it can define motions "that nature does *not* employ." He is concerned, rather, to define the kind of motion that nature *does* employ, and so he supposes a definition that agrees, as he says, "with the essence of naturally accelerated motions." He speaks of the "essence" of such motions, in our understanding, becauses he wishes to abstract from accidental deviations that can arise from impediments and other causes. Although he does not reiterate the expression *ex suppositione* in this sentence, his procedure is equivalent to what we have identified above as supposing a definition of the second or third type, rather than the first. Once this is seen, the only problem that remains is how Galileo can be sure his supposed definition is that actually found in nature. By his own account, although preceded by "continual agitation of mind," he is finally confident of the new supposition on two counts: the first we have referred to as *a posteriori*, because based on natural experiments showing that the properties observed in nature are actually those one should expect as effects of such a definition, and the second we have labelled *a priori*, because based on the simplicity of nature's operation.

For the Aristotelian methodologist, an interesting question arises at this point. Since Galileo's *suppositio* of a definition that agrees with the essence of naturally accelerated motions is to function as a principle or premise of a demonstration, must not this premise be self-evident or immediately confirmable in experience? This is one of the requirements Aristotle lays down for demonstrative knowledge (*Posterior Analytics*, I.2, 72a25–b4), and, as Wisan has shown (1978, pp. 37–45), Galileo frequently insists on such a requirement himself. On the other hand, from reading his justification of the definition in the text analyzed above, one is tempted to see it merely as a hypothetico-deductive argument that is not confirmed apodictically at all, but seems to involve the fallacy of *affirmatio consequentis*. How then can Galileo be sure of his definition, and what is the warrant behind his apparent certitude?

The answer to this question that is implicit in the foregoing essay is that Galileo is assured of the truth of the premised definition because it can be demonstrated *a posteriori* from effects that follow from it and that are directly confirmable by experiment. (The conditional form of the argument does not rule out this interpretation, since any demonstration can be expressed as a conditional, although the converse is not true.) To make this claim, of course, one should have some evidence that Galileo would accept as a premise in a demonstrative syllogism a proposition that is not *per se* evident or based immediately on sense experience, a likelihood that Wisan apparently questions (1978, pp. 42–43). Here reference to Galileo's early logical notes can again be helpful, for one of the questions he raises there is precisely this: "Must the principles of the sciences be so evident [*nota*] that they cannot be proved by any reasoning?" (*Opere* 9: 280). Galileo's answer to this query is in the negative, and completely in accord with the account we have just given of *suppositiones* that are demonstrable. Among the cases of provable premises that he enumerates, in fact, is one wherein principles that would otherwise be unknown are demonstrated *a posteriori* from effects that are more known to us (MS Gal. 27, fol. 6r–v).[8] Thus the methodological procedure we see described in Galileo's treatise "On Naturally Accelerated Motion" is quite consistent with the logical canons he expounded in his earliest extant treatise on scientific method, and which, as we argue in Essays 10 and 14 *infra*, he continued to employ until the end of his life. Not to see this is to involve Galileo in many inconsistencies, and possibly to miss the most significant contribution he wished to make in proposing his *nuova scienza.*

A final observation relates to the author's mention, in the original essay, of Galileo's experiments with free fall as recorded on fol. 116v (p. 146 *supra*) and to his use of the calculus there (note 7, *infra*) to show how Galileo's experimental confirmation effectively had the force of a biconditional argument. Although the calculus enables one to give an elegant proof of this, it is not necessary to employ it, since the same result can be attained by simpler mathematical methods. The point to be proved is that not only does the principle of uniform acceleration (p) imply the times-squared law (q) by mathematical reasoning, but also that the times-squared law (q) can be seen, from experimental evidence, to entail the principle of uniform acceleration (p). To show this, one may start from Galileo's experimental proof of q for motion along an inclined plane, described in the *Two New Sciences* and verified by Settle (1967), which yields the experimentally true result:

$$\frac{s_1}{s_2} = \frac{t_1^2}{t_2^2} . \tag{1}$$

This was known to Galileo shortly after 1604, on the basis of indications on fol. 152r of MS Gal. 72. Additional experiments with the inclined plane, performed around 1608 and recorded on fol. 116v of the same codex, were designed to show that a ball, after descending down an incline set on a table top and being projected horizontally along a line parallel to the table's surface, will travel various distances (D) depending on the height through which it descends (H) before reaching the floor. Galileo's experimental set-up is shown in Fig. 1. He used this to test whether the square of the distance D would vary as the height H according to the relationship:

$$\frac{D_2^2}{D_1^2} = \frac{H_2}{H_1} . \tag{2}$$

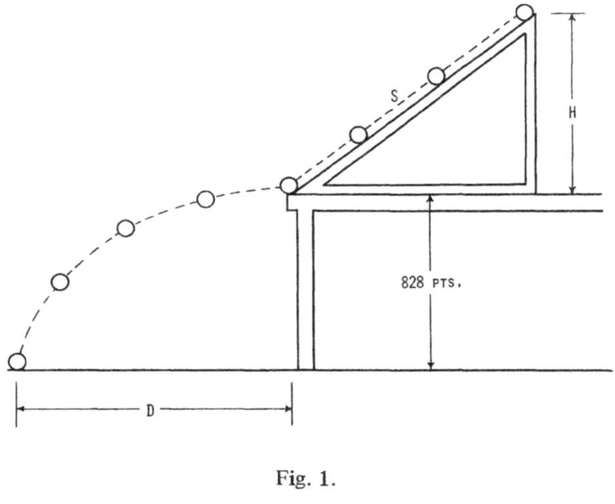

Fig. 1.

Using this proportionality to calculate D's for various H's, he was then able to measure D's experimentally, and found that his measured results, shown in the Table below, confirmed the relationship to a high degree of accuracy.

H	$D_{cal.}$	$D_{meas.}$
300	–	800
600	1131	1172
800	1306	1328
828	1330	1340
1000	1460	1500

Knowing, therefore, that both (1) and (2) are experimentally true to a sufficient degree of approximation, he could reason along the following lines. Since the distance of travel of the ball along the inclined plane, s, is proportional to H, from (1) it is true that

$$\frac{s_1}{s_2} = \frac{t_1^2}{t_2^2} = \frac{H_1}{H_2}. \tag{3}$$

Also, when the ball leaves the table top, since the velocity it acquires during any fall (v_H) is directed horizontally, the horizontal distance of travel by the time it reaches the floor (t_h) – where h is the height of the table – will be

$$D_H = v_H t_h.$$

Since t_h is constant for all the experiments, this is equivalent to saying that

$$\frac{D_1}{D_2} = \frac{v_1}{v_2}. \tag{4}$$

Squaring both sides of (4), and making use of the experimentally verified relationships (2) and (3), one may then write

$$\frac{v_1^2}{v_2^2} = \frac{D_1^2}{D_2^2} = \frac{H_1}{H_2} = \frac{t_1^2}{t_2^2}$$

or taking the square root of the resulting extremes,

$$\frac{v_1}{v_2} = \frac{t_1}{t_2}$$

Q.E.D.[9]

NOTES

[1] See Galileo's *Opere*, A. Favaro (ed.), Vol. 8, p. 207, and Newton's *Principia*, Koyré

and Cohen (eds.), Vol. 2, p. 586. Henceforth all citations to Galileo's writings will be made to Favaro's edition, to which Drake's recent translation of the *Two New Sciences* (1974) is also keyed. Usually Galileo refers to an hypothesis as a *postulato* or as an *ipotesi* (*Opere* 7, p. 29), although sometimes he uses the terms *supposizione* and *ipotesi* interchangeably (e.g., *Opere* 2, p. 212), as will be explained *infra.*

[2] Some details of the methodology this entails are given in my *Causality and Scientific Explanation* (1972, pp. 71–80, 102, 104, and 143; and 1974, pp. 247, 250, 293, and 354). Grosseteste translates the Greek *ex hupotheseōs* as *ex supposicione* in his commentary on the second book of the *Physics*, and Albert the Great, in his paraphrase of the same, renders it variously as *ex suppositione* and *ex conditione*; the Latin text of Averroës' great commentary on the *Physics*, on the other hand, gives the reading *ex positione*. For details of Albert the Great's teaching on reasoning *ex suppositione*, see my 'Albertus Magnus on Suppositional Necessity in the Natural Sciences,' in J. A. Weisheipl, ed., *Albertus Magnus and the Sciences: Commemorative Essays 1980*, Toronto: Pontifical Institute of Mediaeval Studies, 1980, pp. 103–128.

[3] See Aquinas's *In lib. II Physicorum*, lect. 15, n. 2; *In lib. I Posteriorum Analyticorum*, lect. 16, n. 6; and *In lib. II Posteriorum Analyticorum*, lect. 7, n. 2, and lect. 9, n. 11; also *Contra Gentiles*, lib. I, c. 81, and lib. II, cc. 25 and 30; *In lib. I Sententiarum*, dist. 2, q. 1, art. 4, and 3 arg., and *In lib. III Sententiarum*, dist. 20, art. 1, questiuncula 3.

[4] The experiment is described by Galileo as follows: "Let us hang from two strings that are equally long two heavy bodies, for example two musket balls, and let one of the aforementioned strings be attached at the very highest place one can reach and the other at the lowest, assuming that their length is four or five feet. And let there be two observers, one at the highest place and the other at the lowest, and let them pull aside these balls from the perpendicular position so that they begin their free movement at the same instant of time, and then go on counting their swings, continuing through several hundred counts. They will find that their numbers agree to such an extent that, not merely in hundreds but even in thousands, they will be found not to vary by a single swing, an argument concluding necessarily [*argomento necessariamente concludente*] that their falls take place in equal times. And since such falls in the motion along the arcs of a circle are duplicable on the chords drawn from them, there results on earth all that your friend [Fermat] says should happen on inclined planes that are parallel to each other and equally long, one of which is closer to the center of the earth than the other. They fall, I say, exactly in unison [*assolutissimamente*], despite the fact that both are placed outside the surface of the terrestrial globe. And that this might happen between similar planes, one of which were outside the surface of the earth and the other so far inside as to terminate even at the center of the same, I do not wish at the moment to deny, although I have no reason that absolutely convinces me to admit that the movable object that comes to rest at the center would traverse its space in a time shorter than the other movable object traverses it. But to say more, it is apparent to me that it is not well resolved and clear that a heavy movable object would arrive sooner at the center of the earth when leaving from the neighborhood of only a single cubit than a similar body that would depart from a distance a thousand times greater. I do not affirm this but propose it as a paradox, through the solution of which perhaps your friend will have found a demonstration that concludes necessarily [*dimostrazione necessariamente concludente*]." – *Opere* 17, pp. 91–92.

[5] See the more complete citation of this text in the Appendix to this essay. For a fuller

explanation of nature as an efficient cause of falling motion, see Essay 13 in this volume; also the general context provided in Essay 7.

[6] The question suggests itself at this point whether Galileo actually conflated *impedimenti* that prevent ideal mathematical accuracy from being attained with those that prevent perfect and unfailing regularities from being observed in nature. An affirmative answer would seem indicated on the basis of the way Galileo proceeds in the *Two New Sciences*; similar uses of the term *impedimenti* in his earlier writing have been noted by Noretta Koertge (1977), and these likewise suggest a conflation of the two types of cases. With regard to the *Two New Sciences*, as we have seen, when discussing the law of the lever as applied to the balance, Galileo says that this yields results that are only approximate at finite distances, but reasoning *ex suppositione*, i.e., on the supposition that the balance is at an infinite distance from the center of the earth, the results can be perfectly demonstrated, *con assoluta dimostrazione*. In this context he seems to be regarding the physical geometry of the universe as an "impediment" that prevents a mathematical ideal from being realized much as air resistance will prevent an ideal in nature (i.e., uniformly accelerated motion) from being realized. And in both cases it seems that he is employing *ex suppositione* reasoning to make the transit from a real to an ideal case, i.e., from a real balance to an ideal balance, and from the falling motion actually observed in nature to the ideal motion nature is attempting to realize.

[7] If one were to focus on the downward component of the motion alone, here q would stand for the times-squared law (verified approximately) and p for the definition of motion uniformly accelerated with respect to time (now known to be true *ex suppositione* in the order of nature). Then the inference "if q then p" would be verified mathematically by strict implication (and not merely by material implication), since the differentiation of $s \propto t^2$ yields immediately that $v \propto t$. Note that here the reverse inference, "if p then q," would also be true for the same parameters, since one can obtain, by integrating $v = ds/dt \propto t$ with respect to t, the result that $s \propto t^2$; Galileo, as we know, did not see this immediately. Effectively this means that the inference here really has the force of the bi-conditional or equivalence function, since both "if p then q" and "if q then p" are true, and thus the "if's" can be read as "iff's" ("if and only if") in virtue of the mathematics involved. The same, it goes without saying, could not be said of any inference to the proximate cause of uniform acceleration or of the times-squared law, as Galileo was well aware.

[8] Perhaps the clearest statement of the kinds of *suppositiones* that are employed in demonstrative syllogisms is that of Rugerius, in the logic course he gave at the Collegio Romano in 1589–1590. The Latin *reportatio* of a portion of his lecture on this subject reads as follows:

Secundo animadvertendum quod rursus principia complexa alia sunt dignitates, alia positiones. Positiones vero aliae suppositiones, aliae petitiones; suppositiones autem aliae suppositiones simpliciter, aliae suppositiones ad discentem. Explico singulas: dignitates dicuntur communes quaedam animae conceptiones per se notae quae ex sola terminorum cognitione omnibus innotescunt, quibus nemo potest saltem in intellectu non assentiri, ut omne totum est maius sua parte. Petitiones vero sunt quae ab alio discuntur. Et quidem suppositiones simpliciter, quae demonstrari quidem non possunt, sed tamen confirmatione aut explicatione aliqua declarari, vel in aliqua tamen scientia aliqua quidem ratione demonstrari possunt, sed is qui ex illis argumentatur eas supponit tanquam in alia scientia demonstratas. Suntque quasi propositiones immediatae in illa scientia in qua supponuntur tanquam demonstratae in altera, quia in illa non habent medium per quod probari possint. Suppositiones vero ad discentem tamen sunt, quae licet demonstrari possint, tamen non demonstrantur sed apparent vera addiscenti. Petitiones autem sunt quae demonstrari quidem possunt, eas tamen non demonstrat qui argumentatur, sed petit sibi concedi ab eo qui nullam habet de illis opinionem, vel et contrariam, vel quia id suo loco probabitur, vel quia pertinet ad aliam scientiam. De hoc distinctione propositionum lege Aristotelem, primo Posteriorum, 25. – Cod. Bamberg. 62.2, fols. 413v–414r.

Further on Rugerius states that some of these *suppositiones* are such as to be seen immediately on inspection, others require induction or experimentation to be established, and yet others may be proved by demonstration *quia* or *a posteriori*, along the lines we have argued were followed by Galileo. The Latin of this statement is included here as general substantiation of our thesis:

Secunda propositio. Quod dicitur de principiis complexis syllogismi dicendum est etiam de principiis complexis scientiae aut cuiuscunque facultatis. Ob easdem enim rationes praecognita et notissima esse debent, non tamen omnia eodem modo: alia enim ut quae communissima sunt omnibus disciplinis debent esse in omnibus notissima; alia vero in quibusdam, alia vero in superiori aliqua scientia probantur; quae scientia dicitur subalternans, ut deinde sint principia scientiae inferioris, quae dicitur subalternata. Praeterea alia sunt ita nota ut sola cognitione terminorum indigeant ad hoc ut cognoscantur, ut omne totum est maius sua parte, alia vero inductione et experimentis, ut igne esse calidum, reubarbum purgare vinum. Alia etiam per principia quaedam communia aut a superiori scientia accepta probari debent, alia vero aliqua demonstratione quia et a posteriori vel signo, ut probat Aristoteles tria esse rerum naturalium principia, dari primum motorem. Alia etiam sunt omnino prima et immediata alicuius scientiae, quae in tota scientia notissima esse debent, alia vero sunt propria alicuius partis quae nihil repugnat in aliqua alia parte probari, dummodo in illa sint notiora. Sicut etiam principia alicuius demonstrationis poterunt interdum alia demonstratione probari si non sint prima et immediata, dummodo in ipsa demonstratione cuius sunt principia sint nota. Ex quo patet quod in omni demonstratione, ut perfecta cognitio conclusionis habeatur, debent omnia principia vel actu vel saltem habitu cognosci, ex quibus illa cognitio aliquo modo sive mediate sive immediate dependet, unde tandem deveniendum est ad aliqua prima et indemonstrabilia quae ex aliis amplius non dependant. . . . – *ibid*., fols. 414v–415r.

For more details on Rugerius's teaching, see the Appendix to Essay 10, *infra*.

9 The clarifications made in this Appendix are largely the result of correspondence with Dr. Winifred L. Wisan relating to the original essay. The author takes this opportunity to thank her for her interest and her stimulating critique. Following that interchange but unconnected with it, R. H. Naylor (1980) has given an analysis of Galileo's manuscripts that can be used to support the interpretation of his experimentation advanced on pp. 146–147 *supra*.

9. GALILEO AND THE THOMISTS

When Antonio Favaro, the otherwise careful editor of Galileo's *Opere*, came across the name "Caietanus" in Galileo's early notebooks, he assumed that the reference was to Caietanus Thienensis,[1] the Paduan *calculator*; apparently it did not occur to him that the young Galileo would be acquainted with the writings of the Italian Thomist, Thomas de Vio Caietanus.[2] Yet a check of Galileo's citations shows that six of the eight references ascribed by Favaro to Caietanus Thienensis in reality are references to Thomas de Vio Caietanus.[3] Favaro's error here, of course, is more excusable than his failure to identify correctly the more mature Galileo's reliance on the authority of the celebrated Dominican in his letter to the Grand Duchess Christina.[4] There it was a question of the place of the sun in the heavens when Joshua gave his famous command, "Sun, stand thou still," and Thomas de Vio had long been recognized as a competent Scriptural commentator.[5] But De Vio also held distinctive views on matters relating to physical science, as did many of the Thomists of his day, and such views were known and discussed by Galileo in his early writings. Most of these writings, contained in Vol. I of Favaro's National Edition of Galileo's works, show a preoccupation with Aristotle and the problems raised by his philosophy then being discussed in the schools, particularly at Pisa and other Italian universities. Of the 419 pages that go to make up this first volume, in fact, Aristotle is mentioned on 194 pages; the author cited with next greatest frequency is St. Thomas Aquinas, who gets mentioned on 32 pages, followed in order by the Aristotelian commentators Averroës and Simplicius, who are mentioned on 30 and 28 pages respectively.[6] Apart from this recognition of Aquinas as a foremost interpreter of Aristotle, Galileo's early writings reveal also a surprising knowledge of the Thomistic school. On four different pages of the same volume Galileo refers to "the Thomists," and in one of these references he identifies four members of the school. Then, in various individual references, he cites Joannes Capreolus[7] (7 places), Thomas de Vio Caietanus (6 places), Paulus Soncinas[8] (4 places), Ferrariensis[9] (3 places), Hervaeus Natalis[10] (2 places), Dominicus Soto[11] (2 places), and Chrysostomus Javellus[12] (1 place).

This little known acquaintance of the young Galileo with Thomism is worthy of study in its own right, particularly on the part of anyone tracing the impact of the thought of a thirteenth-century scholar like Aquinas on subsequent centuries. From the viewpoint of the history of science, however, there are additional reasons for examining closely Galileo's relationship to the Thomistic school. Although much is known about Galileo, there is a definite *lacuna* in Galilean scholarship in the area of his early writings, and particularly the notes he composed or copied while at the University of Pisa.[13] These notes cover a wide range of topics discussed with a fairly high degree of sophistication, and with a citation of sources that range from classical antiquity through the middle and late scholastic periods to the latter part of the sixteenth century. The resulting mass of material is so refractory to simple analysis that it is not surprising that scholars have contented themselves with rather vague generalities about the early sources of Galileo's ideas. What is needed, as E. A. Moody has already urged,[14] is a detailed study of these early writings, and for this it is necessary to start some place. The present essay is offered as a beginning in this important but hitherto neglected area of scholarship in the hope that it may shed light not only on the development of the Thomistic school but also on its relationships with this celebrated Pisan scientist. In structure it will first describe in general Galileo's citations of St. Thomas and the Thomists, then it will narrow the field of discussion to examine in detail Galileo's understanding of the Thomistic positions he cites on the intension and remission of forms and various teachings on the elements, and finally it will conclude with some observations on the possible sources of these sections of Galileo's early writings.

I. GALILEO'S CITATIONS OF ST. THOMAS AND THOMISTS

Galileo shows little interest in the metaphysical problems that have consistently attracted the attention of Thomistic historians of philosophy, but concentrates instead on the physical problems relating to the heavens and the earth that were to remain a constant concern throughout his life. His citations of St. Thomas in the first volume of the *Opere* are confined to three important references in the treatise *De motu*, with the remaining 29 all occurring in the notes concerned with physical questions. In general these questions treat of two types of problem, the first relating to Aristotle's treatise *De caelo et mundo* and the second to the subject of alteration and the way in which alteration is related to the forms of the elements and their qualities. In the second category there are only eight citations of Aquinas, so that the remaining majority (21 citations) are concerned with the matter of *De caelo et mundo*. Apart from these references in the first volume of the *Opere*, Aquinas is cited seven times in Vol. III, twice each in Vols. IV and V, and once in Vol. XIX, for a grand total of 44 citations, while he is also named, in passing, in at least five other places.[15] Compared to this rather liberal use of Aquinas himself, Galileo's attention to the members of Aquinas's school, "the Thomists," is relatively restricted. Aside from the one citation of Thomas de Vio Cajetan in Vol. V, to which reference has already been made, all of the references to Thomists in the *Opere* are to be found in the physical questions. Cumulatively these amount to 29 references, of which 12 pertain to the matter of *De caelo* and 17 to that of alteration and the elements. The content of all of these citations will now be sketched in a general way, outlining first Galileo's rather extensive use of Aquinas's teaching and then his sparser references to the Thomistic school. This survey will provide the background information necessary for the detailed examination, to be undertaken in Section II of this essay, of the specific understanding and evaluation of the Thomistic tradition revealed in these writings.

A. *Citations of St. Thomas*

The notebooks under discussion purpose to present, in more or less systematic fashion, the essential content of Aristotle's four books *De caelo* and his two books *De generatione*. The treatise is prefaced by two brief questions where Galileo inquires first concerning the subject of Aristotle's *De caelo* (A) and second concerning the order, connection, and titling of these books (B).[16] St. Thomas is mentioned in both questions, first for his view that this subject is the universe according to all its integral parts (A2) and secondly for his insisting on the title *De caelo et mundo*, along with Albertus Magnus and other "Latins," against the Greek tradition represented by Alexander and Simplicius, who would name the books simply *De caelo* from their "more noble part" (B8). Thereupon a further division is made into two treatises, the first concerned with *De mundo* and consisting of four questions and the second concerned with *De caelo* and consisting of six questions.

The treatise *De mundo*, whose title Galileo seems to understand broadly enough to encompass the world or universe as well as the earth, has two preliminary questions which discuss the origins of the universe first as understood by ancient philosophers (C) and second according to the Catholic faith (D), but in neither of these is St. Thomas mentioned. In the third question, however, which treats of the unity and perfection of the universe (E), Aquinas is discussed at some length. He is first invoked in support of Galileo's contention that the universe is one, based on his argument from the order existing in things created by God (E2). Galileo's next conclusion is that the unity of the universe cannot be demonstrated by reason, although it is certain from faith that only one universe exists. Here he raises five queries and offers his own interpretation of St. Thomas to resolve the first three of them. One of Aquinas's arguments seems to maintain that earth's natural motion to a center would preclude there being any other earth than this one, but this is to be understood only of what happens according to nature from God's ordinary power (E7). The second query is whether God can add any species to this universe, or make other worlds that have more perfect species that are essentially different from those found here. Both Scotus and Durandus deny this possibility, but Aquinas and practically everyone else hold that God's infinite power would enable him to make more perfect universes to infinity (E8). The third query is whether God could make creatures more perfect than those he has made in this world, to which Galileo replies that he could make them accidentally more perfect but not essentially so, and here he adduces Aquinas's example of the number four, whose essence cannot be varied, and argues that other essences are like this also (E11).

The fourth question is whether the world could have existed from eternity (F), and here Galileo present three conclusions: that the world did not exist from eternity since it is of faith that it was created in time (F19); that on God's part there is no repugnance that the world could have existed from eternity (F23); and that there is a repugnance, however, on the part of creatures, whether these be corruptible or incorruptible (F24).[17] These conclusions and their proofs are preceded by four opinions, and Aquinas is mentioned in the discussion of three of them. The first opinion is that of Gregory of Rimini and other nominalists, who maintain that the world could have existed from eternity whether it be made up of successive or permanent entities or corruptible and incorruptible ones. This seems to gain some support from St. Thomas, Galileo notes, when he proves that the creation of the universe cannot be demonstrated from reason (F1). Also, Aquinas's maintaining that creation does not involve any action going forth from God to creatures (F9) is used in a quite complex argument in support of the same conclusion. The second opinion is that of Durandus and "many moderns," which holds that there is no repugnance to eternal existence on the part of incorruptible things, whereas there is of corruptible things, and St. Thomas seems also to support this (F11). The fourth opinion, arguing from other *loci* in St. Thomas, Galileo identifies as that of Aquinas himself, which is in agreement with the teachings of Scotus, Ockham, the *Doctores Parisienses*, and Pererius (F18).[18] This would maintain that the world could have existed from eternity on the part of incorruptible things, but that there are problems associated with corruptible things; Durandus points out the absurdities that follow from allowing corruptible things an eternal existence, but these can be solved by admitting infinites in act, or infinites that can be actually traversed, or one infinite that is larger than another.[19]

The second treatise *De caelo* begins with two rather technical questions, the first on the unity of the heavens (G) and the second on the order of the celestial spheres (H), which draw heavily from medieval writers on astronomy but make no mention of Aquinas. In the third question, which inquires whether the heavens are composed of simple bodies (I), Galileo argues that the heavens are a body distinct from the four elements and are not composed from these elements. In discussing the opinion of ancient philosophers before Aristotle, who attributed the same nature to the heavens as to the elements, Galileo documents Aquinas's interpretation of Empedocles's (I8) and Plato's (I9) teachings on this matter. Again, in reply to various objections that are brought against Aristotle's teaching, he culls responses from Aristotelian commentators and among these he cites in some detail Alexander, Simplicius, and Aquinas (I33).[20] The fourth question is whether the heavens are corruptible (J), and here Galileo's main difficulties stem from whether one is to consider the heavens from their intrinsic principles or in relation to the absolute power of God, who can annihilate anything regardless of its natural potencies, provided only that it has an obediential potency (*potentia obedientialis*) to his command.[21] In explaining his solution and the arguments against it, Galileo mentions Aquinas along with Simplicius and Averroës as holding for a twofold alteration, one corruptive and the other perfective (J31). He cites also Aquinas's opinion that, on the Day of Judgment, the heavens will not be corrupted substantially but only with respect to certain of their accidents (J33). The fifth question is a rather lengthy disquisition on whether or not the heavens are composed of matter and form (K), and here Aquinas is given more attention than in any other part of the notebooks. Galileo first identifies St. Thomas's position as being that the heavens are composed of matter and form (K38), but then notes that Aquinas differs from many other Aristotelians in holding that the matter of the heavens is different in kind from the matter here below (K39).[22] Later he cites approvingly an argument taken from Aquinas to prove that the heavens are composed of matter and form and replies to a whole series of objections that have been raised against the Thomistic arguments (K74–78). Again, he quotes the proofs given by Thomas in the commentary on the *Physics* to show that there can be a potency in the heavens, and then gives further arguments to show that this implies that matter is also there, while using the general dichotomy between potency and act as a principle throughout (K92, 94). Galileo is convinced that the heavens are not composed of a matter that is of the same kind as that here below. Here his principal adversary seems to be Giles of Rome, who in turn is arguing against Aquinas (K139), but whose arguments Galileo is at pains to refute. The sixth and final question of this treatise is concerned with the animation of the heavens (L), and here Galileo cites Aquinas as interpreting Aristotle differently in various works. In the *Summa contra gentiles*, as he reads it, Aquinas seems to state that Aristotle held that the intelligences are actually forms of the heavenly bodies (L13), whereas in the *Summa theologiae* and in the question *De spiritualibus creaturis* Aquinas seems to hold that the intelligences merely assist the heavens and are otherwise not their souls (L24).[23] Galileo's conclusion is that, although it might be true that Aristotle regarded the intelligences as actually informing the heavenly bodies, more probably his opinion is that they are merely forms that assist such bodies in their motions, and for this he again invokes the authority of St. Thomas and other scholastics (L29–30).

The second broad division of the physical questions is not so well organized as the first and contains a number of ellipses or omissions that make for difficulty in recognizing its intended structure. It is probable, however, that this was planned to embrace two tractates, the first on alteration (M through O) and the second on the elements (P through Y). The tractate on alteration is the shorter of the two and seemingly was made up of only three questions. The first question is missing except for the last few lines of the text, which state a conclusion suggesting that the question was concerned with the nature of alteration, its subject, and its terminus (M). The second question treats of intension and remission as a species of alteration (N), and here St. Thomas is mentioned at the beginning as one of the authorities in this matter (N2). Otherwise he is not cited, although the members of his school are given close attention, as will become apparent below. The last question discusses the parts or degrees of qualities and consists only of a series of six *praenotamina* relating to the latitudes and degrees of qualities (O). In the fifth of these, Galileo notes that intensification does not come about through addition alone, but in some way requires a greater intensification in the subject, and this is how he thinks St. Thomas can be understood when he holds that intensification results from the eduction of a form in such a way that it becomes more radicated in the subject (O6).[24]

The second tractate is devoted entirely to the elements, and apparently was to consist of four parts, of which only portions of the first two are extant.

After a brief introduction on the nominal definition of an element (P), where Aquinas is cited (P8), the first part is devoted to the nature of the elements, and five questions are allotted to this. St. Thomas is not mentioned in the first two questions, treating respectively of the definitions of the elements (Q) and their material, efficient, and final causes (R). He *is* cited, however, in the third question (S), which inquires into the forms of the elements, for his opinion that these are substantial forms, with which Galileo agrees (S8). The fourth question is whether the forms of the elements undergo intension and remission (T), and here again St. Thomas is cited as an authority along with various members of his school, as will be noted below (T4). The fifth question is incomplete and has no title, but it is concerned with the number and quantity of the elements (U). The question is raised whether elements and natural things have termini of largeness and smallness, and here, among the authorities, St. Thomas is cited for holding that the elements have an intrinsic terminus of smallness but that they have no terminus of largeness (U25). In his solution to this question Galileo holds that, in relation to God, no natural things have maxima and minima, and this despite the fact that some authorities have taught that a quality such as grace cannot be increased intensively to infinity. Other authorities, he says, including some interpreters of St. Thomas, are able to hold the contrary, by distinguishing what can be done by ordinary power and what by the absolute power of God (U34).

The second part of the tractate on the elements is seemingly concerned with their qualities and accidents in general, and is apparently made up of four questions. These discuss: the number of primary qualities (V); whether these are all positive or whether some are privative (W); whether all four qualities are active (X); and what the role of the primary qualities is in activity and resistance (Y). St. Thomas is mentioned only in the third question, where his opinion on what constitutes the passivity of a quality is listed among the notes at the beginning of the question (X15).

This completes the citations of St. Thomas in the physical questions, which are intended to provide the framework for our subsequent analysis. For reasons of completeness we will now mention more summarily the contexts in which Galileo mentions Aquinas in his other writings. The first of these is the group of manuscripts assembled by Favaro in Vol. I of the *Opere* under the general title *De motu*, parts of which have been translated into English and annotated by I. E. Drabkin.[25] In this material there are two direct citations of Aquinas and one marginal notation, associating him with an opinion being discussed in the text. The two direct citations both relate to St. Thomas's distinctive teaching that motion through a vacuum would not take place instantaneously (*Opere* 1: 284.8, 410.21); Galileo is in agreement with the conclusion, but apparently not with the reasoning Aquinas and others use to support it. The marginal note occurs in a chapter where Galileo is discussing the cause of the increased acceleration at the end of a body's fall, and refers the reader to St. Thomas's exposition of Aristotle's reason, namely, "because the weight of the body is more concentrated and strengthened as the body approaches its proper place" (*Opere* 1: 316, n.1). Although these are the only explicit mentions of Aquinas, it should be noted that many of the matters discussed in these manuscripts relating to motion bear on distinctive views of St. Thomas and his school. Even a cursory examination, however, would enlarge this essay beyond reasonable limits, and thus the material relating to motion must be left for further exploration elsewhere.[26]

Of the remaining citations of Aquinas a goodly number occur in Vol. III of the National Edition, where Galileo is discussing mainly astronomical matters. One citation is St. Thomas's elucidation of Aristotle's statement that there is no goodness in mathematics, because mathematicians "abstract from matter, motion, and final causality" (*Opere* 3: 255). Two other references are to Aquinas's teaching on the movement of the heavens, (*ibid.*, 3: 284, 346) and two more to his views on the plurality of worlds as contained in his commentary on the first book of Aristotle's *De caelo* (*ibid.*, 3: 353–354). Yet another reference is to the *Summa contra gentiles*, where Aquinas's authority is invoked to show that it is *ex fide* that the heavens will stop moving at the end of the world (*Opere* 3: 364). These last three citations, it may be noted, refer to matters already discussed in the physical questions and show Galileo's continued preoccupation with topics about which he wrote in his early notebooks. Four other citations invoke St. Thomas's assistance in the interpretation of Sacred Scripture, two of these occurring in Galileo's letter to the Grand Duchess Christina already mentioned in connection with Cajetan (*Opere* 3: 290; 5: 333–334). Finally, in Vol. IV, Aquinas is mentioned twice, along with others, for his opinion that shape is not the cause of motion, but of its being slower of faster (*Opere* 4: 424, 738). These particular references are worthy of detailed examination in the context of fuller discussions of the manuscripts *De motu*, as has already been observed.

B. *Citations of Thomists*

With this general overview established it is now possible to deal at greater depth with points in St. Thomas's teaching that were taken up and developed by his followers. Although there are fewer references in the physical questions to Thomists than there are to St. Thomas, the former occur where points of doctrine are being subjected to closer scrutiny and thus shed as much light on Galileo's knowledge of Thomism as do the more extensive references to Aquinas.

That Galileo was aware of the existence of a Thomistic school seems incontestable in the light of his reference to "Thomists." In the first place where he uses this term he identifies four members of the school, all Dominicans and easily recognized for their professed loyalty to St. Thomas. The context is in a discussion whether the heavens are a composed body, understanding this in the sense of composed of matter and form, and here he includes Capreolus, Cajetan, Soncinas, and Ferrariensis among "all Thomists," who, with Aquinas, answer in the affirmative (K38).[27]

Before and after this text there is a fairly comprehensive citation of authors ranging from Plato and the Stoics to sixteenth-century philosophers. Following a rather extensive discussion of the matter-form composition of the heavens and the kind of matter found in them, Galileo raises the question how the various matters might be said to differ. He writes: "You inquire here, on what basis do these matters differ? Capreolus thinks they differ by different forms, Cajetan in themselves alone" (K159).[28] In his answer Galileo seems to side more with Cajetan than with Capreolus, although allowing that "the matters differ likewise in relation to forms" (K160). Then, following a disquisition on the implications of this for the corruptibility of the heavens, he raises another question. This concerns the matter of the celestial spheres, and whether it is one or many, just as the spheres are many (K170). In answering, Galileo sides with Cajetan, as against Capreolus, Soncinas, "and others,"[29] that a different matter is to be found in each of the heavenly spheres. The reason he gives is that if there were a single matter in the heavenly bodies and their forms were to differ specifically, that matter would be in potency to several forms, and then, when it exists under one form it would be deprived of another and be in potency to it; therefore there would be an intrinsic principle of corruption in the heavens. Having stated this, he notes the reply of "Capreolus and others" that this would not happen, because heavenly matter is not an apt subject of privation and the form of the heavenly body so informs the matter that it exhausts the matter's potency entirely (K173).[30] Galileo responds to this, in turn, that it is open to all the objections that he has made against the arguments of Giles of Rome, which connect the sameness of the matter with the sameness of its potency. Here again he answers an objection of Capreolus that this argument cannot be applied to the matter of the heavenly bodies, because although sublunary matter can admit privation, celestial matter cannot (K174). To the contrary, writes Galileo, for if the matter of the heavens is in potency to several heavenly forms, it can be the subject of privation, and secondly, if this is not the reason why matter is the subject of privation, Giles could say that matter that is not of the same definition as an inferior type is not the subject of privation from its nature but from the fact that if receives a form with its contrary (K174–175). After further extended discussion, Galileo concludes that since the matters and the forms of the heavenly spheres are different, one planet is essentially different from any other, since the planets are to be identified with their spheres, whereas all of the stars of the firmament have the same species (K177–178).

Of the various Thomists mentioned in the physical questions, Ferrariensis receives the least detailed consideration, being mentioned only for the three opinions: that the world could have existed from eternity, where he is said to agree with John Canonicus and "many moderns" (F1); that the heavens are composed, as in the text mentioned above (K38); and that it is the teaching of Aristotle that the intelligences are forms actually informing the heavens, where he is listed for his commentary on the text of St. Thomas (L13).

From what has been said up to now, Galileo can be regarded as being quite sympathetic to the teachings of St. Thomas and his school, while being cognizant in the cases mentioned of differences between Cajetan and Capreolus, and generally siding with Cajetan. When, however, we come to the next series of topics relating to the intension and remission of forms and special problems pertaining to the elements, we find that his attitude towards the Thomistic school becomes more critical, and that the Thomistic conclusions and the arguments in their support are generally contested.

Galileo begins his treatment of intension and remission by stating generally that this is the process by which a quality is varied according to more or less, and by noting the importance of the topic, namely, that, since it is found in almost every alteration, one cannot understand how alteration comes about if he does not understand intension and remission; and, since every action comes about through intension and remission, this process must be understood if one is to understand how any body acts on another (N1). Immediately following this he lists the authorities from which he will draw his arguments, and these include St. Thomas, Capreolus, Herveus, and Soncinas, among others (N2).[31] He then lays down three *praenotamina*, in the second of which he observes that both alteration and intension are successive, although he notes that it is possible that some alterations take place instantaneously, and cites the example of water freezing (N5–6).[32] The third of the *praenotamina* contains his classification of the various ways in which intension can take place in qualities (N7–8).[33]

Immediately following this preliminary material, Galileo states his own conclusions and proceeds to argue in their favor. The first of these is that intension does not take place through an extrinsic change alone, nor does it take place through the expulsion of the contrary or the disposition of the subject, with the quality itself remaining indivisibly the same (N9). Five different arguments are given in support of this position, in the third of which the interesting statement is made that "motion is nothing more than a *forma fluens*" (N12).[34] Then space is left for a second conclusion, which, however, is missing from the manuscript. Following this Galileo states his third conclusion, namely, that in intension the prior part of the quality does not perish (N18).[35] Five major arguments are again given in support of this, the last of which has many supporting syllogisms and distinctions. At the conclusion of all these Galileo makes a summary statement that, as it turns out, is transitional to the next conclusion he wishes to draw. He writes:

Therefore one must conclude that intension comes about through the production of a new quality in such a way that, when the later part comes, the earlier remains. And this again can take place in two ways: first, the later degrees that are added are produced in single instants, and in this way the intension would be discrete, as the Thomists prefer; second, in such a way that the intension comes about successively by a type of continued action (N28).[36]

Note here that "the Thomists" are referred to as a group and that Galileo merely lists their opinion without commenting on it in one way or another.

The fourth conclusion, however, addresses this point directly and sides against the Thomistic school. Galileo holds that intension comes about continuously, and supports his conclusion with two proofs: the first, that otherwise the alteration would not constitute a single motion, and the second, that the postulation of successive instants involves a host of inconsistencies and absurdities (N29–30).[37] Up to this point there has been no discussion of the Thomistic opinion, but now Galileo proceeds:

Nor can you say with the Thomists that heat having been produced, at the same instant it is then extended to the other parts of the subject: for the first part in which heat is already produced is closer to the agent, and so the agent, having finished with that, will not act on the more distant part. Add to this: it would follow that the agent would act no less on the closer parts than on the most distant, nor would it initially induce the first degree in the closest part rather than in the most distant. Confirmation: because a form is produced successively by reason of the resistance of the contrary, and so it cannot be produced through instants alone, for it would otherwise follow that each part would be produced in an instant and, as a consequence, without resistance and all at once. Therefore one must conclude with Simplicius on the eighth *Physics*, 12, at text 23, Giles on the first *De generatione* at text 20, Jandun on the eighth *Physics*, question 8, and others, that intension and remission come about continuously (N31).[38]

Here, then, is the first explicit rejection of a teaching of the Thomistic school, although it is noteworthy that the particular Thomists are not mentioned *nominatim*, nor is there any indication of differences of opinion that might exist within Thomism on this particular conclusion.

Having finished this exposition of the intension and remission of qualities, Galileo next turns his attention to special questions relating to the elements, among which are found the question "Whether the forms of elements undergo intension and remission?" (T) Here he divides the authorities into two groups, the first including the Averroists and the Scotists, among others (T2), and the second including Avicenna, St. Thomas and the Thomistic school, nominalists, and others (T4). His account of the second grouping is the following:

The second opinion is that of others who deny that forms undergo intension and remission. This is Avicenna's in the first *Sufficientia*, chapters 10 and 11, and in the First of the First, doctrine 3, chapter 1; Averroës, however, opposes him. The same is the opinion of St. Thomas in the opusculum *De mixtione*, in the second *Sentences*, distinction 15, and in the First Part, question 76, article 4, and there also Cajetan; Capreolus in the second *Sentences*, distinction 15, question 1, conclusion 2, and in the solutions to the arguments against it, Soncinas in the eighth *Metaphysics*, question 25 and question 26, the tenth *Metaphysics*, question 27, and elsewhere; Gregory in the second *Sentences*, distinction 15, question 1; Ockham, *Quodlibet* 3, question 4; Marsilius, first *De generatione*, question 22; Themistius, second *De anima*, text 4; and Philoponus, second *De generatione*, comment 33. These, however, say only this, that the forms of the elements do not remain actual in the compound, but from this the other conclusion follows. Moreover, the same opinion is defended by Durandus in the first *Sentences*, distinction 17, question 6, Henry, *Quodlibet* 3, question 2, Nobilius, in chapter 3, Buccaferrus in text 18, many of the commentators on the first *Microtechni*, comment 15, Hervaeus in the tract *De unitate formarum*, Giles in the first *De generatione*, question 18, Albert on the first *Techni*, chapter 25, and Javelli, eighth *Metaphysics*, question 5 (T4).[39]

Note in this citation that the Thomists are put in different places; first there is St. Thomas, with whom are mentioned Cajetan, Capreolus, and Soncinas, and then, after an enumeration of eight other thinkers, Herveus, followed three names later by Javelli. From the enumeration of the Thomists already noted (K38), it is obvious that Galileo recognized Cajetan, Capreolus, and Soncinas as Thomists, and from the authorities listed at the beginning of the section on intension and remission, Herveus is grouped with Capreolus and Soncinas (N2); with regard to Javelli, this is the only explicit mention of him in the physical questions, but there is indirect evidence that he was used for summaries of Thomistic teaching.

After this presentation of the various opinions, Galileo appears to side with those in the second grouping, for he states immediately after the text just cited: "This second, true opinion is proved by the following arguments" (T5). He then goes on to list four different arguments in some detail, proving that the forms of the elements do not undergo intension and remission. Following this, however, he lists a number of arguments *sed contra* that have been held by proponents of the first opinion. Unfortunately the question is not complete, and does not proceed beyond the listing of four objections. These are not answered, but it is probable that had the question been completed the resolutions of their arguments would have been given in a way that would safeguard the proofs offered in favor of the second opinion.

The last question wherein the opinion of Thomists are referenced is that devoted to the problem, "Whether elements and other natural things have termini of largeness and smallness?" (U9). As in some previous questions, the preliminaries include the various definitions and distinctions that will be employed in the discussion, and these make up eight *praenotamina*. These are followed by a listing of four opinions, of which the first is of some interest:

> The first opinion is that of those saying that all natural things, elements excepted, have intrinsic termini of largeness and smallness; elements, on the other hand, have an intrinsic terminus of smallness but none of largeness: so St. Thomas in the first *Physics*, texts 36, 38, *De generatione*, text 41, and the First Part, question 7, article 3; Capreolus, in the second *Sentences*, distinction 19; Soto, in the first *Physics*, question 4; and all Thomists (U25).[40]

Note here the significant inclusion of Soto's name after that of Capreolus, with the implicit acknowledgment that he is to be enumerated in the Thomistic school. Soto is the only Spanish Dominican mentioned by Galileo, although he is aware of other Spanish writers such as Pererius[41] and Vallesius.[42] Spanish Dominicans differed from Italian Dominicans in their treatment of some questions discussed in the physical questions, and thus this supplies an interesting point of comparison for judging Galileo's knowledge of variations within the Thomism of his day.

Following the foregoing enumeration, five arguments are offered in support of the first opinion (U25–29). The remaining three opinions that are listed, immediately following these arguments, are those of Averroës and the Averroists, the opinion of Paul of Venice, and finally the opinion of Scotus, Ockham, and Pererius (U30–32). Then Galileo lists ten assertions or conclusions which he wishes to establish, and in the first and seventh of these he makes mention of individual Thomists. The first conclusion and its proof read as follows:

I say, first: it seems certain, whatever others may think, that no things have maxima and minima in relation to God – not in the sense that they can go to infinity, for concerning this elsewhere – but only God can, by his absolute power, increase and diminish all created things forever and ever. The proof of this for living things: these require quantity, as something extrinsic, for their operation and conservation; but God can supply for the concursus of any extrinsic cause; therefore [living things have no maxima and minima in relation to God].

Concerning qualities, on the other hand, Scotus and Durandus in the third *Sentences*, distinction 13, Richard and Giles in the first *Sentences*, distinction 17, Henry, *Quodlibet* 5, question 22, Cajetan, on the Third Part, question 7 and question 10, article 4, and on the Second Part of the Second, question 24, article 7, speaking of the quality of grace, deny that a quality can be increased to infinity intensively. For, since qualities other than grace are created and limited, the properties of the essence must have a fixed limit, granted intrinsic, in intension. Capreolus, however, in the third *Sentences*, distinction 13, question 1, and in the first *Sentences*, distinction 17, question 4, Almainus and Gregory, same place, Ockham, same place and in the third *Sentences*, distinction 13, question 7, and Soto, in the first *Physics*, question 4, article 2 – where he shows that this is the opinion of St. Thomas in *De veritate*, question 29, article 3, and that if, in the Third Part, question 7, article 12, he seems to say the contrary, this should be understood of ordinary law – these all hold that, although quality of itself has a fixed terminus in intension, nonetheless it can be increased by absolute power. And this argument can be given: because qualities are not so intrinsically the intruments of forms that they essentially include the latitude owed to the form itself. Bonaventure and Cartarius are in agreement with this opinion.

And so I say, first: no quality of itself, abstracting from an order to a subject or an agent, has a fixed terminus in intension, and yet it does not tend to infinity simply – for this is incompatible with a created nature – but to infinity syncategorematically. For this reason I say, second: the same quality, so considered, can by God's power always increase and decrease continually without a terminus. The reason: because a quality, as a quality, in itself and abstracting from the subject, does not itself require a fixed terminus. Add to this also that the opinion of Capreolus is very probable (U33–35).[43]

The text goes on to enumerate various objections and the replies that may be given to them, and the exposition reveals Galileo's respectable knowledge of the discussions current among nominalists and late scholastics generally.

The seventh conclusion is obviously directed against the Thomistic opinion, for it reads: "I say, seventh, that elements and homogeneous compounds of themselves have no termini of largeness or smallness, either extrinsic or intrinsic" (U59).[44] The proof is divided into two parts, the first consisting of four arguments to show that the elements have no termini of smallness, or *minima*. After this Galileo continues:

These arguments prove that there is no minimum. Much easier is it to prove that there is no maximum, especially since this is denied by no one expect Cajetan, on the First Part, question 7, article 3. For Aristotle says, second *De anima*, chapter 41, that fire can increase to infinity. And this is obvious: for, if straw is added to the maximum fire, it will certainly increase; for to say that the straw is not going to burn, or, if it does burn, that in such an event the fire would turn into air, seems plainly ridiculous (U64).[45]

This refutation of Cajetan is the only argument against *maxima*, and Galileo then turns to his remaining conclusions.

As to the rest of the question, there is no indication by Favaro that the treatment is incomplete, but one may suspect that it is from the fact that the solutions of the arguments that relate to the conclusions are limited to a refutation of the five arguments that have been given in support of the first opinion, that, namely, of the Thomistic school. It may be, however, that Galileo felt that the variations introduced in the second, third, and fourth opinions had been sufficiently accounted for in his own conclusions and in the arguments given in their support. If such is the case, then it would seem that the point of the entire question is to refute this particular Thomistic teaching.[46]

This, then, completes the exposition of the teaching of St. Thomas and the various Thomists that are contained in Galileo's physical questions. The matter that they provide would seem to be sufficient to form some judgment of the knowledge of Thomism that is manifested in them and the sources from which it may be drawn, to which topics we now turn.

II. THOMISM AND THE SOURCES OF THE NOTEBOOKS

As can be seen from the foregoing, the notebooks contain a wealth of information that sheds light on Galileo's intellectual formation at Pisa and his general sympathies regarding philosophical issues that were being debated at the time. Here we shall have to restrict ourselves to those few points where his notes bear directly on St. Thomas and the Thomistic school, leaving for later study his relationships to nominalist and other positions. From the texts that have been cited, however, and the annotations of their likely sources that have already been given, it is possible to draw some tentative conclusions regarding Galileo's views at Pisa and the sources from which either he or his mentors drew their inspiration.

A. *The Thomism of the Notebooks*

In general, it seems fairly safe to conclude that Galileo is correctly informed on the teaching of St. Thomas and the Thomists whom he cites, and is sympathetic to their conclusions and the main lines of argument in their support. He is somewhat eclectic, however, and is not always accurate in his citation of the *loci* he purports to use. In one instance, while disagreeing with those whom he identifies as "the Thomists," he actually defends a position that was urged by Domingo Bañez, writing about that time, as the authentic Thomistic position.[47] The argumentation on this particular point was current when Galileo was composing the notes and shows an up-to-date knowledge of the literature that had recently appeared and debates going on in Spanish and Italian universities.

These general conclusions will now be substantiated, first from the viewpoint of Galileo's knowledge of, and agreement with, the teaching of St. Thomas and, second, with his comparable relationship to the Thomistic school.

Not unexpectedly, considering the atmosphere in sixteenth-century Italy, St. Thomas is held in great respect by Galileo as the foremost among Catholic Doctors and is treated with deference; this extends even to points on which he disagrees with St. Thomas but utilizes his argumentation benignly in support of his own position. The opinion or argumentation of St. Thomas is cited in ten of the questions discussed above, and in eight of these Galileo sides with Aquinas's conclusions; in the remaining two, while disagreeing with one teaching or another, he does not directly oppose any of Aquinas's statements but prefers rather to argue against the positions of his commentators. The following is a listing of these questions, with a comment on the extent of his agreement or disagreement with St. Thomas:

1. *The unity and perfection of the universe*. Galileo agrees with St. Thomas that there is only one universe (E2), and urges Aquinas's support for his teaching that God could have made the universe more perfect in an accidental way, but not essentially so (E11).

2. *Whether the world could have existed from eternity*? Galileo here presents three conclusions, the third of which he attributes to St. Thomas, although this is rejected by Thomists as not being the authentic teaching of Aquinas (F15).

3. *Whether the heavens are one of the simple bodies, or composed of them*? Galileo's conclusion is that the heavens are a body distinct from the four elements and not composed of them; this he recognizes as also the teaching of St. Thomas (I33).

4. *Whether the heavens are corruptible*? Galileo proposes a twofold conclusion: that they are probably incorruptible by nature, but not incorruptible in

such a strict sense as to limit God's power in corrupting them if he so wished, since he alone is *ens necessarium*. He invokes St. Thomas and other theologians in support of this double conclusion (J33).

5. *Whether the heavens are composed of matter and form*? Galileo again proposes a twofold conclusion: that they are so composed, but that the matter of the heavens is not the same as the matter here below (K57, 130). He adduces St. Thomas's support for both elements of this teaching (K38, 139).

6. *Whether the heavens are animated*? Here Galileo concludes that the heavens are not animated by a vegetative or by a sensitive soul, and that the problem of their being animated by an intellective soul can best be solved by holding that the intelligences are merely the movers of the heavens; in thus functioning they assist the heavenly bodies rather than inform them the way in which a soul informs a body. This he again presents as the teaching of St. Thomas (L30).

7. *On intension and remission*. Here Galileo concludes that the intensification of a quality requires more than an extrinsic change in the quality, that in such intensification the prior part of the quality does not perish, and that such intensification takes place in continuous fashion. St. Thomas is mentioned only as an authority on this topic (N2), and is not otherwise discussed, although the opinion of "the Thomists" that intensification is not continuous is rejected. (Thomists after Bañez, as has been noted, could agree with all three of Galileo's conclusions).

8. *What are the forms of the elements*? Here Galileo offers a threefold conclusion: that the forms of the elements are not proper alterative qualities and that they are not motive qualities, but that each element has its proper substantial form distinct from all others. This he proposes as the teaching of St. Thomas (S8).

9. *Whether the forms of the elements undergo intension and remission*? Here Galileo answers in the negative, giving the arguments of St. Thomas and the Thomistic school in support of his conclusion (T4–13).

10. *Whether elements and other natural things have termini of largeness and smallness*? Here Galileo offers ten conclusions, most of which resolve arguments raised in the nominalist schools that flourished after St. Thomas's death. Of the three conclusions that relate to Aquinas's thought, he is in agreement with two, one regarding God's power with respect to maxima and minima (U33), and the other regarding the termini of living things and heterogeneous compounds (U47), and rejects the third, regarding the termini of elements and homogeneous compounds (U59).

Of the 24 sub-conclusions that go to make up the answers to these ten questions relating to Aquinas's teaching, therefore, only two or three conclusions, depending on how one views the continuity of intensification as Thomas's teaching, imply a rejection of the Angelic Doctor. These two or three are basically Scotistic conclusions; and all three, perhaps by coincidence but noteworthy nonetheless, were taught by Vallesius, whereas the two less arguable ones were taught also by Pererius.[48]

The situation with respect to the Thomists falls into the same general pattern as that with respect to St. Thomas, except that in the former case Galileo is less restrained in expressing his disagreement with one or other member of the school. As we have seen, he cites seven Thomists: Herveus, Capreolus, Soncinas, Cajetan, Ferrariensis, Javelli, and Soto. Among these, there is little evidence of any serious study of Herveus or Ferrariensis, whereas for the remaining five it appears that their texts were analyzed with some degree of care. On one question Galileo prefers Cajetan's teaching to that of Capreolus, on another, Capreolus's over that of Cajetan; he is definitely opposed to Soncinas's teaching on intensification; he uses Soto extensively for Thomistic views on maxima and minima, while disagreeing with some of his conclusions; Javelli he seems to find useful as a compendium of Thomistic teachings, particularly as a guide to the thought of Capreolus and Soncinas.

A more detailed characterization of Galileo's agreements and disagreements with the Thomistic school will become apparent from the following listing of the topics in relation to which their teachings are discussed:

1. *The nature of celestial matter.* Galileo agrees with the Thomistic teaching that the heavens are composed of matter and form (K57) and that their matter differs from the earthly matter of the sublunary region (K130). He prefers Cajetan's explanation to Capreolus's as to the nature of this heavenly matter (K159), and also sides with Cajetan (against Capreolus and Soncinas) in holding that there is a different type of matter in each of the heavenly spheres (K170).

2. *The continuity of intensification.* Here Galileo is directly at variance with the teaching of Herveus, Capreolus, Soncinas, and Javelli (N2, 31). Soto has nothing explicit on this thesis, although he does lay the groundwork for a different interpretation of intensification in his tract on natural minima, from which Bañez was to draw his inspiration for what was to become the accepted Thomistic teaching.[49] Galileo was acquainted with this section of Soto's questionary on Aristotle's *Physics* (U25, 34), but apparently did not grasp its connection with the problem of the intensification of qualities.

3. *The intensification of elemental forms.* Here Galileo's solution is

identical with that of the Thomistic school, among whom he enumerates the teaching of Herveus, Capreolus, Soncinas, Cajetan, and Javelli on this point (T4). His own arguments are directed against Averroës and his followers, and also against Scotus and his school.

4. *Problems associated with maxima and minima*. Among the ten conclusions reached by Galileo in his attempt to resolve these problems, only two bear on the teachings of the Thomistic school. The first is that, in relation to God, things do not have maxima and minima in the sense that God can always make them larger and smaller, but not in the sense that they will become actually infinite in a categorematic sense.[50] In stating this conclusion, he opposes himself to Cajetan and aligns himself with Capreolus, while using Soto's citation of a text of Aquinas in support of Capreolus's interpretation (U34–35). The second conclusion relates to the termini of largeness and smallness as found in elements and homogeneous compounds, and here Galileo rejects Cajetan's opinion that elements have an intrinsic terminus of largeness – an opinion that is rejected as unintelligible by Soto[51] and is not common to the school. Galileo rejects also the teaching of Thomists generally that elements have an intrinsic terminus of smallness (U59), using mainly Soto's arguments in his rejection of this conclusion, and opting rather for the Scotistic solution.

To summarize, then, on two of the four topics (n. 1 and n. 3), Galileo is in complete agreement with the Thomistic school, and on one topic (n. 1) engages in an intra-mural dispute wherein he favors the more recent opinion (Cajetan's) over an earlier teaching (Capreolus's). On a third topic (n. 4), he agrees with the more theological conclusion of the Thomistic school (God's power), while disagreeing with its thesis bearing directly on natural philosophy (elemental minima); his intramural arguments here favor Capreolus over Cajetan. On a fourth topic (n. 2), while disagreeing with a Thomistic teaching of his time, he himself argues for a position that was later to be accepted as the authentic teaching of Aquinas. When these results are taken together with Galileo's attitude towards St. Thomas himself, they support the conclusion that he is well acquainted with the teaching of Thomists and is generally sympathetic to this school.

B. *The Sources of the Notebooks*

This brings us to our final consideration, namely, that of the implications of this study for a better understanding of the sources of Galileo's early writings. Here only a brief indication will be given with respect to four different hypotheses: (1) that the notebooks are original with Galileo, representing his own work with primary sources: (2) that they were copied by Galileo from printed secondary sources; (3) that they were Galileo's class notes based on the lectures of one or more of his professors at the University of Pisa; and (4) that they were a summary Galileo prepared for himself with an ulterior motive in mind by borrowing or cribbing from the handwritten notes of others, possibly *reportationes* of lectures he himself had not attended.

The first hypothesis, as has been pointed out by Favaro (*Opere* 1: 10–12), is quite unlikely considering the neatness of the autograph and clues of its having been transcribed from another source. These include several spaces left vacant in the original writing and then filled in later either with cramped lettering or with sentences flowing over into the margins. These are also expressions that are written down, then crossed out as not making sense in the context, only to appear a line or two below in their proper place. Yet the transcription, if such it is, was not made slavishly, as there are also evidences of words being changed and expressions being altered to convey a clearer and more consistent sense. Moreover, it could be that Galileo was actually recopying his own poorly written notes; in this connection, there is certainly no a priori reason for excluding the possibility that such notes were the composition of even a twenty-year old university student. The work on Thomistic theses that we have analyzed is not appreciably superior to what a bright twenty-year old Italian or Spanish seminarian might do even in our own day. In the century before Galileo, Giovanni Pico della Mirandola had, at the age of 23, challenged all comers to debate 900 selected theses in philosophy, theology, and science, and it should be remembered that Galileo was as much a genius as Pico in his own right. The question as to what primary sources might have been available to Galileo for such a composition poses no serious problem. Among the authors we have discussed the most recent would be the works of Pererius (Rome 1576), Vallesius (Alcala 1556), Javellus (Venice 1555), and Ferrariensis's commentary on the *Physics* (Venice 1573). Although Soto's commentary and questionary on the *Physics* was not published in Italy until the Venice 1582 edition, at least six Spanish editions had appeared before this date, beginning *c.* 1545. The author has found copies of Spanish editions of Soto's works in Italian libraries, and suspects that Vallesius's work was similarly available in Italy shortly after its publication. Thus there is nothing to exclude the possibility that Galileo could have worked with these sources and made his own synthesis from them. The strongest argument against this hypothesis, of course, is that the notes contain too many references. Galileo was notorious for having read *very* little, and although a bright young man of those days could have controlled a hundred books of the kind cited, it is extremely unlikely Galileo really did so.

The hypothesis that the notebooks were copied from secondary sources is still tenable in the light of our study. It should be noted, however, that there is no evidence of direct copying from any of the Dominican authors mentioned in this study. The only printed works that give evidence of being studied by Galileo, and that could have been used in the composition of the notebooks, are those of Clavius, Toletus, and Pererius discussed in the following essay in this volume. None of the remaining sixteenth-century printed works listed in that essay (Table II) shows recognizable similarities with Galileo's text.

The third hypothesis, that the notebooks are Galileo's class notes based on lectures given at the University of Pisa, presents a problem when one attempts to identify the professor who could have delivered the lectures. To the author's knowledge only two candidates have been proposed to date, Francesco Buonamici and Flaminio Nobili. With regard to Buonamici, the results of this study would be adverse to his identification as the professor involved. A perusal of Buonamici's *De motu* shows a quite different citation of authors from that in the notebooks, with strong emphasis placed on Averroës and classical commentators such as Alexander of Aphrodisias and Simplicius.[52] Buonamici cites St. Thomas with about half the frequency of Averroës, and occasionally mentions *Thomistae*, but he does not identify any Thomists nor does he discuss their teachings in any detail. He accepts the Averroist teaching on the forms of the elements, and has not progressed beyond Walter Burley and James of Forli in his discussion of the intension and remission of qualities.[53] Thus the general tenor of his thought is quite different from that contained in the physical questions. Certainly Buonamici would not be identified as a Thomist or as one sympathetic to the teachings of this school.

Because of the difficulty of tracing any correspondence between the thought patterns of Buonamici and the young Galileo, Eugenio Garin has called attention to Galileo's "quotations from the lectures of Flaminio Nobili."[54] From the viewpoint of this study, however, there are difficulties with this suggestion also. Whereas Buonamici is not cited in the notebooks, Nobili is cited three times, and in two of these the author is quite harsh in his rejection of Nobili's teaching. The first rejection is with regard to the question, "What are the forms of the elements?" Nobili's teaching is listed as the first opinion, "that the forms of the elements are something made up of primary qualities and a kind of substantial form" (S2). Galileo goes on: "But this is unintelligible, nor does he [Nobili] seem sufficiently to understand the nature of a substantial form" (*ibid.*). The second rejection occurs in the context of a discussion as to how the primary qualities are related to resistance. After giving his own conclusions, Galileo writes: "From this is apparent the error of Nobili, who, in the first *De generatione*, doubt 11 in chap. 7, distinguishes a twofold resistance: one of animals, which would consist in a certain effort that is a kind of action; another in other things, which he reduces to an impotency for receiving; here, as you can see, he takes the extrinsic cause of resistance for resistance formally, when they should be differentiated." (Y8). The third citation is merely an enumeration of Nobili along with St. Thomas and others as holding that the forms of the elements do not undergo intension and remission (T4). All three citations of Nobili refer to his *De generatione*; even from these brief indications of its contents, one could hardly hold that Nobili was the professor hidden behind the notebooks.

This leaves the fourth hypothesis, which, at the original writing of this essay, gained its plausibility mainly from the exclusion of the other three. At that time (1971), the author had concluded that, should Galileo's manuscript have been based on the handwritten notes of others, these would derive from a professor or professors who had a consistently good knowledge of Thomism. Other characteristics then attributed to this hypothesized source include the following: (1) an eclectic Aristotelian with scholastic leanings, quite well acquainted with and sympathetic to the teachings of the Thomistic school, but accepting nonetheless some Scotistic theses and a few interpretations deriving from Averroës; (2) one knowledgeable with respect to the classical Greek commentaries, who also knew and appreciated the nominalist arguments that were common among the "Latins"; and (3) a person well acquainted with developments at both Spanish and Italian universities, and sympathetic to the writings of two members of the newly-formed Society of Jesus, Pererius and Toletus (Wallace, 1974b, p. 327).[55]

Subsequent research, as detailed in the three following essays and in *Galileo's Early Notebooks* (1977), has revealed that the physical questions are based, almost in their entirely, on *reportationes* of lectures given by Jesuit professors at the Collegio Romano around the year 1590. Some of the *reportationes* the author has studied date from as early as 1565–1566, others as late as 1596–1597, but the materials taught in the years 1589–1591 show the closest agreement with Galileo's text. The lecturers involved, together with the dates of their "reported" lectures, are the following: Benedictus Pererius, 1565–1566; Hieronymus de Gregoriis, 1567–1568; Antonius Menu, 1577–1579; Paulus Valla, 1588–1589; Mutius Vitelleschi, 1589–1590;

Ludovicus Rugerius, 1590–1591; Robertus Jones, 1592–1593; and Stephanus del Bufalo, 1596–1597.

For purposes of completeness, all of the references relating to St. Thomas and the Thomists reported in the present essay are listed in the Table below. These are keyed to Galileo's text by the sigla explained in note 16, next to which are indicated degrees of correspondence to the *reportationes* of the lecturers listed in the previous paragraph. Three degrees of correspondence are noted: *Excellent* signifies that the agreement is close enough to suggest almost verbatim copying; *Good* means that the expression and content are quite similar; and *Fair* means that the content is roughly the same, though the manner of expressing it differs. Fuller paleographical and historical details relating to the composition of the notebooks are given in the Introduction and Commentary contained in *Galileo's Eary Notebooks.*

TABLE I

Paragraphs in Galileo's Text	Degree of Correspondence With Jesuit Lecturers at the Collegio Romano[56]
A2	*Good*: De Gregoriis, Vitelleschi, Rugerius
B8	*Good*: Vitelleschi, Rugerius
E2	*Good*: De Gregoriis, Menu, Vitelleschi
7	*Good*: Menu, Vitelleschi
8	*Good*: Menu
11	*Good*: Menu, Vitelleschi
F1	*Excellent*: Pererius (p); *Good*: Menu, Vitelleschi; *Fair*: Rugerius
9	*Excellent*: Pererius (p); *Fair*: Vitelleschi
11	*Excellent*: Pererius (p); *Good*: Menu, Vitelleschi, Rugerius, Toletus (p)
15	*Excellent*: Pererius (p); *Good*: Vitelleschi, Rugerius, Toletus (p)
18	*Excellent*: Pererius (p); *Good*: Menu, Vitelleschi, Rugerius
19	*Good*: Vitelleschi; *Fair*: Rugerius
23	*Excellent*: Rugerius; *Good*: Vitelleschi
24	*Good*: Vitelleschi
I 8	*Good*: Vitelleschi, Rugerius; *Fair*: Menu
9	*Good*: Menu, Vitelleschi, Rugerius, Pererius (p)
33	*Good*: Menu
J 31	*Good*: Vitelleschi; *Fair*: Rugerius
33	*Good*: Vitelleschi
K38	*Good*: Menu, Vitelleschi; *Fair*: Rugerius
39	*Good*: Menu, Vitelleschi; *Fair*: Rugerius
57	*Good*: Menu, Vitelleschi, Rugerius, Pererius (p)
74	*Good*: Menu, Vitelleschi

TABLE I Continued

Paragraphs in Galileo's Text	Degree of Correspondence With Jesuit Lecturers at the Collegio Romano[56]
75	*Fair*: Menu
76	*Good*: Vitelleschi
77	No recognizable agreement
78	*Good*: Menu
92	*Fair*: Menu
94	*Good*: Menu
K130	*Good*: Menu, Vitelleschi, Pererius (p); *Fair*: Rugerius
139	No recognizable agreement
159	*Good*: Menu
160	*Fair*: Menu
170	*Good*: Menu; *Fair*: Pererius (p)
172	*Fair*: Menu
173	*Good*: Menu
174	*Good*: Menu
L13	*Good*: Menu, Vitelleschi, Rugerius; *Fair*: Pererius
24	*Good*: Menu, Vitelleschi, Rugerius
29	*Good*: Pererius, Vitelleschi
30	*Good*: Vitelleschi, Rugerius
N1	*Good*: Rugerius; *Fair*: Vitelleschi
2	*Good*: Menu, Vitelleschi, Rugerius
9	*Good*: Vitelleschi; *Fair*: Pererius, Menu, Rugerius
12	*Good*: Vitelleschi, Rugerius
18	*Good*: Menu, Vitelleschi, Rugerius
28	*Good*: Rugerius; *Fair*: Pererius, Vitelleschi
29	*Good*: Rugerius; *Fair*: Pererius, Menu, Vitelleschi
30	*Good*: Pererius, Vitelleschi, Rugerius
31	*Good*: Pererius, Rugerius; *Fair*: Vitelleschi
O6	*Good*: Vitelleschi, Rugerius; *Fair*: Pererius
P8	*Good*: Menu, Valla, Vitelleschi, Rugerius
S8	*Good*: Pererius, Menu, Valla, Rugerius
T2	*Excellent*: Valla; *Good*: Menu, Rugerius
4	*Good*: Menu, Valla, Rugerius
5	*Good*: Rugerius
U9	*Good*: Valla; *Fair*: Pererius, Manu, Vitelleschi
25	*Excellent*: Valla; *Good*: Menu, Vitelleschi, Rugerius
26	*Good*: Menu, Valla, Vitelleschi
27	*Good*: Menu, Valla; *Fair*: Vitelleschi
28	*Good*: Valla, Vitelleschi; *Fair*: Pererius
29	*Good*: Menu, Valla, Vitelleschi
33	*Good*: Pererius, Valla, Vitelleschi, Rugerius
34	*Excellent*: Vitelleschi; *Good*: Valla, Toletus (p)
35	*Good*: Vitelleschi; *Fair*: Rugerius

TABLE I Continued

Paragraphs in Galileo's Text	Degree of Correspondence with Jesuit Lecturers at the Collegio Romano[56]
47	*Good*: Pererius, Valla, Vitelleschi, Rugerius, Pererius (p), Toletus (p)
59	*Good*: Menu, Valla, Vitelleschi, Rugerius; *Fair*: Pererius
64	*Good*: Menu, Valla, Vitelleschi; *Fair*: Pererius
76	*Good*: Menu, Valla, Vitelleschi
77	*Good*: Valla, Vitelleschi; *Fair*: Menu
78	*Good*: Vitelleschi
79	*Good*: Valla, Vitelleschi; *Fair*: Menu
80	*Good*: Menu, Valla
X15	*Good*: Menu, Valla, Vitelleschi; *Fair*: Rugerius

NOTES

1 Gaetano da Thiene (1387–1465), professor at Padua from 1422 to his death, and author of important commentaries on William Heytesbury and on the *Physics* of Aristotle.

2 Tommaso de Vio, of Gaeta (1469–1534), cardinal, commonly known as Gaetanus or Caietanus from his birthplace; master general of the Dominican Order and principal commentator on the *Summa theologiae* of St. Thomas.

3 See A. Favaro, ed., *Le Opere di Galileo Galilei*, Edizione Nazionale, 20 vols., (Florence 1890–1909), Vol. I, 422; henceforth cited as *Opere* 1: 422. Caietanus Thienensis is cited only on pp. 72 and 172, whereas Thomas de Vio Caietanus is cited on pp. 76, 96, 101, 133, 146, and 153. Favaro ascribed all these references to Thienensis.

4 See *Opere* 5: 347 and index for the entire work, 20: 339.

5 The letter, written in 1615, has been translated into English by Stillman Drake, *Discoveries and Opinions of Galileo* (New York 1957). For the reference to Thomas de Vio and its context, see p. 214; Drake correctly identifies the Italian Dominican.

6 These data are taken from a simple count of Favaro's entries in the "Indice degli autori citati," *Opere* 1: 421–423. Other authors who are frequently cited include Ptolemy (on 20 pages), Alexander of Aphrodisias (on 19 pages), Plato (on 18 pages), John Philoponus (on 17 pages), Albertus Magnus (on 15 pages), Giles of Rome (on 14 pages), and Archimedes (on 10 pages). Omitted from the index entirely are the Calculator, Richard Swineshead, cited on p. 172, and the Conciliator, Pietro d'Abano, cited on p. 36.

7 John Capreolus, (*c.* 1380–1444), French Dominican and principal expounder and defender of Thomistic doctrine in the century before Cajetan; known especially for his *Defensiones*, cast in the form of a commentary on the *Sentences*, and directed against Duns Scotus and Henry of Ghent, among others.

8 Paul Soncinas (d. 1494), Italian Dominican and admirer of Capreolus, who published a compendium of the latter's work as well as a lengthy Thomistic exposition of the *Metaphysics* of Aristotle.

9 Francesco Silvestri of Ferrara (*c.* 1474–1528), Italian Dominican noted for his

commentary on the *Summa contra gentiles* of St. Thomas; among his works is a commentary on the *Physics* of Aristotle.

10 Harvey or Hervé Nedellec (*c.* 1255–1323), master general of the Dominican Order and author of numerous *Quaestiones* and *Quodlibets*; his polemics were frequently directed against Henry of Ghent.

11 Domingo de Soto (1494–1560), Spanish Dominican, theologian and political theorist, but also a commentator on Aristotle's logic and physics; he was regarded by Pierre Duhem as a precursor of Galileo on the basis of his adumbration of the law of falling bodies *c.* 1545. For details, see Essays 3 through 6 *supra.*

12 Giovanni Crisostomo Javelli (*c.* 1470–*c.* 1538), Italian Dominican, student of Cajetan and colleague of Ferrariensis, who commented on the works of Aristotle and defended a Thomistic interpretation against the Averroists of his day.

13 Since preparing the original version of this essay, the author has translated and commented on the portion of these notes transcribed by Favaro in *Opere* 1: 15–177. (See his *Galileo's Early Notebooks: The Physical Questions.* A Translation from the Latin, with Historical and Paleographical Commentary. Notre Dame: University of Notre Dame Press, 1977.) Favaro entitled his transcription the *Juvenilia*, or youthful writings, on the basis that they were written in 1584 while Galileo was yet a student at Pisa. The author now questions this dating, and argues that they were written *c.* 1590, when Galileo was already teaching at the University of Pisa. In the original version of this essay, he accepted uncritically both Favaro's dating of the notes and his titling of them as *Juvenilia.*

14 Treating of the sources of Galileo's ideas in 'Galileo's Precursors,' *Galileo Reappraised*, ed. C. L. Golino (Berkeley 1966), 41, Moody writes: "In what form, or through what books, these ideas were conveyed to Galileo, are questions that, if answerable in whole or in part, would cast a good deal of light on the way Galileo's thinking developed. If historians of science had given more time to historical research on this problem, instead of engaging in a priori debates over the validity or invalidity of Duhem's thesis, better insight into the nature of Galileo's scientific achievements might well have been gained."

15 The references in passing occur in *Opere* 4: 421; 12: 265; 14: 260; and 19: 298, 319. The remaining citations are located, verified and in most cases explained in detail in what follows.

16 The letters A, B, C, etc., refer to the sigla used by the author in *Galileo's Early Notebooks* to designate the questions of which the treatise is composed. Arabic numerals following these letters designate paragraph numbers within these questions as they are found in the same work.

17 The more common Thomistic opinion would agree with the first two conclusions but would disagree with the third, maintaining that there is no repugnance on the part either of the act of production or of the thing produced (whether corruptible or incorruptible) that creatures with a stable nature should have existed from eternity. Apart from St. Thomas, this was held by Capreolus, Cajetan, Ferrariensis, Soncinas, Javelli, Soto, and Bañez. St. Bonaventure, on the other hand, held the third conclusion as stated by the author; so did Philoponus, Henry of Ghent and, among the late sixteenth-century authors, Toletus and Vallesius. Giles of Rome, Scotus, and Pererius argued dialectically on both sides of the question and came to no firm conclusion. Here Galileo, surprisingly enough, identifies the opinion of St. Thomas with that of Scotus and Pererius (see note 19 *infra*).

[18] On the *Doctores Parisienses*, see Essay 10 *infra*.

[19] St. Thomas does not state the conclusion in this way in any of the foregoing references nor would most Thomists say that it is his conclusion (see note 17). In *De aeternitate mundi* he does allude to the question of infinites in act, but this is merely to state that the question has not yet been solved (Marietti ed., n. 310, p. 108). The terminology used by the author is to be found, however, in later scholastics and particularly in Soto's *Questiones* on the *Physics* of Aristotle, Bk. 3, qq. 3–4, and Bk. 8, qq. 1–2.

[20] Citing Aquinas's commentary on the *De caelo*, Bk. 3 [lect. 8] and the *Metaphysics*, Bk. 7 [lect. 1]. Galileo here appears to be quoting Aquinas, whereas in actuality he is summarizing and interpreting his argument.

[21] The expression "potentia obedientialis" derives from Aquinas (*De virtutibus in communi*, q. un., a. 10 ad 13) and gained acceptance among theologians in later centuries; other medievals, such as Alexander of Hales and St. Bonaventure, used the related expression "potentia obedientiae."

[22] The best exposition and analysis of Aquinas's teaching on this subject is Thomas Litt, *Les Corps célestes dans l'univers de saint Thomas d'Aquin* (Louvain 1963), pp. 54–90.

[23] This inconsistency is discussed at length by Ferrariensis in his commentary on the *Contra gentiles*, Bk. 2, c. 70, n. 3, where he offers reasons why St. Thomas may have wished to leave his interpretation of Averroës open on this point.

[24] Thomas speaks of a greater participation of the form by the subject (1a2ae, q. 52, a. 2; 2a2ae, q. 24, a. 5), but within his school it became common to speak of the form being more "radicated" in the subject.

[25] Galileo Galilei, *On Motion* and *On Mechanics*. Comprising *De Motu* (*c*. 1590), translated with Introduction and Notes by I. E. Drabkin, and *Le Meccaniche* (*c*. 1600), translated with Introduction and Notes by Stillman Drake (Madison 1960); see p. 1–12.

[26] Some further details are given in Essays 14 and 15 of this volume.

[27] In this and other texts quoted in this section, unless otherwise indicated the author has verified all citations of St. Thomas and Thomists and noted any errors on Galileo's part; he has not done this, however, for other authorities cited. For St. Thomas he has used the standard Leonine and Marietti editions, together with the Mandonnet-Moos edition of the commentary on the *Sentences*, while for the Thomists he has used the following: Capreolus, *Defensiones theologiae divi Thomae Aquinatis*, ed. C. Paban and T. Pègues, 7 vols., (Tours 1900–1907); Cajetanus, *Commentaria in Summam Theologiae Sancti Thomae* as printed in the Leonine edition of the *Summa*, and *Commentaria in De anima Aristotelis*, ed. P. I. Coquelle (Rome 1938); Soncinas, *Acutissime questiones metaphysicales* (Venice 1505); and Ferrariensis, *Commentaria in Summam Contra Gentiles*, as printed in the Leonine edition of the *Contra Gentiles*.

[28] The reference to Capreolus is *In 2 sent*., d. 12, concl. 2a; Cajetan's argument is in *In Iam*, q. 66, a. 2, n. 7, and is directed against Giles of Rome and Scotus.

[29] Here Galileo cites Soncinas erroneously as the tenth book of his *Metaphysics*, q. 10, whereas this teaching is actually found in his twelfth book, q. 10. The "others" possibly refers to Javelli, *In 10 meta*., q. 22. For Javelli the author has used *Totius rationalis, naturalis, divinae ac moralis philosophiae compendium . . . his adjecimus in libros physicorum, de anima, metaphysicorum ejusdem questiones . . .* , 2 vols. (Lyons 1568).

[30] In the place cited (K159), Capreolus is arguing against Durandus, Aureoli, and Giles of Rome. The "others" probably refers to Soncinas, *In 12 meta*., q. 7, and Javelli, *In 10 meta*., q. 22.

31 In verifying citations from Herveus the author has used his *Quodlibeta undecim cum octo . . . tractatibus* (Venice 1513), which includes the treatise *De unitate formarum.*

32 Most of the distinctions on which this discussion is based are to be found in Soncinas, *In 8 meta.*, q. 22, where he is inquiring whether the intensification of forms is a continuous motion. The discussion of freezing (*congelatio*) is quite similar to that in Toletus, or Francisco de Toledo (1533–1596), *Commentaria in libros de generatione et corruptione Aristotelis* (Venice 1575), Bk. 1, q. 6, 4 concl.

33 Galileo's particular mode of classification is not found in any of the Thomistic authors cited, although it is consistent with Soncinas's discussion in *In 8 meta.* q. 22, and with Capreolus's sixth conclusion in *In 1 sent.*, d. 17, q. 2, a. 1.

34 This statement would seem to align Galileo with the nominalist view of motion, rather than with the realist view which identified motion as a "fluxus formae." For a discussion of the distinction between "forma fluens" and "fluxus formae," see E. J. Dijksterhuis, *The Mechanization of the World Picture*, tr. C. Dikshoorn (Oxford 1961), 174–175. It is noteworthy, however, that Herveus Natalis speaks of motion in the same manner: " . . . illud non potest esse aliud nisi forma fluens que communiter ponitur esse ipse motus . . . " *Quodlibet* 2, q. 13 (Venice 1513), fol. 59ra. Herveus's arguments in this locus are similar to those discussed in the physical questions.

35 The position that the prior part of the quality perishes in intensification is identified by Javelli (*In 8 meta.*, q. 6) as that of Walter Burley [in his *De intensione et remissione*] and of Gregory of Rimini, *In 1 sent.*, d. 17, q. 4, a. 2. Gregory, in this locus, attributes the position to Burley and to Godfrey of Fontaines.

36 The Thomists to whom Galileo refers seem to be Herveus, *Quodlibet* 2, q. 13, and Soncinas, *In 8 meta.*, q. 22. According to Soncinas, the intensification of forms can be continuous in three senses: (1) in the Heraclitean sense of continually going on; (2) in the sense of deriving continuity from the parts of the subject being altered, either without the corruption of a previous part as in heating or with corruption of a previous part as in illuminating; and (3) in the sense of being continuous on the part of the form being intensified. Soncinas holds that St. Thomas denies that the intension of forms is continuous in the third of these senses. He writes: "St. Thomas holds the contrary opinion, namely, that the intension of forms is not a motion that is strictly and completely continuous; in fact, an intermediate rest intervenes, since the altering body, if it is sufficiently close to the patient, begins to act on the patient causing in it as perfect a form as it can; and afterwards the body acted upon rests for some time under that form, and is uniformly disposed according to it until, either because of a greater disposition of the subject or because of more power in the agent or from some other cause, it can make that form more perfect" (fol. 109rb). This teaching, it should be noted, was not accepted by Spanish writers under the influence of Domingo de Soto, being rejected as not authentically Thomistic by Toletus and then, in more definite fashion, by Domingo Bañez in his *Commentaria et Quaestiones in duos Aristotelis de generatione et corruptione libros* (Salamanca 1585), a work approximately contemporaneous with the writing of the physical questions. See note 47 *infra.*

37 The arguments Galileo adduces are given by Javelli, *In 8 phys.*, q. 7, as objections to the Thomistic position, although he connects no names with the objections, speaking only of "via nostra" and the "multi" who are opposed to it (ed. cit., pp. 602–604). Diego de Astudillo, O. P., in his *Questiones super libros de generatione et corruptione* (Valladolid 1532), Bk. 1, q. 10 traces the opposition back to Burley and "all the

nominalists," who hold that the motion of intensification is "simpliciter continuus"; his own position he states as follows: "motus intensionis et remissionis non est continuus sed successivus . . . multae mutationes instantanee sibi succedentes" (fol. 21ra). Toletus, while arguing in the same vein as Galileo, does not identify the source of his arguments, except to mention a special teaching of Giles of Rome, that alteration is discrete with respect to the parts of the subject but continuous with respect to intensification, and to note that John of Jandun seems to follow Giles. The position that alteration is continuous in both of Giles's respects is identified by Toletus as "fere communis" (ed. cit., fol. 16vb).

38 The arguments here are drawn from Soncinas, *In 8 meta.*, q. 22, fol. 110rb; Galileo's mention of Giles and Jandun, together with the form of his arguments, suggests some dependence on Toletus, *In 1 de gen.*, q. 6, la concl., as mentioned in the previous note. Immediately following the text cited, in N32, Galileo goes on to answer an objection; this too is taken from Soncinas, loc. cit., fol. 109va.

39 The *De mixtione* of St. Thomas has the fuller title, *De mixtione elementorum ad magistrum Philippum*, and is printed in the Marietti ed. of the *Opuscula philosophica* (Rome 1954), pp. 155–156; see especially n. 433, p. 155, and nn. 436–437, p. 156. Galileo's reference to Aquinas's *In 2 Sent.*, d. 15, may be based on conjecture, since Capreolus has a long discussion of the subject in this *locus*; St. Thomas has only a tangential reference to it in q. 2, a. 1. Cajetan's arguments are directed against Scotus, Henry of Ghent, and the Averroists; see Vol. 5 of the Leonine ed., n. 19, p. 227. Javelli discusses the intension and remission of substantial forms in both q. 4 and q. 6, but not in q. 5, although he does state in q. 6, "hanc opinionem pertractavimus in q[uaestione] praecedenti," whereas the treatment is actually to be found in q. 4; perhaps this is the source of Galileo's miscitation.

40 Note that Galileo's method of citing Aquinas's commentary on Aristotle here (and in S8) is not the usual one, and probably indicates that he took the citations from a secondary source. See the Table at the end of this essay. To verify the reference to Soto the author has used the 2d ed. of his *Quaestiones super octo libro physicorum Aristotelis* (Salamanca 1555) and have checked this against the Venice 1582 edition, which may have been available to Galileo. Among "all Thomists" one would have to include Cajetan, Ferrariensis, and Javelli; other Thomists who taught this doctrine but are not mentioned in the notebooks include Gratiadei (John of Ascoli), Peter Crokart of Brussels, and Diego de Astudillo; Toletus also followed the Thomistic teaching on this subject.

41 Benedictus Pererius, or Benito Pereyra (*c.* 1535–1610), Spanish Jesuit philosopher and Scriptural exegete who spent most of his teaching life in Rome. His *De communibus omnium rerum naturalium principiis et affectionibus* (Rome 1576) is cited five times in Vol. I of Galileo's *Opere*, approvingly for its treatment of questions relating to the eternity of the universe and critically for its discussion of falling bodies. Pererius is somewhat eclectic but in his preface accords St. Thomas a place of honor among the philosophers, while favoring at times Scotistic or nominalist teachings himself. The Thomists he cites include Herveus, Capreolus, Cajetan, Ferrariensis, and Soncinas (see pp. 173, 197, 223, 227, 265).

42 Francisco Vallés (1524–1592), Spanish philosopher and physician who composed an exposition and Commentary on Aristotle's *Physics* (Alcala 1562), to which he appended his *Controversiarum naturalium ad tyrones pars prima, continens eas quae spectant ad octo libros Aristotelis de physica doctrina* (Alcala 1563); the latter is cited by Galileo

(Y1). Vallés taught at Alcala, where he was a friend of Gaspar Cardillo de Villalpando (1527–1581), a classical Aristotelian who also commented on the *Physics* (Alcala 1566), but whose texts were soon replaced at Alcala by Soto's more scholastic commentaries. Vallés mentions Soto in his *Controversiae* but is generally opposed to his teaching.

43 Cajetan is aware that he is disagreeing with the majority Thomistic opinion on this matter; mentioning the *Thomistae*, he adds "inter quos forte ego aliquando fui" [*In 2am 2ae*, q. 24, a. 7, n. 3 (Leonine ed., Vol. 8, p. 183)]. Capreolus's arguments are directed against Aureolus, Durandus, Ockham, and Scotus, among others. Soto is aware of Cajetan's arguments, cites two of the three *loci* given by Galileo, and gives a fairly lengthy refutation of Cajetan's teaching.

44 This is the thesis of Vallesius, Villalpandus, and Pererius; it is opposed to the teaching of Soto, Javelli, Astudillo, and Toletus, among others, including the Averroist John of Jandun.

45 Cajetan's arguments are directed against Scotus [Leonine ed., Vol. IV, pp. 76–79, esp. n. 8 (p. 77) and n. 12 (p. 77)]. These arguments are discussed at length by Soto, *In 1 phys.*, q. 4, a. 2.

46 The five arguments are given in U25–29, and their refutation in U76–80. The Thomistic sources on which they are based are difficult to verify, but the following represents the preliminary results of the author's research into these aspects of Thomistic teaching:

U25: The argumentation seems to be drawn from Soto, *In 1 phys.*, q. 4, aa. 2 and 3.

U26: The argumentation is probably drawn from Soto, *In 1 phys.*, q. 4, a. 2, and from Javelli, *In 1 phys.*, q. 19, who gives essentially the same reasoning as Soto. Others who earlier argued in similar terms include Astudillo, *In 1 phys.*, q. 8, and Peter Crokart of Brussels, *In 1 phys.*, q. 2, a. 5. These last-named little known Dominicans were influential in the formation of Francisco Vittoria and Soto and thus indirectly influenced the rise of "second scholasticism." Astudillo's commentary and questions on the *Physics* was finished on July 4, 1530, at five o'clock in the morning, as he states in the colophon, and is bound with his questions on the *De generatione* (Valladolid 1532); Peter Crokart's arguments are to be found in *Argutissime, subtiles et fecunde questiones phisicales magistri Petri de Bruxellis, alias Crokart* (Paris 1521).

U27: This brief argument is expanded at considerable length by Soto, Javelli and Crokart in the *loci* cited, and by Cajetan in his commentary on the *De anima*, chap. 4, q. 4, and *In Iam.*, q. 7, a. 3, nn. 5, 6, and 9.

U28: This is based on Soto, *In 1 phys.*, q. 4, a. 3. The argument is also implicit in Javelli, *In 1 phys.*, q. 19, in his refutation of the Scotistic position.

U29: This type of argument is to be found in Peter Crokart, *In 1 phys.*, q. 2, a. 5, but otherwise is not to be found in the Thomists mentioned in the physical questions, apart from a passing remark in Soto, *In 1 phys.*, q. 4, a. 3, which might be interpreted in this fashion.

U76: The author has not been able to locate the source of these arguments; they are somewhat similar to the exegesis of Aristotle offered by Vallesius in his *Controversiae*, nn. 4 and 5, and are possibly based on Pererius or Villalpandus.

U77: reading *requirat* for *requirant* in line 34. The line of reasoning here seems to be Scotistic and is similar but not identical to that discussed and refuted by Javelli, *In 1 phys.*, q. 19. It is also stated and answered by Peter Crokart, *In 1 phys.*, q. 2, a. 5.

U78: The author has not been able to locate this argument in precise form, although it is somewhat similar to that offered by Vallesius, *Controversiae*, n. 5 (fol. 7r) and may be based on Pererius or Villalpandus. An intimation of the argument is also to be found in Crokart, loc. cit.

U79: A Scotistic argument, similar to those refuted by Javelli, loc. cit.

U80: This line of reasoning is found as early as the mid-fourteenth century in Albert of Saxony's questions on the *Physics* (Venice 1516), Bk. 1, q. 9, "Utrum cognitio totius dependeat ex cognitione suarum partium?" It is repeated by authors in the nominalist tradition and by those arguing against them, such as Crokart, *In 1 phys.*, q. 2, a. 5.

[47] See note 36 *supra*. Bañez devotes seven questions in his commentary on Bk. 1 of *De generatione* to the subject of alteration (ed. cit., pp. 50–82); of these, q. 6 is entitled "Utrum alteratio sive intensio sit motus continuus?" (pp. 73–77). In the first article of this question Bañez analyzes the teachings of Thomists, including Capreolus, Soncinas, Javelli, Ferrariensis, and Astudillo, and concludes that their view is erroneous and not consonant with St. Thomas's teaching. His second article is devoted to the question "Utrum detur minimum in accidentibus que intenduntur?", and here he analyzes the roots of his teaching in Soto's questions on the first book of the *Physics*, q. 4, a. 3, and refutes the interpretation of "quidam novus philosophus" who is undoubtedly Toletus.

[48] Valesius taught that creatures could not have been produced from eternity in his commentary on the eighth book of the *Physics*, and that intensification is continuous and that elements have no natural intrinsic terminus of largeness or smallness in n. 5 of his *Controversiae*. Pererius held that one cannot demonstrate, one way or another, that corruptible creatures could have existed from eternity in Bk. 15, c. 13 of his *De communibus*, and that elements have no natural minima in Bk. 10, c. 23, of the same.

[49] On Bañez, see his *In 1 de gen.*, q. 6, a. 2 (ed. cit., pp. 74–77). For a brief sketch of the history of this thesis in Thomism, see A. M. Pirotta, O.P., *Summa philosophiae Aristotelico-Thomisticae* (Turin 1936), II, pp. 315–329, esp. n. 456, pp. 317–318.

[50] See U35 and U44, where Galileo allows the possibility of an infinite in a syncategorematic sense. The terms categorematic and syncategorematic derive from the nominalist controversies over infinity in the fourteenth century.

[51] Soto states: "Caietanus adducens illic auctoritatem Aristotelis contra S. Thomam effingit nescio quam distinctionem, certe, ut pace doctissimi authoris dixerim, parum physicam." *In 1 phys.*, q. 4, a. 2 (ed. cit., fol. 14rb).

[52] For details concerning Buonamici's life and works, see the author's article on him in the *Dictionary of Scientific Biography* (New York 1970), 2: 590–591. The author has used the Florence 1591 folio edition of Buonamici's *De motu* and has counted the number of citations of authors in Bks. 4, 5, and 8 of this work; these contain the chapters that most relate to the elements, local motion, and alteration.

[53] On the forms of the elements, see ed. cit., pp. 745–753; on intension and remission, see pp. 759–763.

[54] Eugenio Garin, *Science and Civic Life in the Italian Renaissance*, translated by Peter Munz (New York 1969), 140; see also 97–113.

[55] The Society was officially confirmed in 1540. Toletus became the first Jesuit cardinal; shortly after being received into the Society, he was called to Rome from Spain in 1559 to teach philosophy and later theology at the Collegio Romano. He himself had studied theology at Salamanca under Soto, who regarded him as a favored disciple.

Pererius entered the Society as a youth in 1552 and likewise taught philosophy and theology, including Scripture, at the Collegio Romano.

56 All names cited refer to *reportationes* of their respective lectures, unless the letter (p) follows, in which case the reference is to a printed text.

10. GALILEO AND THE *DOCTORES PARISIENSES*

The title of this study translates that of a note published by Antonio Favaro in 1918 in the transactions of the Accademia dei Lincei, wherein he presented his considered opinion of the value of Pierre Duhem's researches into the 'Parisian precursors of Galileo.'[1] Earlier, in 1916, only a few years after the appearance of Duhem's three-volume *Études sur Léonard de Vinci*, Favaro had reviewed the work in *Scientia* and had expressed some reservations about the thesis there advanced, which advocated a strong bond of continuity between medieval and modern science.[2] In the 1918 note he returned to this topic and developed a number of arguments against the continuity thesis, some of which are strikingly similar to those offered in present-day debates. Since Favaro, as the editor of the National Edition of Galileo's works, had a superlative knowledge of Galileo's manuscripts – one that remains unequalled in extent and in detail to the present day – it will be profitable to review his arguments and evaluate them in the light of recent researches into the manuscript sources of Galileo's early notebooks. Such is the intent of this essay.

I. FAVARO'S CRITIQUE OF THE DUHEM THESIS

The purpose of Favaro's 1916 review, as shown by its title, was to question whether Leonardo da Vinci really had exerted an influence on Galileo and his school. To answer this Favaro focussed on the concluding portion of Duhem's final tome, wherein Duhem had cited the first volume of Favaro's National Edition containing the previously unedited text of Galileo's Pisan notebooks. Duhem had noted in Favaro's transcription of the manuscript containing the notes the mention of Heytesbury and the 'Calculator,' the discussion of degrees of qualities in which Galileo uses the expression *uniformiter difformis*, and explicit references to the *Doctores Parisienses*.[3] For Duhem these citations, and particularly the last, clinched the long argument he had been developing throughout his three volumes. Galileo himself had given clear indication of his Parisian precursors: they were Jean Buridan, Albert of Saxony, and Themo

Judaeus – and here Duhem even ventured the volume from which Galileo had extracted his information, the collection of George Lokert, published at Paris in 1516 and again in 1518.[4] So the continuity between medieval and modern science was there for all to see, spelled out in Galileo's youthful handwriting.

Favaro, however, was not convinced. He saw Duhem's effort as partial justification for a general thesis to which he himself did not subscribe, namely, as he put it,

> that the history of science shows scientific development to be subject to laws of continuity, that great discoveries are almost always the fruits of a slow and involved preparation worked out over centuries, that. . .[they] result from the accumulated efforts of a crowd of obscure investigators. . .[5]

For Favaro such a thesis was too disparaging to the truly great men, to Newton, Descartes, his own Galileo; it accorded insufficient credit to these fathers of modern science.

But Favaro's refutation of Duhem was not argued at this level. Instead, proper historian that he was, Favaro turned to the manuscript itself. Admittedly it was in Galileo's handwriting, and Favaro was the first to concede that, but this proved only that it was the work of Galileo's hand, not necessarily the work of his head. The notes were extremely neat, spaces were occasionally left to be filled in later, some of the later additions did not fit and spilled over into the margins, there were signs of copying – phrases crossed out and reappearing a line or two later – all indications that this was not an original composition. Moreover, the notes were very sophisticated, they had a magisterial air about them, quoting authorities and opinions extensively – in a word, they manifested considerably more learning than one would expect of the young Galileo.[6] And 'young' Galileo truly was when these notes were written. Internal evidence could serve to date them, as Favaro had already indicated in his introduction to the edited text. "Without any doubt," he had written, "Galileo wrote these notes during the year 1584."[7] That is why Favaro himself had entitled them *Juvenilia*, or youthful writings, perhaps stretching the term a bit, for his 'youth' was by then twenty years of age. Though in Galileo's hand, therefore, they were but his work as a student at the University of Pisa, probably copied from one of his professors there – a likely candidate would be Francesco Buonamici, who later published a ponderous tome of over a thousand folio pages discussing just these problems.[8] So one ought not make too much of references to the *Doctores*

Parisienses. The *Juvenilia* are not Galileo's expression of his own thought, they do not represent his work with original sources, not even with Duhem's 1516 edition of Parisian writings. They are materials copied at second or third hand, the scholastic exercises of a reluctant scholar who probably had little taste for their contents, and indeed would soon repudiate the Aristotelianism their writing seemed to imply.[9]

Favaro's refutation of the Duhem thesis, so stated, is difficult to counter, and it is not surprising that it has remained virtually unchallenged for over fifty years. This in spite of two facts that go against Favaro's account, to wit: (1) there is no mention of the *Doctores Parisienses* in Buonamici's text, or indeed, in the writings of any of Galileo's professors who taught at Pisa around 1584, nor do any of them discuss the precise matters treated in the notes, and (2) the curator who, considerably before Favaro, first collected Galileo's Pisan manuscripts and bound them in their present form had made the cryptic notation: "The examination of Aristotle's work *De caelo* made by Galileo around the year 1590."[10] Now in 1590, as we know, Galileo was by no accounting a youth; he was 26 years of age, already teaching mathematics and astronomy at the University of Pisa. So Favaro's critique leaves two questions unanswered: (1) if Galileo copied the notes, from whom did he copy them, particularly the references to the *Doctores Parisienses*; and (2) must they be the notes of an unappreciative student, even one twenty years of age, or could they be the later work of an aspiring professor who understood the arguments contained in them and possibly made them his own? Depending on how we answer these questions we may have to revise Favaro's judgment on the value of the notes, and in particular their bearing on his critique of the Duhem thesis.

II. MS GAL 46: CONTENTS AND CITATIONS

Before attempting a reply it will be helpful to review the contents of the notebooks that have been labelled for so long as Galileo's *Juvenilia*. Actually they are made up of two sets of Latin notes, the first containing questions relating to Aristotle's *De caelo et mundo* and the second containing questions relating to Aristotle's *De generatione et corruptione*. These questions are dealt with in a stylized scholastic manner, first listing various opinions that have been held, then responding with a series of conclusions and arguments in their support, and finally solving difficulties that have been raised in the opinions of those who hold the contrary. Each of Galileo's sets of notes contains twelve questions that follow this general pattern, but with some variations. The first set begins with two introductory questions, then has four questions that make up its *Tractatio de mundo*, followed by six questions of a *Tractatio de caelo*, the last question of which is incomplete. The second set of notes, like the first, is incomplete at the end, but it is also incomplete at the beginning. It comprises three questions from a *Tractatus de alteratione* that discuss intensive changes in qualities, and nine questions of a *Tractatus de elementis*, five of which treat the elements of which the universe is composed and the remaining four the primary qualities usually associated with these elements.[11] Apart from the fact that both sets of notes are patently incomplete, there are numerous internal indications that they either are, or were intended to be, parts of a complete course in natural philosophy that would begin with questions on all the books of Aristotle's *Physics*, go through the *De caelo* and the *De generatione*, and terminate with a series of questions on the *Meteorology*.[12] Whether this complete set of notes was actually written by Galileo and subsequently lost, or whether what has been preserved constitutes the whole of his writing on these subjects, must remain problematical.

Unlike much of Galileo's later composition, these notes are replete with citations – a fortunate circumstance, for such citations are frequently of help in determining the provenance of manuscripts of that type. In this particular manuscript Galileo cites 147 authors, with short titles of many of their works, as authorities for the opinions and arguments he adduces. Many of these are writers of classical antiquity, the authors of primary sources such as Plato, Aristotle, and Ptolemy. But a goodly number of medieval and Renaissance authors are mentioned also, and, based on frequency of citation, it was to these later sources that Galileo seems to have had recourse more generally. Indeed, it is instructive to list the principal authors he cites in the order of frequency of citation, as shown in Table I, since this reveals some curious facts. Note in this list, for example, that the author who is most frequently referenced is one of Aristotle's medieval commentators, Thomas Aquinas, who is mentioned 43 times. After him comes Averroës, 38 times, and then a group of Dominicans whom Galileo refers to as 'the Thomists' (*Tomistae*), whose citations singly and collectively total 29.[13] After them come classical

TABLE I

A Selection of Authors Cited in MS Gal 46
Including All the most Recent Authors and
Listed in Order of Frequency of Citation

Aquinas		43	Democritus	5
Averroes		38	Marsilius of Inghen	5
Thomistae	4	29	Pietro d'Abano	5
Capreolus	7			
Caietanus	6		Alfonsus	4
Soncinas	4		Henry of Ghent	4
Ferrariensis	3		Hippocrates	4
Hervaeus	2		John of Sacrobosco	4
Soto	2		Mirandulanus	4
Javelli	1		Paul of Venice	4
Simplicius		28	Plotinus	4
			Pomponatius	4
Plato		20	Taurus	4
Ptolemy		20		
Philoponus		19	Biel	3
Albertus Magnus		17	Buccaferreus	3
Alexander Aphrodisias		16	Cardanus	3
			Contarenus	3
Galen		15	Diogenes Laertius	3
Aegidius Romanus		12	Nobili	3
Anaxagoras		11	Pererius	3
Avicenna		10	Pliny	3
Duns Scotus		10	Pythagoras	3
Plutarch		10	Scaliger	3
Durandus		9	Alfraganus	2
Regiomontanus		9	Algazel	2
Augustinus		8	Alpetragius	2
Empedocles		8	Antonius Andreae	2
Achillini		7	Avempace	2
Jandunus		7	Avicebron	2
Proclus		7	Burley	2
Themistius		7	Copernicus	2
Ockham		6	Doctores Parisienses	2
Porphyry		6	Gaetano da Thiene	2
Zimara		6	Iamblichus	2
			John Canonicus	2
Albategni		5	Nifo	2
Bonaventura		5		

Albumasar	1	Pavesius	1
Aristarchus	1	Philo	1
Balduinus	1	Philolaus	1
Carpentarius	1	Puerbach	1
Cartarius	1	Sixtus Senensis	1
Clavius	1	Strabo	1
Crinitus	1	Swineshead	1
Heytesbury	1	Taiapetra	1
Lychetus	1	Valeriola	1
Marlianus	1	Vallesius	1
Nominales	1		

commentators, such as Simplicius and Philoponus, Plato and Ptolemy with 20 citations apiece, Galen with 15, Scotus with 10, Ockham with 6, Pomponazzi with only 4, the *Doctores Parisienses* with 2, and over a score of other authors with only 1 apiece.

In looking over this list, one might wonder whether printed editions of any of these authors were available to Galileo, for if these could be identified and located, they might provide a clue to dating the notes, or even turn out to be the source or sources from which the notebooks were compiled. Pursuing this lead, one finds that, of the most recent titles cited by Galileo, 27 were works printed in the sixteenth century, all of which are listed in Table II. Some of

TABLE II

Sixteenth-Century Printed Sources Cited in MS Gal 46*

Balduinus, Hieronymus, *Quaesita. . .et logica et naturalia*, Venice 1563
Buccaferreus, Ludovicus, *In libros de generatione*, Venice 1571
Caietanus, Thomas de Vio, *In summam theologicam*, Lyons 1562
Cardanus, Hieronymus, *De subtilitate*, Nuremberg 1550
Carpentarius, Jacobus, *Descriptio universae naturae*, Paris 1560
Cartarius, Joannes Ludovicus, *Lectiones super Aristotelis proemio in libris de physico auditu*, Perugia 1572
Clavius, Christophorus, *In sphaeram Ioannis de Sacrobosco*, Rome 1581†
Contarenus, Gasparus, *De elementis*, Paris 1548
Copernicus, Nicolaus, *De revolutionibus*, Nuremberg 1543
Crinitus, Petrus, *De honesta disciplina*, Lyons 1554
Ferrariensis, Franciscus Sylvester, *In libros physicorum*, Venice 1573
Ferrariensis, Franciscus Sylvester, *In summam contra gentiles*, Paris 1552
Javellus, Chrysostomus, *Totius. . .philosophiae compendium*, Venice 1555

Lychetus, Franciscus, *In sententiarum libros Scoti*, Venice 1520
Mirandulanus, Bernardus Antonius, *Eversionis singularis certaminis libri*, Basel [1562?]
Niphus, Augustinus, *In libros de generatione*, Venice 1526
Nobilius, Flaminius, *De generatione*, Lucca 1567
Pavesius, Joannes Jacobus, *De accretione*, Venice 1566
Pererius, Benedictus, *De communibus omnium rerum naturalium principiis*, Rome 1576††
Pomponatius, Petrus, *De reactione*, Bologna 1514
Scaliger, Julius Caesar, *Exercitationes de subtilitate. . .ad Cardanum*, Paris 1557
Sixtus Senensis, *Bibliotheca sancta*, Venice 1566
Sotus, Dominicus, *Quaestiones in physicam Aristotelis*, Salamanca 1551
Taiapetra, Hieronymus, *Summa divinarum et naturalium quaestionum*, Venice 1506
Valleriola, Franciscus, *Commentarium in libros Galeni*, Venice 1548
Vallesius, Franciscus, *Controversiarum medicarum et philosophicarum libri*, Alcala 1556
Zimara, Marcus Antonius, *Tabula delucidationum in dictis Aristotelis et Averrois*, Venice 1537

* The list cites the earliest known printed edition, except for the works of Clavius and Pererius.
† See note 20, *infra*.
†† See note 22, *infra*.

these are by authors well known to historians of science, such as Cardanus, whose *De subtilitate* was printed at Nuremberg in 1550, and Copernicus, whose *De revolutionibus* was printed there in 1543. Similarly well known are the works of Nifo, Pomponazzi, and Zimara, all from earlier decades of the sixteenth century. Other works are less well known, for example, Nobili's *De generatione*, which was printed at Lucca in 1567. But only one book on the list, as it turns out, was published in the 1580's. That is Christopher Clavius's *In sphaeram Ioannis de Sacrobosco commentarius*, the second edition of which was printed at Rome in 1581. Clavius, of course, was the famous Jesuit, 'the Euclid of the sixteenth century,' who taught mathematics and astronomy at the Collegio Romano, the influential Jesuit university in Rome. And, by coincidence, the next most recently published work on the list is that of another Jesuit professor at the Collegio Romano, Benedictus Pererius, an edition of whose *De communibus omnium rerum naturalium principiis* was published in Rome also, but five years earlier, in 1576.

Moreover, if one looks at a tabulation of the places and dates of earliest imprints of these volumes, as shown in Table III, some interesting results emerge. Most of the books, as might be expected, were printed at Venice, and

TABLE III

Sixteenth-Century Printed Sources Cited in MS Gal 46

Places of earliest imprints	
Venice	11
Paris	4
Rome	2
Lyons	2
Nuremberg	2
Bologna	1
Perugia	1
Lucca	1
Basel	1
Salamanca	1
Alcala	1
Dates of earliest imprints	
1580's	1
1570's	4
1560's	7
1550's	7
1540's	3
1530's	1
1520's	2
1510's	1
1500's	1

Italian publishing houses predominate, although there are some French, German, and Spanish also. With regard to the dates of imprint, books published in the 1550's and 1560's are in the majority, with some from the 1570's and one from the 1580's. If Galileo composed these notes from printed sources, therefore, he must have had access to a good library with a wide variety of works that were kept fairly up to date. More important, if he composed them from printed sources, he must have done so *after* 1581 – a point to be recalled later.

Since Clavius and Pererius are the most recent authors cited in MS Gal 46, it would appear worthwhile to consult their works for any evidences of copying that would tie Galileo to these sources. During the past few years this has been done by Alistair Crombie, Adriano Carugo, and the writer, with results that are extremely gratifying. Crombie has shown that practically all of the astronomy contained in MS Gal 46, in fact, could have been taken almost verbatim from Clavius's commentary on the *Sphere* of Sacrobosco, and Carugo, on Crombie's report, has shown that other sections that treat of the composition of the heavens and the intension and remission of forms could have derived from Pererius's *De communibus omnium rerum naturalium principiis.*[14] Taken together, by my estimate the materials contained in these two Jesuit authors can account for about 15% of the entire contents of MS Gal 46. Moreover, though it is possible that Galileo could have copied the notes from some intermediate author who in turn based them on Clavius and Pererius, there is *prima facie* evidence that Galileo himself used these two authors when composing the notes. And whether he did so or not, Favaro's suspicion is partially confirmed by these discoveries: Galileo clearly did not compose his notes from primary sources alone, although he supplies references to such sources. Rather he freely utilized the citations found in more recent works such as those of Clavius and Pererius, or of some even later writer whose excerpts from these authors were somehow made available to him.

III. MS GAL 46: CLAVIUS AND PERERIUS

Since the relationships between Galileo's notes and the printed works of these two Jesuits will assume considerable importance in what follows, it will be well to review here the *prima facie* evidence that supports the view that Galileo had access to these two books and actually copied from them, rather than from some intermediate source.

Looking first at Crombie's evidence, the portion of Galileo's manuscript that seems to be based on Clavius consists of 10 folios of the 97 that make up the two sets of notes. They constitute the first two questions of the *Tractatio de caelo* already referred to, which treat respectively of the number and the ordering of the celestial orbs. Galileo begins the initial question of this treatise by listing opinions, the first of which he cites as follows:

> The first opinion was that of certain ancient philosophers, whom St. Chrysostom and some moderns follow, holding that there is only one heaven. Proof of this opinion: all of our knowledge arises from the senses; yet, when we raise our eyes to heaven we do not perceive several heavens, for the sun and the other stars appear to be in one heaven; therefore [there is only one heaven]. Nor do the heavens fall under any sense other than sight.[15]

To show the comparison of texts that suggests copying, the Latin versions are reproduced below in parallel column, with the transcription of Galileo's hand on the right and the relevant portions of Clavius from which the notes were apparently taken on the left:

CLAVIUS	GALILEO
In *Sphaeram*. . .	Tractatio de caelo
Commentum in primum caput	Quaestio prima
De numero orbium caelestium	An unum tantum sit caelum?
Antiquorum[1] philosophorum nonnulli[2] unicum duntaxat caelum esse affirmabant, quos pauci[3] admodum ex recentioribus[4] imitantur[5]. . .Omnis scientia[6] nostra. . .a sensu oritur.[7] Cum igitur, quotiescunque[8] ad caelum oculos attolimus, non percipiamus[9] visu multitudinem caelorum, Sol enim[10]. . .et reliquae. . .stellae[11] in uno. . .caelo videntur[12] existere, caelumque ipsum sub nullum[13] alium sensum praeter[14] visum, cadere possit[15]. . .	Prima opinio fuit veterum[1] quorumdam[2] philosophorum, quos secutus est[5] D. Chrysostomus et aliqui[3] recentiores,[4] sententium unicum esse caelum. Probatur haec opinio: omnis nostra cognitio[6] ortum habet[7] a sensu: sed, cum[8] attolimus oculos ad caelum, non percipimus[9] multitudinem caelorum, cum[10] sol et reliqua astra[11] in uno caelo videantur[12] existere: ergo [etc.]. Neque[13] vero sub alium[13] sensum quam[14] sub visum cadunt.[15]

Note here that, except for Galileo's reference to St. John Chrysostom, about which more will be said later, all of the information contained in Galileo's text is already in Clavius, although expressed in slightly different words. Note also that instances of word changes are shown in the parallel columns by superscript numbers: for example, the superscript 1 indicates that where Clavius has *antiquorum* for 'ancient' Galileo has the synonym *veterum*; superscript 2 shows the substitution of *quorumdam* for *nonulli*, and so on. Now it is possible to proceed in this fashion through the ėntire 10 folios that contain these two questions and to find places in the 1581 edition of Clavius from which practically every sentence in Galileo's notes could have been obtained.[16] Clavius, of course, has a more detailed treatment, since in addition to treating the number and the ordering of the spheres he also discusses their motions and periodicities, the trepidation of the fixed stars, and the earth's location at the center of the universe.[17] Even when restricting attention to the number and ordering of the spheres, moreover, Galileo's text summarizing the relevant passages contains roughly half the words used by Clavius to explain the same material.[18] Significantly, Galileo's text is not only abbreviated but in some places incorporates phrases and even entire clauses from Clavius's commentary. Usually, however, the sentence structure is changed, the inflection of words is altered, and synonyms are employed, as seen in the foregoing sample with the numbered superscripts. Withal the abbreviation is done skillfully, and in particular the order of Clavius's arguments is rearranged in a consistent pattern of exposition, so that the work is clearly that of a person who knows the subject matter well and is attempting to present it as synthetically as possible.

To check the stylistic characteristics of the author who worked from Clavius's text, presumably Galileo, a detailed analysis was made of all the word preferences and changes of inflection that occur in the noted 10 folios of MS Gal 46 vis-à-vis the 31 pages of Clavius's text on which they seem to be based. The system of superscript numbering shown in the above sample was continued for the entire two questions, and it was found that in all there were about 630 numbered expressions indicating conversions, i.e., that 630 of the 5650 words, or some 11%, were changed from the way they appear in Clavius. Of these total conversions, about 180 (or 35%) were mere changes of inflection, the word-stem remaining the same, whereas in the remaining 450 cases the word had been replaced by a synonym. By preparing index cards for each of the words changed, and counting the number of conversions to the word and from the word, it was possible to calculate an index of the author's stylistic and word preferences. The results of these calculations are presented in Table IV entitled 'List of Conversions,' with the 'likes' (those for which the conversions *to* the word exceed those *from* the word) shown on the left and the 'dislikes' (wherein the conversions *from* the word exceed those *to* the word) on the right. The changes are categorized according to the parts of speech and inflected forms, and only changes with a significant conversion index are included. A comparison of the 'likes' with the 'dislikes' listed in Table IV shows that the author's latinity is unsophisticated, that he usually prefers the simple word, and exhibits none of the classical variety of Clavius's excellent Latin prose. Many of the changes of inflection can be explained by the writer's attempt to abbreviate and synthesize, as in the preference for specific relative pronouns, but generally his changes in verbs, nouns, and

TABLE IV

List of Conversions

Conversion Index = no. of conversions *to* the word (+)/no. of conversions *from* the word (–)

LIKES			DISLIKES	
PARTICLES (187 total conversions):				
	nam	+22/–2	enim	+0/–25
	et	+17/–4	quare	+0/–7
	ut	+14/–2	quoniam	+0/–6
	cum	+12/–3	deinde	+0/–5
	sed	+6/–0	denique	+0/–4
	neque	+5/–0	unde	+0/–3
	quia	+4/–0		
INFLECTED FORMS (179 total conversions):				
Change of Conjugation (98 total conversions):				
	Subjunctive	+17/–0	Infinitive	+18/–22
Change of Declension (81 total conversions):				
	Accusative	+26/–14	Nominative	+21/–35
VERBS (122 total conversions):				
	movere	+9/–0	collocare	+0/–7
	patet	+7/–0	statuere	+0/–6
	constituere	+5/–0	deprehendere	+0/–5
	esse	+10/–5	efficere	+0/–4
	posse	+4/–0	dicere	+5/–9
	debere	+4/–0	colligere	+0/–3
NOUNS (58 total conversions):				
	orbis	+8/–3	caelum	+4/–7
	occasum	+4/–1	occidens	+1/–4
	cursus	+3/–0	circulus	+0/–3
	corpus caeleste	+2/–0	stella	+2/–4
	astronomus	+2/–0	sphaera	+4/–5
PRONOUNS (33 total conversions):				
	qui, quae, quod	+12/–2	is, ea, id	+0/–7
	ille, illa, illud	+12/–0		
	hic, haec, hoc	+6/–1		
ADJECTIVES (36 total conversions):				
	-- nothing significant			
VARIA (16 total conversions):				
	-- nothing significant			

adjectives are so content-determined that they cannot be used as stylistic indicators. The particles, however, and particularly the conjunctions, show strong preferences for simple words like *nam*, *et*, *ut*, and *cum*, and almost equally strong rejections of slightly longer connectives such as *enim*, *quare*, *quoniam*, and *deinde*.

Since the foregoing use of conjunctions would be a likely characteristic of Galileo's somewhat simple style – his Latin by his own admission was never good[19] – a further study was made of the frequency of occurrence of the conjunctions shown in the 'likes' and 'dislikes' columns of Table IV to see how these occur throughout the entire 97 folios of MS Gal 46. The results of this analysis are shown in Table V, where the same consistent pattern of preference can be seen. The two middle columns of this table show the comparison between the use of these conjunctions in the 10 folios based on

TABLE V

Frequency of Occurrence of Conjunctions in Galileo's Early Latin MSS
(Number of occurrences per thousand words)

PREFERRED CONJUNCTIONS	MS Gal 27*	MS Gal 46† (16v–26r)	MS Gal 46† (entire)	MS Gal 71†
nam	2.3	6.4	4.1	2.4
et	24.8	31.5	37.9	24.9
ut	9.2	12.0	11.6	12.4
cum	3.8	6.9	6.2	7.7
sed	6.6	3.9	5.9	6.3
neque	3.2	1.8	2.3	1.6
quia	16.9	3.9	11.8	4.9
NON-PREFERRED CONJUNCTIONS				
enim	1.1	1.2	1.8	7.7
quare	1.1	0.0	0.8	1.8
quoniam	0.0	0.0	0.0	0.0
deinde	0.3	0.9	0.6	1.3
denique	0.1	0.5	0.2	0.0
unde	0.4	0.2	1.1	0.6

* Word counts are based on a transcription of the text by Adriano Carugo.
† Word counts are based on Favaro's reading in *Opere* 1.

Clavius and in the entire composition, and the agreement here is very good. Indeed, the only serious departure is in the less frequent use of *quia*, for which Galileo seems to have a more decided preference than is obvious from his conversion index for this word based on Clavius. Clavius, in fact, prefers more sophisticated modes of expression, and so does not use the simple *quia* when giving reasons – a factor that apparently inhibited Galileo's use of it in this portion of his composition.

When the analysis is extended to Galileo's other Latin manuscripts dating from his Pisan period, moreover, additional confirmation for a characteristic style is given. These results are presented in the two outer columns of Table V. MS Gal 27 is a set of notes in Galileo's hand not unlike the two sets contained in MS Gal 46, except that it is concerned with matters relating to Aristotle's *Posterior Analytics*; MS Gal 71, on the other hand, includes all of Galileo's early writings on motion, his rather extensive *De motu antiquiora*, commonly dated by historians between 1590 and 1592. The latter work, in particular, has always been regarded as Galileo's own composition, and since essentially the same preferences are regulative throughout all of these works, it seems on face value that Galileo was the author who put them all together.

These observations, based as they are on word counts, at best have suasive force, but they provide some basis for maintaining that all three Pisan manuscripts are the work of Galileo's head as well as his hand, and that they probably were composed by him, as will be argued in more detail below, in connection with his Pisan professorship from 1589 to 1591.

To return, then, to the portion of MS Gal 46 based on Clavius, other *prima facie* evidences can be adduced to support the view that Galileo himself actually used Clavius's text when composing these notes, and that they are not therefore the work of some intermediate author merely copied by Galileo as a scholastic exercise. Clavius's commentary went through five editions, the first appearing in 1570, but there are indications that Galileo made use of the second edition of 1581 or possibly a later printing.[20] A likely account of how Galileo worked is suggested by Plate I, which shows folios 17r and 18r of Galileo's manuscript juxtaposed with pp. 62, 63, and 46 of the 1581 edition. On this illustration passages have been identified that show errors in copying, where, for example, Galileo copied from the wrong line, crossed out his mistake, and then wrote the passage correctly, or changed his copy to agree with the text. While it is possible that Galileo could have done this from a

PLATE I

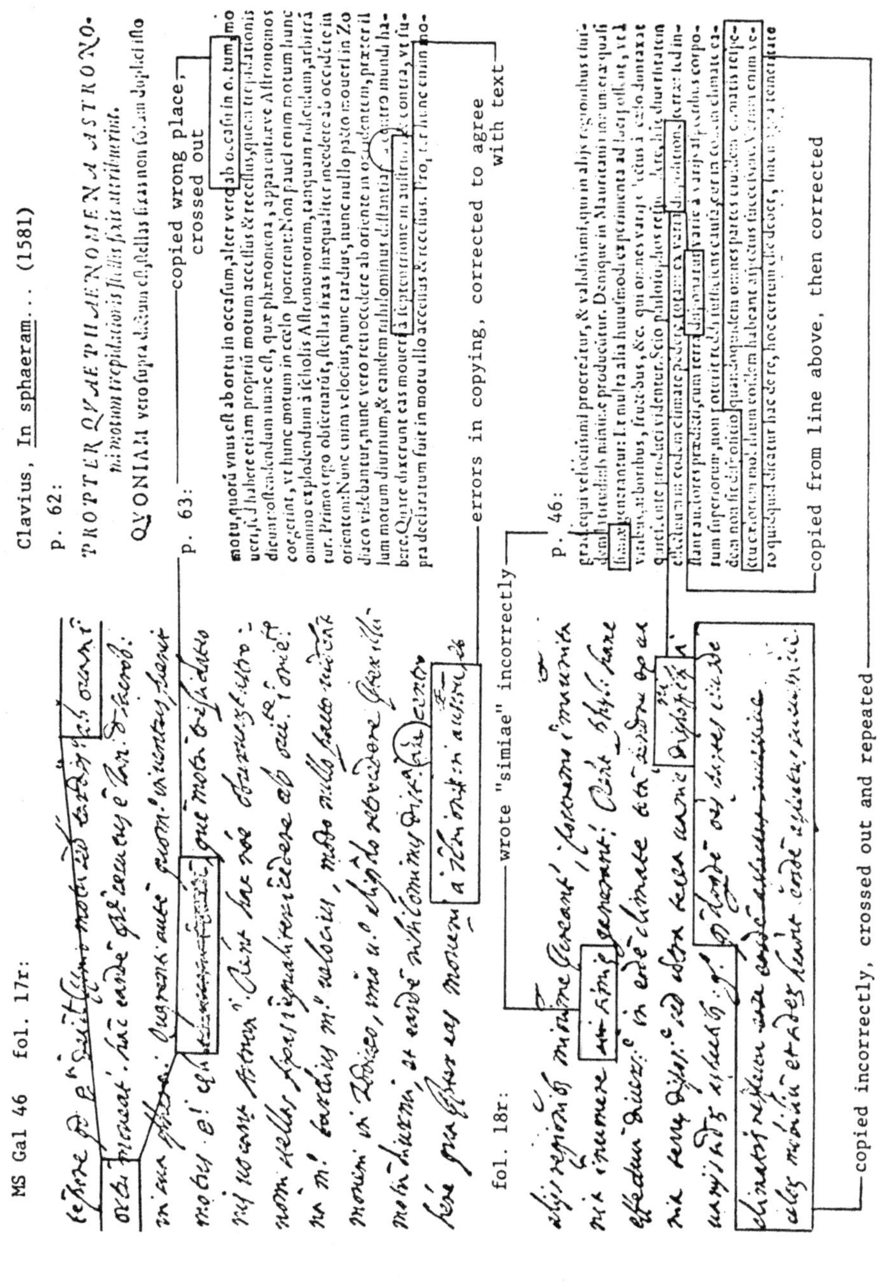
MS Gal 46 fol. 17r:
fol. 18r:
Clavius, In sphaeram... (1581)
p. 62:
p. 63:
p. 46:
copied wrong place, crossed out
errors in copying, corrected to agree with text
wrote "simiae" incorrectly
copied from line above, then corrected
copied incorrectly, crossed out and repeated

manuscript source rather than from the printed text here illustrated, on face value one would be disposed to see Clavius's commentary as his direct source.[21]

To come now to the dependence on Pererius, suspected by myself and independently identified by Carugo, this is somewhat similar to that on Clavius but it is not so striking – the passages used are neither as extensive nor as concentrated as those used for the 10 folios. At one place in MS Gal 46, however, Galileo makes explicit reference to Pererius's text; this is at the end of the first question in the *Tractatio de mundo*, which discusses the opinions of ancient philosophers concerning the universe. There Galileo concludes his exposition with the words:

> The opinion of Aristotle is opposed to the truth – for his arguments and those of Proclus, Averroës, and others supporting the eternity of the world, together with the answers, read Pererius, Book 15.[22]

Checking this citation, one finds in Book 15 of *De communibus* a considerable discussion paralleling closely the material found in the fourth question of Galileo's *Tractatio de mundo*, entitled 'Whether the world could have existed from eternity?' This occurs in Chapter 10 of that book, also entitled 'Whether the world could have existed from eternity,' which begins with a 'first opinion' that is quite similar to the 'third opinion' Galileo gives in answer to the same query.[23] Below are reproduced in parallel column excerpts from the two Latin texts, to show the resemblance between them:

PERERIUS	GALILEO
De communibus. . .principiis	MS Gal 46
Lib. 15. De motus et mundi aeternitate	Tractatio prima de mundo
Caput decimum	Quaestio quarta
An mundus potuerit esse ab aeterno?	An mundus potuerit esse ab aeterno?
Prima opinio est Henrici de Gandavo, Quod. 1, quaest. 7, et Philoponi in eo libro quo respondet argumentis Procli pro aeternitate mundi. . .	Tertia opinio est Philoponi in libro quo respondet argumentis Procli pro aeternitate mundi, Gandavensis in quodlibeto p°, . . .
Quod probant his rationibus, 1. creatio	Probatur haec opinio. . .2° quia creatio

est productio ex nihilo. . .necesse est in creatione non ens et nihil praecedere ipsum esse rei quae creatur. . .	est productio ex nihilo: quo fit ut non ens necessario debeat praecedere rem creatam. . .
2. De omni producto verum est dicere ipsum produci, . . .sed de creatura non est verum dicere ipsam semper produci; alioquin nulla creatura haberet esse permanens, sed tantum esse successivum. . . ergo si in aliquo instanti producitur, non semper, nec ab aeterno habet esse.	Confirmatur: quia de eo quod producitur verum est dicere producitur: sed creatura quae producitur non potest semper produci, alioqui [sic] non haberet esse suum permanens sed successivum. . .ergo in instanti producitur, et, ut est consequens, non potest ab aeterno esse.
3. Creatura habet esse acquisitum a Deo, ergo habet esse post non esse, nam quod acquirit aliquid, id non semper habuit. . . .	Confirmatur adhuc: quia creatura habet esse acquisitum a Deo; ergo habet illud post non esse; quia non acquiritur quod habetur, sed quod non habetur. . . .

Here a careful study of the Latin texts shows a close dependence of MS Gal 46 on Pererius's text completely analogous to the dependence noted earlier on Clavius. The arguments have been rearranged, some of the words have been changed and the inflections have been altered, but the content is essentially the same. Thus the conclusions that have been reached above with regard to Clavius's text would seem to apply *a pari* to Pererius's. The works of the two Jesuits explicitly cited in Galileo's notes, the two most recent books in all the works he there references, would therefore appear to have supplied a significant portion of the information they contain.

IV. MS GAL 46: MANUSCRIPT SOURCES

The foregoing researches are of paramount importance; not only do they throw unexpected light on possible sources of Galileo's early notebooks, but they register a considerable advance over Favaro's conjecture that the notes were based on the teaching of Buonamici – whose writings contains nothing comparable to the materials just discussed. On one matter, however, they are not particularly helpful, and this is Galileo's references to the *Doctores Parisienses.* Now it is significant that Galileo mentions the *Parisienses* in only two places in MS Gal 46, and that one of these citations occurs in precisely the question just discussed that seems to be based on Pererius. A careful search of Pererius's text, on the other hand, shows no mention of the *Doctores*

Parisienses in this or in any other locus. Where, therefore, did Galileo get this citation? Or did he in fact use Pererius as a guide and actually return to primary sources himself, as Duhem claimed in the first place? Allied to these questions are others relating to the composition of MS Gal 46 as a whole. Already noted is the fact that the printed works of Clavius and Pererius can account for 15% of the contents of these notes. That leaves 85% of the notes unaccounted for. What was *their* source? Or could Galileo possibly have composed them himself?

Fortunately, to answer this last question there is available one of Galileo's Latin compositions known beyond all doubt to be original, the autograph of the famous *Nuncius sidereus* written by him in 1609. Although the subject matter of this work was quite familiar to Galileo, its actual writing in Latin proved somewhat tortuous: as can be seen on Plate II, which duplicates the first page of his composition, almost every line has cross-outs and corrections, there are frequent marginal inserts, and the manuscript as a whole is decidedly messy. When this Paduan composition is compared with the Pisan manuscripts, in fact, the latter are found to be quite neat and clean. They have their evidences of copying, as already remarked, and on most pages there are a few words crossed out, others inserted, frequent superscript or subscript modifications of inflection, minor deletions, and so forth, but nothing of the magnitude of the marking-up found in Galileo's original Latin composition. Favaro's paleographical comments thus appear to be correct: it seems probable that *all* of MS Gal 46 is derivative, although not in the precise way Favaro thought. To verify this a careful check of all 97 folios was made, noting the number of corrections on each page and the evidences of copying they reveal, to see if these were concentrated in places for which the source was known or were distributed uniformly throughout the entire composition. Again the results, summarized on Plate III, are surprising. The 10 folios based on Clavius and the others based on Pererius are not markedly different from any other folios; indeed those toward the end of the manuscript, not yet discussed, give even more indications of copying than do the earlier ones. Moreover, some of the cross-outs and changes in this part of the manuscript raise the question whether Galileo had difficulty not merely with the subject matter, but even with reading the text from which he was working. This query further suggests unfamiliar handwriting: perhaps Galileo based these notes not on printed sources alone but on partially illegible manuscript sources as well?

SIDEREUS NUNCIUS -- AUTOGRAPH OF FIRST PAGE:

PLATE II

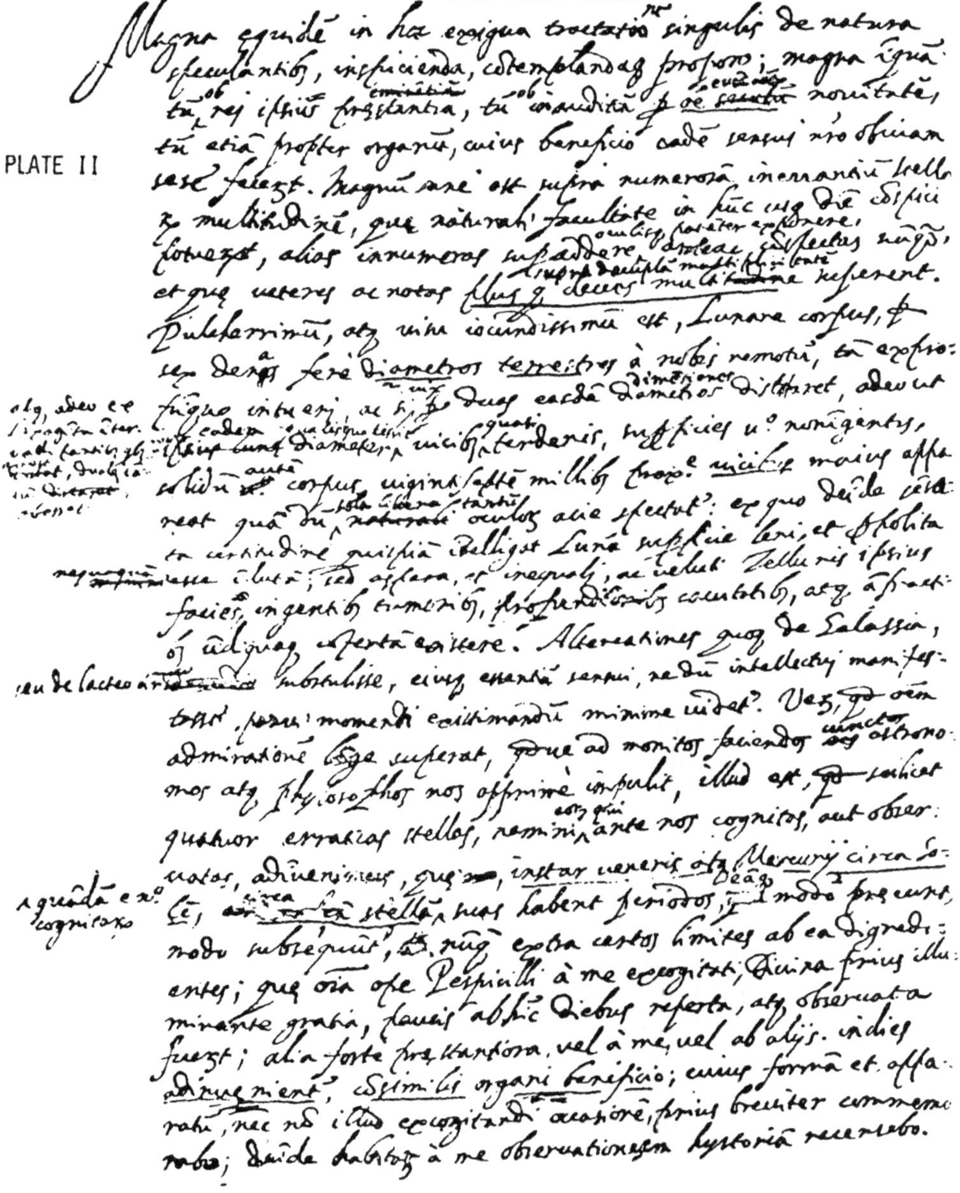

PLATE III

CORRECTIONS AND EVIDENCES OF COPYING IN MS GAL 46

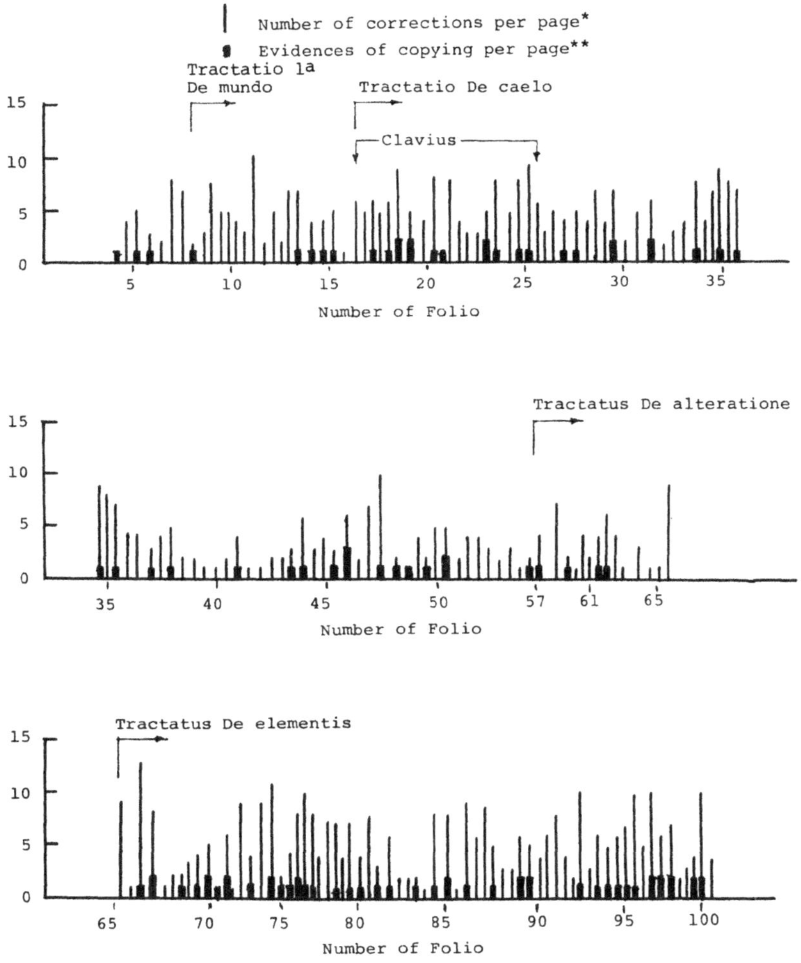

*Total = 945 **Total = 88

Further investigation, following up this line of reasoning, showed that Pererius had taught the entire course of natural philosophy at the Collegio Romano in 1565–1566, and that handwritten copies or reports (*reportationes*) of parts of his course during that academic year are still extant. Not only this, but a considerable collection of similar lecture notes from the Collegio have been preserved, spanning the interval from the 1560's through most of the seventeenth century. It turns out that the Jesuit professors who taught natural philosophy at the Collegio generally did so for only one year, although occasionally they repeated the course; in most instances, moreover, they had taught the entire logic course the year previous, ending with the *Posterior Analytics.*[24] Many of those who taught these courses prepared questionaries on the relevant works of Aristotle; professors of natural philosophy did so for the *Physics*, the *De caelo*, and the *De generatione*, and in some cases they even reached the *Meteorology* – precisely the span of subject matter envisaged in Galileo's enterprise. Unfortunately not all of these questionaries have been preserved; thus far the writer has located only the following from the period of Galileo's lifetime:

Year	Professor	Notes
1565–1566	Benedictus Pererius	*Physics, De caelo, De generatione*
1567–1568	Hieronymus de Gregorio	*Physics, De caelo, De generatione*
1588–1589	Paulus Valla	*Meteorology*
1589–1590	Mutius Vitelleschi	*De caelo, De generatione, Meteorology*
1597–1598	Andreas Eudaemon-Ioannes	*Physics, De caelo, De generatione*
1603–1604	Terentius Alciati	*De cometis*
1623–1624	Fabius Ambrosius Spinola	*Physics, De generatione*
1629–1630	Antonius Casiglio	*Physics, De caelo, De generatione*

Further research will undoubtedly turn up other Collegio *reportationes* for courses taught in the years between 1560 and 1642, for they are known to be diffused widely through European manuscript collections. For the time being, however, it should be noted that all of the first four authors on the list above have teachings that can be found in MS Gal 46, and that particularly the notes of the third and fourth, i.e., Paulus Valla and Mutius Vitelleschi, contain substantial amounts of Galileo's material. Indeed, when these manuscript sources are taken in conjunction with the printed sources already discussed, practically the entire contents of MS Gal 46 can be seen to derive, in one way or another, from courses taught at the Collegio Romano up to the year 1590. This, then, supplies the clue to Galileo's otherwise odd citation of sources as given in Table I: trained both scholastically and humanistically, the Jesuit professors from whose writings Galileo worked were well acquainted with medieval sources, and particularly with Thomas Aquinas and 'the Thomists,' but they also knew the Averroist thought then being propagated in the universities of Northern Italy, and of course they were no strangers to the primary sources of classical antiquity.

To give some idea of these manuscript sources, there are reproduced below transcriptions of portions of two of the *reportationes* that could well be the loci from which Galileo obtained his references to the *Doctores Parisienses*. The first is that of Vitelleschi, who lectured on the *De caelo* during the academic year 1589–1590. The question wherein the citation occurs is the same as that already seen in Galileo and Pererius, namely, whether the world could have existed from eternity. Galileo gives four different opinions on the question, whereas Vitelleschi gives only three; a close examination of the first of Vitelleschi's opinions, however, shows that it can be divided into two interpretations, and it is a simple matter to make a fourth opinion out of one of the alternatives. Portions of the Latin text of all four of Galileo's opinions are given below on the right, and in parallel column on the left are shown excerpts from Vitelleschi's notes, rearranged where necessary to show similarities in the text:

VITELLESCHI	GALILEO
APUG/FC 392 (no foliation)[25]	BNF MS Gal 46, fol. 13r–15r
Tractatio prima. De mundo.	Tractatio prima. De mundo.
Disputatio quarta.	Quaestio quarta.
An mundus potuerit esse ab aeterno?	An mundus potuerit esse ab aeterno?
Prima sententia {ita Gregorius in 2° dist. p^{a} quest. 3^{a} art. p^{o} et 2°, Occam ibidem. . .Gabriel. . .hi tamen duo defendunt hanc sententiam tamquam probabiliorem, et volunt contrariam etiam probabiliter posse defendi; idem sentiunt Burlaeus et Venetus in t. 15 8^{i} Phys.,	Prima opinio est Gregorii Ariminensis, in 2° dist. p^{a} q^{e} 3 art. p^{o} et 2°, Gabrielis et Occam ibidem; quamvis Occam non ita mordicus tuetur hanc sententiam quin etiam asserat contrariam esse probabilem. Hos secuntur Ferrariensis, in 8 Phys. qu. 15, et Ioannes Canonicus, in

Canonicus ibidem. . .Eandem sententiam sequuntur Ferrariensis 8 Phys. quest. 3^{a}, . . .Soncinas. . .Herveus. . .Capreolus }[26] vult mundum. . .secundum omnia entia que continet, tam permanentia quam successiva, tam corruptibilia quam incorruptibilia, potuisset esse ab aeterno.	8 Phys. q. p^{a}, et plerique recentiorum, existimantium mundum potuisse esse ab aeterno, tam secundum entia successiva quam secundum entia permanentia, tam secundum corruptibilia quam secundum incorruptibilia.
Secunda sententia { ita Durandus in 2° dist. p^{a} quest. 2^{a} et 3^{a}, quem sequuntur plurimi recentiores; Sotus vero 8 phys. quest. 2^{a}. . . } concedit res incorruptibiles potuisse esse ab aeterno, negat tamen res corruptibiles. . .	Secunda opinio est Durandi in 2° dist. p^{a}, quem secuntur permulti recentiorum; et videtur etiam esse D. Thomae, in p^{a} parte q. 46 art. 2°, sentientis aeternitatem repugnare quidem corruptibilibus, non tamen incorruptibilibus. . .
Tertia sententia { ita Philoponus contra Proclum sepe. . .Enricus quodlibeto p^{o} . . .Marsilius in 2° dist. p^{a} quest. 2^{a} art. p^{o}, S. Bonaventura ibidem parte p^{a} art. p^{o} quest. 2^{a}. . .plurimi S. Patres quos infra citabo } . . .	Tertia opinio est Philoponi in libro quo respondet argumentis Procli pro aeternitate mundi, Gandavensis in quodlibeto p^{o}, D. Bonaventura in 2° dist. p^{a} quo^{e} 2^{a}, Marsilii in 2° dist. p^{a} art. 2°, Burlei in 8 Phys. in quo^{e} hac de re super t. 15; et Sanctorum Patrum. . .
[Prima sententia:] Cum hac sententia convenit S. Thomas ut colligitur ex opusculo de hac re et. . .alibi { Scotus autem in 2° dist. p^{a} quest. 3^{a} neutram partem determinate sequitur sed putat utramque esse probabilem. Quod etiam *docent Parisienses 8 Phys. quest. p^{a}*. . . }	Quarta opinio est D. Thomae, in q^{e} 3^{a} De potentia art. 14, et in opusculo De aeternitate mundi, et in 2° Contra gentiles c. 38, et in p^{a} parte q. 46, Scoti in 2° dist. p^{a} q. 3, Occam in quodlibeto 2° art. 5, *doctorum Parisiensium 8 Phys. q. p^{a}*, Pererii in suo 15, et aliorum. . .

What is most striking about these parallels, of course, is that practically every authority and locus cited by Vitelleschi is also in Galileo, usually in abbreviated form. And, most important of all, the *Parisienses* are mentioned explicitly by Vitelleschi, and in a completely similar context, although for Galileo this occurs in his fourth opinion whereas for Vitelleschi it is in a variation of his first. More than that, although Galileo cites Pererius immediately after the *Parisienses*, Galileo's enumeration of opinions does not follow Pererius's enumeration but rather Vitelleschi's.[27] In light of this discovery one may now question whether Galileo did copy from Pererius after all, or whether his proximate source was a set of notes such as Vitelleschi's, which in turn could have taken into account the exposition of Pererius.[28]

Galileo's other mention of the *Doctores Parisienses* occurs in the second set of notes that make up MS Gal 46, more specifically in his Treatise on the Elements, where he makes the brief statement:

We inquire, second, concerning the size and shape of the elements. Aristotle in the third *De caelo*, 47, and in the first *Meteores*, first summary, third chapter, followed by the *Doctores Parisienses*, establishes a tenfold ratio in the size of the elements, i.e., water is ten times larger than air, and so on for each. Actually, however, whether this is understood of the extensive magnitude of their mass or of the magnitude and portion of their matter, I will show elsewhere by mathematical demonstration that it is false. . . .[29]

Now Vitelleschi in his lecture notes on *De generatione* mentions the problem of the size and shape of the elements, but says that he treats it elsewhere,[30] and so he is of no help on this particular citation. But Paulus Valla, who taught the course in natural philosophy the year before Vitelleschi, i.e., in 1588–1589, wrote an extensive commentary on Aristotle's *Meteorology* to which he appended a Treatise on the Elements, and in the latter there is a passage that is similar to Galileo's. Again the two are reproduced below in parallel column, with Valla's text on the left and Galileo's on the right:

VALLA	GALILEO
APUG/FC 1710 (no foliation)	BNF/MS Gal 46, fol. 76r
Tractatus de elementis	Tractatus de elementis
De quantitate elementorum	De magnitudine et figura elementorum
Aliqui existimant elementa habere inter se decuplam proportionem, ita ut aqua sit decuplo maior terra, aer decuplo maior aqua, [etc.]. . .ita tenent *Doctores Parisienses*, p° Meteororum, q. 3, . . .Probatur ex Aristotele 3° Caeli, t. 47, . . .p° Meteororum, summa pª, cap. 3. . .Contrariam sententiam habent communiter omnes mathematici et multi etiam Peripatetici non dari scilicet in elementis ullam determinatam proportionem in quantitate.	Aristoteles in 3° Caeli, 47, et p° Meteororum, summa pª, capᵉ 3°, quem secuti sunt *Doctores Parisienses*, in magnitudine elementorum constituit proportionem decuplam: idest, aqua sit decies maior quam terra; et sic de singulis. Verum, hoc, sive intelligatur de magnitudine molis extensiva, vel de magnitudine et portione materiae illorum, demonstrationibus mathematicis alibi ostendam id esse falsam.

Note here exactly the same reference to the *Doctores Parisienses*, as well as the same peculiar way of citing Aristotle's *Meteorology*. Also interesting is Galileo's statement that he is going to disprove the *Parisienses*' opinion elsewhere 'by mathematical demonstration,' which in fact he does not do, at least not in the portions of his notes that have come down to us; Valla, on the other hand, gives all the arguments of the 'mathematicians' that go counter to the *Parisienses*, and thus could have been a source for Galileo's projected demonstrations.[31] And needless to say, if Galileo had access to Valla's notes or to others like them, he could easily have written what he did without ever having looked at these authors from fourteenth-century Paris.[32]

Before leaving the subject of manuscript sources for MS Gal 46, it may be well to raise the question of possible intermediates for the portions of the notes apparently based on Clavius. Galileo's presentation differs most markedly from Clavius's in the simple fact that it locates many details of Ptolemaic astronomy, not in an exposition of the *Sphere* of Sacrobosco, as does Clavius, but rather in notes on Aristotle's *De caelo*. Less striking is an odd interpolation apparently introduced by Galileo in this material and already mentioned above, i.e., his citation of St. John Chrysostom in the first opinion on the number of the heavens. These differences suggest two apparently innocuous questions: (1) Did Galileo have any precedent for treating the number and ordering of the heavenly orbs in questions on the *De caelo*; and (2) Where did Galileo obtain his reference to Chrysostom, a Church Father about whom he would not be expected to know much in his own right, when such a reference is not found in any edition of Clavius? Now suggestions for answers to these may be found in the questionaries of Hieronymus de Gregorio, who taught the course in natural philosophy at the Collegio in 1568. Among his questions on the second book of the *De caelo* are two entitled respectively *De numero orbium caelestium* and *De ordine orbium caelestium*. Note here that the first title is exactly the same as that in Clavius, whereas the second differs by only one word: Clavius's title reads *De ordine sphaerarum caelestium*, i.e., it substitutes *sphaerarum* for *orbium*. Returning now to Galileo's notes, we find that his second question in the *Tractatio de caelo* follows De Gregorio's reading, not Clavius's. And more remarkable still, among the many opinions mentioned by De Gregorio at the outset of his treatment of the first topic is to be found the name of John Chrysostom. De Gregorio's notes, of course, are very early, 1568, composed apparently before the first edition of Clavius,[33] and one need not maintain that Galileo actually used these *reportationes* when composing his own notes. But it could well be that a tradition of treating some astronomy in the course *De caelo* already existed at the Collegio Romano, and that another Jesuit (yet unknown) had previously culled from

Clavius's *Sphaera* the type of material contained in MS Gal 46, thus serving as the proximate source of Galileo's note-taking.[34] This still leaves open, to be sure, the question of multiple sources for MS Gal 46, and it also allows room for Galileo's distinctive style of Latin composition to show through the final result, as explained above.[35]

V. MS GAL 46: THE PROBLEM OF DATING

All of this evidence makes highly plausible one aspect of Favaro's critique of Duhem, i.e., that MS Gal 46 derives from secondary rather than from primary sources. In another respect, however, Favaro might have erred, and this in dating the manuscript's composition in 1584, particularly considering the late date of the Valla and Vitelleschi materials, written as these were in 1589 and 1590 respectively. More pointedly, perhaps the earlier curator was closer to the truth when indicating that the notes were composed 'around 1590.' If so, in dating them too early Favaro could well have underestimated their role in Galileo's intellectual development, and so overlooked an important element of truth in Duhem's continuity thesis. On this score it is important to turn now to the problems of dating, to the evidence Favaro considered indisputable for the year 1584, and to a possible reinterpretation of that evidence.

First it should be remarked that it is practically impossible to date MS Gal 46 as a juvenile work on the basis of Galileo's handwriting alone. To the writer's knowledge only five Latin autographs of Galileo are still extant in various collections, and portions of each of these are duplicated on Plate IV. At the top are the opening lines of the draft of *Sidereus nuncius* written by Galileo – and this is the only date known for certain – in 1609, at the age of 45. Just under this is an extract from the notes on motion, the *De motu antiquiora* of MS Gal 71, usually dated between 1590 and 1592. Under this again are a few lines from MS Gal 46, whose writing was put by Favaro in 1584. Under this yet again is an extract from the smaller set of notes in MS Gal 27 relating to Aristotle's *Posterior Analytics*, which seem to be earlier than those in MS Gal 46; because of spelling errors they contain Favaro regarded them as Galileo's first attempt at Latin scholastic composition.[36] The bottom sample, finally, contains a few lines from a brief exercise of nine folios (at the back of MS Gal 71), wherein Galileo is translating from Greek to Latin a passage attributed to the Greek rhetorician Isocrates. These

PLATE IV

GALILEO'S LATIN HANDWRITING 1579?-1609

1609
MS 48
fol.
8r

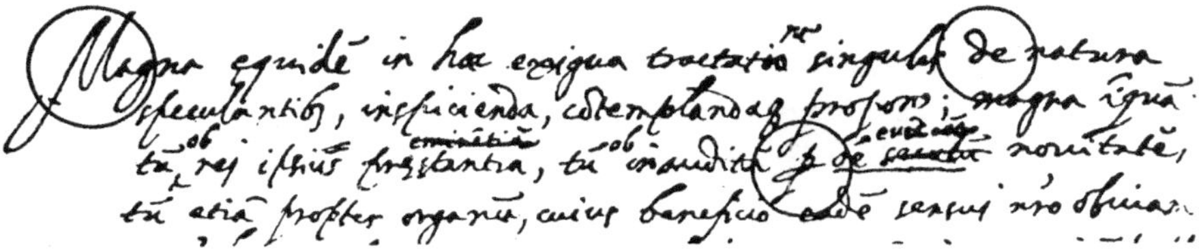

MS 71
fol.
3v

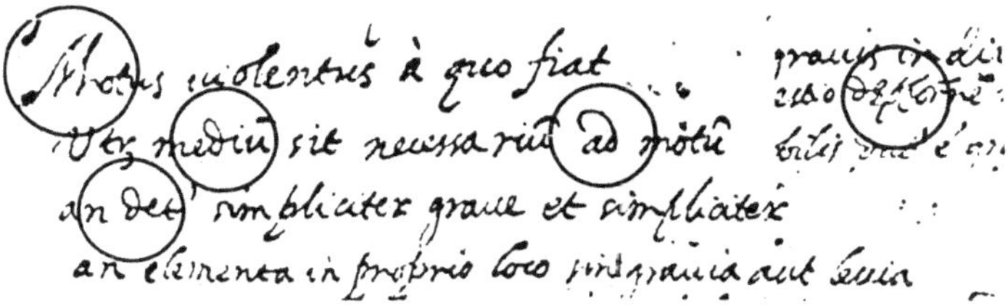

MS 46
fol.
33r

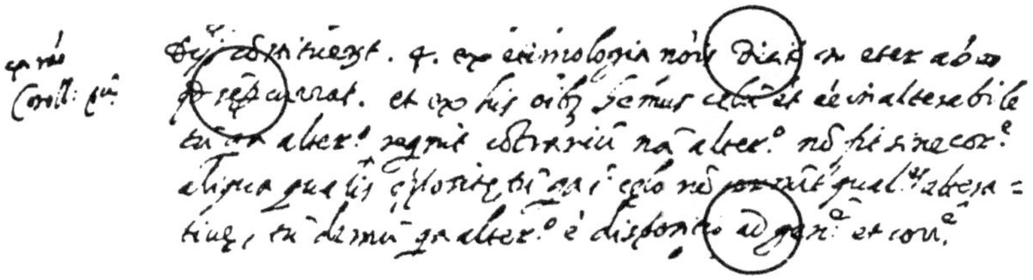

MS 27
fol.
4r

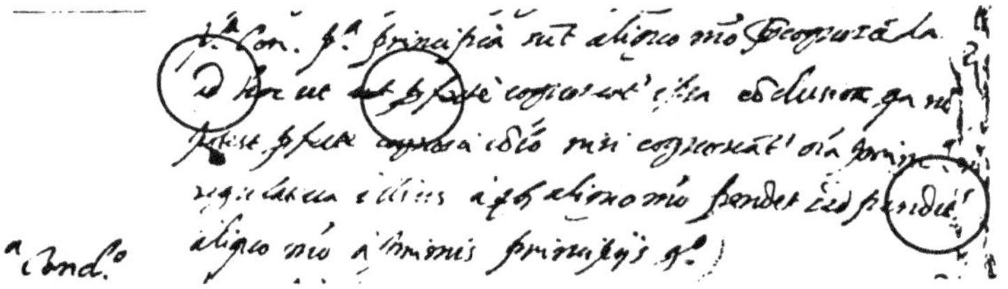

1579?
MS 71
fols.
132r
132v

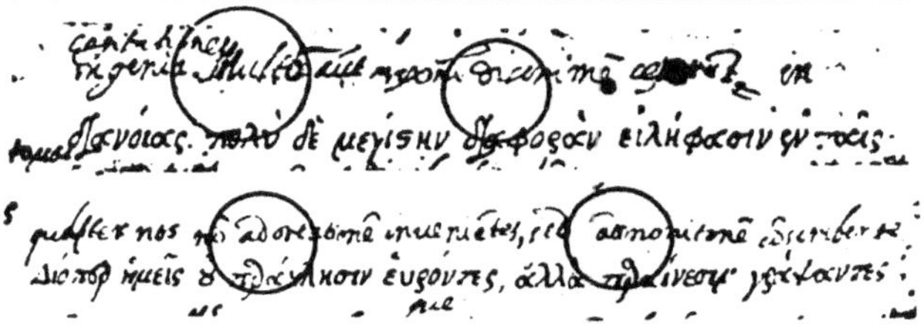

particular folios show evidence of an older hand correcting Galileo's translation, a fairly good indication that they are a student composition.[37] Presumably Galileo wrote this exercise during his humanistic studies, before he entered the University of Pisa in 1581; it could date from as early as 1579, when he was only 15 years of age.

In all of these samples of Galileo's Latin hand, and they span a period of almost 30 years, certain similarities are recognizable. For example, his capital 'M' remains pretty much the same; the way he links the 'd-e' and the 'd-i' is constant; the abbreviation for 'p-e-r' does not change; his manner of writing 'a-d' varies only slightly. Galileo's hand is consistent and in no way idiosyncratic; indeed, it resembles many other Tuscan hands of the late *cinquecento*. And, from a comparative point of view, the hand that wrote MS Gal 46 is no more patently juvenile than the one that penned the opening lines of the *Sidereus nuncius* or the extract from the *De motu antiquiora*. Thus it is not transparently clear that these notes can be written off as a youthful exercise, at least on the basis of the hand that wrote them.

To come now to Favaro's internal evidence, in the second question of the *Tractatio de mundo* Galileo considers the age of the universe. This particular question arose because of Aristotle's teaching that the universe is actually eternal – not the same problem as that already mentioned, whether it *could be* eternal, which questions the possibility rather than the actuality of its eternal existence. Galileo's answer is orthodox, in conformity with the Church's teaching on creation: Aristotle is wrong, the universe was created in time. He goes on:

> To anyone inquiring how much time has elapsed from the beginning of the universe, I reply, though Sixtus of Siena in his *Biblioteca* enumerates various calculations of the years from the world's beginning, the figure we give is most probable and accepted by almost all educated men. The world was created five thousand seven hundred and forty-eight years ago, as is gathered from Holy Scripture. For, from Adam to the Flood, one thousand six hundred and fifty-six years intervened; from the Flood to the birth of Abraham, 322; from the birth of Abraham to the Exodus of the Jews from Egypt, 505; from the Exodus of the Jews from Egypt to the building of the Temple of Solomon, 621; from the building of the Temple of Solomon to the captivity of Sedechia, 430; from the captivity to its dissolution by Cyrus, 70; from Cyrus, who began to reign in the 54th Olympiad, to the birth of Christ, who was born in the 191st Olympiad, 560; the years from the birth of Christ to the destruction of Jerusalem, 74; from then up to the present time, 1510.[38]

Now that is a fascinating chronology, and one can well imagine Favaro's excitement when he transcribed this passage and saw in it a definitive way of dating the composition of these notes. All we need do, said he, is focus on the last two figures Galileo has given, for these, when added together, give the interval from the birth of Christ to Galileo's writing. So add 74 to 1510, and we obtain the result desired, *Anno Domini* 1584. That is the year in which the notes were obviously composed.[39]

This, of course, is a piece of evidence to be reckoned with. But Favaro's interpretation of it is not the only one possible. To permit a different reckoning the crucial point to notice is that the foregoing chronology does not supply a single absolute date; all that it records are intervals between events. Now, if one can accurately date any event in Galileo's account, then of course he can calculate other dates from these intervals. For example, if one knows the year in which the destruction of Jerusalem took place, then by adding 1510 to that date he will determine the year referred to in the notes as 'the present time.' As it just happens, the best known date of all the events mentioned, and one confirmed in secular history, is the date of the destruction of Jerusalem, which took place in A.D. 70.[40] So, add 1510 to 70 and the result is 1580, not 1584, as the date referred to as 'the present time.' Moreover, if Galileo wrote these notes in 1580, he was then a mere sixteen years of age, not yet a student at the University of Pisa, and Favaro's title would become even more appropriate than he thought – they would be *juvenilia* beyond all question and doubt.

What this observation should serve to highlight is a more problematic dating that lies behind Favaro's calculation, namely, determining the year in which Christ was born. Favaro simply assumed that Christ's birth was in A.D. 0, but scholars who were writing chronologies in the 1580's would permit themselves no such assumption. Most of their attention was devoted to calculating the years that transpired between the creation of the world and the birth of Christ so as to establish the time of Christ's birth in relation to the age of the universe. Going back, then, to Galileo's epochs and summing the intervals he gives from Adam to the birth of Christ, one obtains a total of 4164 years. Now, it is an interesting fact that throughout the course of history hundreds of calculations of this particular sum have been made, and despite Galileo's saying that his figures are "most probable and accepted by almost all educated men," no recent historian has uncovered any chronology that gives Galileo's implicit sum, 4164 years from the world's creation to the birth of Christ. The chronologer William Hales, in his *New Analysis of Chronology*, printed in 1830, gives more than 120 different results for this computation, and says that his list "might be swelled to 300" without any difficulty.[41] Some of his figures will be of interest to historians of science: they range from the calculation of Alphonsus King of Castile, in 1252, who gives 6984 as the total number of years between creation and Christ's birth, to that of Rabbi Gerson, whose sum is only 3754. The calculators include some eminent figures in science's history: Maestlin, who gives 4079 years; Riccioli, who computes 4062; Reinhold, who has 4020; Kepler, who gives 3993; and Newton, who computes only 3988.[42] Such calculations, needless to say, are very complex and require a detailed knowledge of Scripture to be carried out. It is quite unlikely that Galileo would have been able to do this himself; more probably he copied his result from another source. But again, the identity of that source has been a persistent enigma, for no chronology in Hales's list or elsewhere yields Galileo's precise result.[43]

Here again a possible answer can be found by employing the procedure that worked with the *Doctores Parisienses*, i.e., by turning to the Jesuits of the Collegio Romano. We have already discussed Benedictus Pererius, who taught the course in physics at the Collegio between 1558 and 1566, at which time he disappeared from the *rotulus* of philosophy professors and began teaching theology.[44] In 1576, moreover, he was the professor of Scripture at the Collegio, a post he held until 1590, and again in 1596–97.[45] Only fragments of his Scripture notes survive, but fortunately for our purposes he published two Scriptural commentaries, one on the Book of Daniel printed at Rome in 1587, the other on the Book of Genesis printed there in 1589.[46] When these are studied for evidences of a chronology of creation it is found that the chronology recorded by Galileo, with the exception of only a single interval, employs the figures given by Pererius. Thus this computation, like practically all else in MS Gal 46, derives from work done at the Collegio Romano and, in this case, recorded in a printed source not available until the year 1589.

The details of these calculations are given in Table VI and its accompanying notes, but a few additional observations may help locate these materials in context. The problem with intervals of the type shown on Table VI is that not all Scripture scholars choose precisely the same events for their computations; depending on their knowledge of particular books of the Bible they find some intervals easier to compute than others. But a more or less standard chronology, standard in the sense that it was widely diffused and frequently cited, is that given by Joseph Scaliger in his *De emendatione temporis*, first published in 1583 and subsequently reprinted many times.[47] Now, if one excludes from Galileo's list the last two intervals, which have already been discussed, and focuses on those between Adam and the birth of Christ, he can see that Galileo's epochs are basically Scaliger's. Actually Scaliger records one more event, the call of Abraham, which he interpolates between Abraham's birth and the Exodus from Egypt, but otherwise he has all the intervals recorded by Galileo, although his figures for them are generally different.

Pererius, on the other hand, has two chronologies, one in his commentary on Daniel and the other in his commentary on Genesis; their difference lies in the fact that they interpolate the reigns of David and of Achaz respectively into the epochs computed.[48] Although Pererius's epochs are thus different from both Scaliger's and Galileo's, with a little computation they can be converted into equivalent epochs, and when this is done it is found that, with the exception of one interval, Pererius's figures are precisely Galileo's. The one exception is the interval between the Exodus from Egypt and the building of the Temple, for which Galileo gives 621 years whereas Pererius, and Scaliger also, give only 480; the difference is that the larger figure (621) is that calculated by the profane historian Josephus, whereas the smaller figure (480) is that computed from the Hebrew text of the Bible.[49] Thus it seems likely that Galileo's figure derives at least indirectly from Pererius and probably comes from a Collegio Romano source like the rest of MS Gal 46.[50]

The question remains as to how such a computation can be reconciled with a composition of the notes 'around 1590,' when the figures themselves seem to indicate 1580, or 1510 years after the destruction of Jerusalem in A.D. 70. A plausible answer is that Galileo made a simple mistake in recording, a mistake that involves only one digit. It seems unlikely that he changed the intervals of the Biblical epochs; whatever his source, he probably took down what was there recorded. But when he came to the interval between the destruction of Jerusalem and 'the present time,' which could be his own computation, he wrote down 1510 instead of 1520, and thus came out ten years short in his entire sum. Now it is not unprecedented that Galileo makes an error of one digit in this way; in fact, in the same MS Gal 46, a few folios

TABLE VI

Possible Sources of the Chronology of Creation in MS Gal 46

Event	Galileo MS Gal. 46	Joseph Scaliger 1583	Pererius (Daniel) 1587	Pererius (Genesis) 1589	Galileo's Chronology Again (1)
Adam					
	1656	1656	1656	1656	1656 (2)
Flood					
	322	292	322	322	322 (3)
Birth of Abraham					
		75			
	505		505		505 (4)
[Call of Abraham] *					
		430		942	
Exodus from Egypt					
	621	480	480		621 (5)
[David]					
Building of Temple					
				473	
			283		
[Achaz]	430	427			430 (6)
Captivity of Sedechia					
	70	59	776	[630?] 730	630 (7)
Dissolution by Cyrus					
	560	529			
Birth of Christ					
SUBTOTAL	4164	3948	4022	4123 [4023?]	4164 (8)
	74				
Destruction of Jerusalem					
	1510 (9)				
Present					

*Entries enclosed in square brackets are not listed in Galileo's chronology.

TABLE VI

Possible Sources of the Chronology of Creation in MS Gal 46

Notes to Table VI

(1) Galileo's epochs are essentially those of Scaliger.

(2) Same in all sources.

(3) Pererius adds 30 years to Scaliger's figure to allow for the generation of Cain; Galileo follows Pererius.

(4) Same for Scaliger (75 + 430) and for Pererius on Daniel.

(5) Both Scaliger and Pererius on Daniel computed this interval from the Hebrew text (= 480 years); Galileo gives the longer interval found in Josephus (= 621 years) – see William Hales, *New Analysis of Chronology and Geography, History and Prophecy*, (London: 1880), p. 217.

(6) Galileo's figure is consistent with Pererius's two chronologies and can be calculated from them:

$$942 + 473 - 505 - 480 = 430$$

(7) Galileo's figure for this interval is the sum of 70 and 560 or 630; this is about the same as the sum given by Pererius on Daniel:

$$776 + 283 - 430 = 629 \simeq 630$$

It is probable that the figure given by Pererius on Genesis (730) is a misprint and should read 630, as given by Galileo.

(8) Galileo's sum is the same as Pererius's except for one particular, i.e., the interval between the Exodus from Egypt and the building of the Temple – see (5) above; it differs from Scaliger's sum in five particulars, viz, the intervals between the Flood and Abraham, between the Exodus and the Temple, between the Temple and the Captivity, between the Captivity and the Dissolution, and between the Dissolution and the Birth of Christ.

(9) The figure 1510 puts the writing of the notes in A.D. 1580, because the Destruction of Jerusalem took place in A.D. 70, which is 74 years after the Birth of Christ (4 B.C.). If Galileo wrote the wrong figure here, i.e., 1510 instead of 1520, it is necessary to change only one digit to make the date of composition 1590, which agrees with the original notation on the notebook. The date of 1580 for the time of composition would seem to be ruled out by Galileo's extreme youth (16 yrs.) at that time, in light of the sophistication of the notes and their use of materials only available later.

beyond the passage being discussed, when again giving the total number of years from creation to the present time, Galileo there recorded 6748 years, whereas earlier he had given 5748 for that identical sum.[51] There is simply no way of explaining such a difference (here a difference of one digit in the thousands column, rather than of one digit in the tens column as just postulated) except ascribing it to an error on Galileo's part. And it would seem much simpler to admit such a slip than to hold that Galileo did the copying in 1580, when he was a mere sixteen years of age, before he had even begun his studies at the University of Pisa.

Moreover, if one persists in the 1580 or 1584 dating, there is no way of accounting for the sophistication of the notes, the lateness of the sources on which they seem to be based, and other historical evidences that will be adduced in the next section. The 1580 dating, which otherwise would be the more plausible, is particularly untenable in light of the dependence of the notes on the 1581 edition of Clavius's *Sphaera*, for, as noted earlier, the copying from Clavius (either directly or through an intermediate) had to be done *after* 1581. On the other hand, if one admits the appreciably later composition consistent with the dates of the *reportationes* of Valla and Vitelleschi (1589 and 1590) and with that of Pererius's commentary on Genesis (1589), then he need no longer view the notes as *Juvenilia*, the writings of a reluctant scholar or of an uncomprehending student, but rather will be disposed to see them as the serious work of an aspiring young professor. From such a viewpoint, of course, the Duhem thesis itself will bear re-examination – not in the form originally suggested by Duhem, to be sure, but in a modified form that takes fuller account of the Collegio Romano materials and their possible influence on Galileo.

VI. MS GAL 46: PROVENANCE AND PURPOSE

Before suggesting such a qualified continuity thesis, it will be helpful to speculate briefly as to how Galileo could have gotten his hands on the sources already discussed, and what use he intended to make of them. In the writer's opinion Clavius remains the key to a proper understanding of these notes. Galileo had met the great Jesuit mathematician during a visit to Rome in 1587, as is known from a letter written by Galileo to Clavius in 1588; thereafter they remained on friendly terms until Clavius's death in 1612.[52] What better explanation for Galileo's possession of all this Collegio Romano material than that it was made available to him by Clavius himself. In Galileo's position at Pisa from 1589 to 1591, and in the openings for a mathematician-astronomer for which Galileo applied at Bologna in 1588 and at Padua in 1592, it would certainly be desirable to have someone with a good knowledge of natural philosophy, who would be conversant, in particular, with matters relating to Aristotle's *De caelo* and *De generatione*. It makes sense to suppose that Clavius would help his young friend, whose mathematical ability he regarded highly, by supplying him with materials from his own Jesuit colleagues that would serve to fill out this desirable background.

Again, among the Jesuits at that time there was a movement to integrate courses in mathematics with those in natural philosophy,[53] and even at Pisa Galileo's predecessor in the chair of mathematics, Filippo Fantoni, had written a treatise *De motu* that was essentially philosophical.[54] Consider this in conjunction with the fact that, both at Pisa and at Padua, professors of philosophy were paid considerably more than professors of mathematics.[55] Could it not be that Galileo was aspiring to a position wherein he could gradually work into philosophy and thus earn for himself a more substantial salary? In this event it would be advantageous for him to have a set of notes, preferably in Latin, that would show his competence in such matters philosophical. What better expedient for him, in these circumstances, than to seek out a good set of lecture notes and from these compose his own. Such composition, in those days, was not regarded as copying and certainly not as plagiarism in the modern sense; it was rather the expected thing, and everyone did it.[56] That would explain Galileo's rearrangement of the arguments, his simplified Latin style, a certain uniformity of presentation not to be found in the varied sources on which the notes seem to be based. And it gives much more meaning to this rather intelligent set of notes, than it does to maintain that they were trite scholastic exercises copied by an uninterested student from a professor whose later publication bears no resemblance to the materials they contain.

Thus far largely internal evidences have been cited that converge toward the year 1590 as the likely time of writing the notes contained in MS Gal 46. Apart from these there are some external evidences confirmatory of this date that may shed some light on the purpose for which the notes were composed. Already mentioned is the dependence of the first set of notes on Clavius's

Sphaera. To this should now be added that, in the second set devoted mainly to *De elementis*, Galileo refers often to the Greek physician Galen, mentioning his name no less than 15 times, and that the *reportatio* of Valla's *Tractatus de elementis* similarly contains extensive references to Galen. Consider then, in relation to these apparently disconnected facts, a letter written by Galileo to his father on November 15, 1590, which reads as follows:

> Dear Father: I have at this moment a letter of yours in which you tell me that you are sending me the Galen, the suit, and the *Sfera*, which things I have not as yet received, but may still have them this evening. The Galen does not have to be other than the seven volumes, if that remains all right. I am very well, and applying myself to study and learning from Signor Mazzoni, who sends you his greeting. And not having anything else to say, I close. . . Your loving son. . .[57]

The reference to the Galen volumes, in light of what has just been said, could well be significant. Perhaps Galileo, having seen in a secondary source so many citations of Galen, an author with whom he would have been familiar from his medical training, wished to verify some of them for himself.[58] Again, the reference to the *Sfera* is probably not to Sacrobosco's original work but rather to the text as commented on by Clavius. In view of Galileo's personal contact with, and admiration for, the Jesuit mathematician, it would seem that he should turn to the latter's exposition of the *Sphere* rather than to another's. Further, one can suspect from details of Galileo's later controversy with the Jesuits over the comets of 1618 that he was quite familiar with Clavius's commentary.[59] And finally, in Galileo's last published work, the *Two New Sciences* of 1638, he has Sagredo make the intriguing statement, "I remember with particular pleasure having seen this demonstration when I was studying the *Sphere* of Sacrobosco with the aid of a learned commentary."[60] Whose 'learned commentary' would this be, if not that of the celebrated Christopher Clavius?

Again in relation to Clavius, it is known that while at Pisa Galileo taught Euclid's *Elements*, and significantly that he expounded Book V, entitled *De proportionibus*, in the year 1590.[61] This is an unusual book for an introductory course, but it assumes great importance in applied mechanics for treatises such as *De proportionibus motuum*. If Galileo lectured in Latin, it is possible that he used Clavius's commentary on Euclid for this course, for the second edition of that work appeared the year previously, in 1589, and has an extensive treatment of ratios, precisely the subject with which the fifth book is concerned.[62]

Yet again, in 1591 Galileo taught at Pisa the 'hypothesis of the celestial motions,' and then at Padua a course entitled the *Sphera* in 1593, 1599, and 1603, and another course on the *Astronomiae elementa* in 1609.[63] His lecture notes for the Paduan courses have survived in five Italian versions, [64] none of them autographs, and these notes seem to be little more than a popular summary of the main points in Clavius's commentary on Sacrobosco's *Sphaera*. They show little resemblance, on the other hand, to Oronce Finé's commentary on that work, which was used as a text for this course at Pisa as late as 1588.[65] In one of the versions of Galileo's Paduan lecture notes, moreover, there is reproduced a 'Table of Climes According to the Moderns,' lacking in the other versions, but taken verbatim from Clavius's edition of 1581 or later.[66] Indeed, when the *Trattato della Sfera* is compared with the summary of Clavius's *Sphaera* in MS Gal 46, the latter is found to be more sophisticated and richer in technical detail. It could be, therefore, that the notes contained in MS Gal 46 represent Galileo's first attempt at class preparation, and that later courses based on the *Sphaera* degenerated with repeated teaching – a phenomenon not unprecedented in professorial ranks.

Another possible confirmation is Galileo's remark in the first essay version of the *De motu antiquiora* to the effect that his "commentaries on the *Almagest* of Ptolemy. . .will be published in a short time."[67] In these commentaries, he says, he explains why objects immersed in a vessel of water appear larger than when viewed directly. No commentary of Galileo on Ptolemy has yet been found, but significantly this particular phenomenon is discussed in Clavius's commentary on the *Sphere*.[68] The latter also effectively epitomizes the *Almagest* and so could well be the source Galileo had in mind for his projected summary.

All of this evidence, therefore, gives credence to the thesis emerging out of this study, namely, that the materials recorded in MS Gal 46 need no longer be regarded simply as *Juvenilia*, as the exercises of a beginning student; perhaps more plausibly can they to be seen as lecture notes or other evidences of scholarship composed by Galileo in connection with his Pisan professorship from 1589 to 1591.[69] As such they then merit serious consideration, not merely for the insight they furnish into Galileo's intellectual formation, but for identifying the philosophy with which he operated during the first stages of his teaching career.

VII. CONTINUITY REVISITED: GALILEO AND THE *PARISIENSES*

Now back to the thorny problem of continuity. Favaro, as is known from his critique, did not possess the facts here presented; so impressed was he by the novelty of Galileo's contribution, moreover, that he was not disposed to discern any law of continuity operating on the thought of his master. Duhem, at the other extreme, was disposed to see continuity even where none existed. Duhem's own philosophy of science was decidedly positivistic, placing great emphasis on "saving the appearances," according no realist value to scientific theories.[70] In his eyes the nominalists of the fourteenth century, the *Doctores Parisienses*, had the correct view, and he wanted to connect Galileo directly with them. The afore-mentioned evidences, unfortunately for Duhem, will not support such an immediate relationship.[71] And yet they do suggest some connection of early modern science with medieval science *via* the writings of sixteenth-century authors, along lines that will now be sketched.

In his studies on Leonardo da Vinci, Duhem correctly traced nominalist influences from the *Doctores Parisienses* all the way to Domingo de Soto, the Spanish Dominican who taught at Salamanca in the 1540's and 1550's.[72] Now it is an interesting fact, in connection with Soto, that many of the early professors at the Collegio Romano were either Spanish Jesuits, such as Pererius, or they were Jesuits of other nationalities who had completed some of their studies in Spain or Portugal, such as Clavius, who had studied mathematics at Coimbra under Pedro Nuñez. Warm relationships also existed between Pedro de Soto (a blood relative of Domingo and also a Dominican) and the Roman Jesuits, as Villoslada notes in his history of the Collegio Romano.[73] Admittedly, Duhem pursued his theme a long way, but as it has turned out he did not pursue it far enough. In the passage from fourteenth-century Paris to sixteenth-century Spain, moreover, Duhem failed to note that many nominalist theses had given way to moderate realist theses, even though the techniques of the *calculatores* continued to be used in their exposition. Duhem never did study the further development, the passage from Spain to Italy, when other changes in methodological orientation took place, particularly among the Jesuits. The nominalist emphasis on the logic of consequences, for example, quickly ceded to a realist interest in Aristotle's *Posterior Analytics* and in the methodology explained therein for both physical and physico-mathematical sciences. The Averroist atmosphere in

Italian universities also led to other emphases, to a consideration of problems different from those discussed in fourteenth-century Paris. The movement that gave rise to Galileo's science, therefore, may well have had its origin, its *terminus a quo*, among the *Doctores Parisienses*, but its *terminus ad quem* can be called nominalist only in a much attenuated sense. More accurately can it be described as an eclectic scholastic Aristotelianism deriving predominantly from Aquinas, Scotus, and Averroës, and considerably less from Ockham and Buridan, although it was broadly enough based to accommodate even the thought of such nominalist writers.

Exemplification of this qualified continuity thesis can be given in terms of Galileo's treatment of topics in natural philosophy and in logic that are usually said to have nominalist connotations. Among the former could be listed, for example, creation, the shapes of the elements, maxima and minima, uniformly difform motion, impetus, the intension and remission of forms, degrees of qualities, and analyses of infinities. For purposes here only one such topic need be discussed, viz., the reality of local motion, for this can serve to illustrate some of the conceptual changes that took place in the centuries between Ockham and Galileo.

For William of Ockham, as is well known, local motion was not an entity in its own right but could be identified simply with the object moved. To account for local motion, therefore, all one need do is have recourse to the moving body and its successive states; the phenomenon, as Ockham said, "can be saved by the fact that a body is in distinct places successively, and not at rest in any."[74] Since this is so it is not necessary to search for any cause or proximate mover in the case of local motion: "local motion is not a new effect. . .it is nothing but that a mobile coexist in different parts of space."[75]

In contrast to Ockham, as is also well known, Jean Buridan subscribed to the traditional analysis of local motion, with consequences that were quite significant for the history of science. In his commentary on Aristotle's *Physics* Buridan inquired whether local motion is really distinct from the object moved, and answered that while the "later moderns" (*posteriores moderni* – an obvious reference to Ockham) hold that it is not, he himself holds that it is, and went on to justify his conclusion with six different arguments.[76] Thus, for Buridan and other Parisians, local motion was decidedly a "new effect"; it was this conviction that led them to study it *quoad causes* and *quoad effectus* and, as part of the former investigation, to develop their theories of impetus, the proximate forerunner of the modern concept of inertia.

As has been indicated elsewhere, a development similar to Buridan's among sixteenth-century commentators such as Juan de Celaya and Domingo de Soto provided the moderate realist background against which Galileo's *De motu antiquiora* must be understood.[77] Soto was not adverse to calculatory techniques; in fact he was the first to apply them consistently to real motions found in nature, such as that of free fall, and so could adumbrate Galileo's 'law of falling bodies' some 80 years before the *Two Chief World Systems* appeared.[78] Moreover, Jesuits such as Clavius saw the value of applying mathematical techniques to the study of the world of nature. In a paper written for the Society of Jesus justifying courses of mathematics in Jesuit *studia*, Clavius argued that without mathematics "physics cannot be correctly understood," particularly not matters relating to astronomy, to the structure of the continuum, to meteorological phenomena such as the rainbow, and to "the ratios of motions, qualities, actions, and reactions, on which topics the *calculatores* have written much."[79] And it is significant that the *reportatio* of Vitelleschi discussed above concludes with the words: "And thus much concerning the elements, for matters that pertain to their shape and size partly have been explained by us elsewhere, partly are presupposed from the *Sfera*"[80] – an indication that by his time at the Collegio the mathematics course was already a prerequisite to the lectures on the *De caelo* and *De generatione*.

Duhem's thesis with regard to the study of motion, therefore, requires considerable modification, and this along lines that would accent realist, as opposed to nominalist, thought. With regard to logic and methodology, on the other hand, his thesis runs into more serious obstacles. Although the nominalist logic of consequences was known to the Jesuits, and Bellarmine preferred to express himself in its terms in his letter to Foscarini,[81] it was never adopted by them as a scientific methodology. Clavius, in particular, ruled it out as productive of true science in astronomy. So he argued that if one were consistently to apply the principle *ex falso sequitur verum*,

> then the whole of natural philosophy is doomed. For in the same way, whenever someone draws a conclusion from an observed effect, I shall say, "that is not really its cause; it is not true because a true conclusion can be drawn from a false premise." And so all the natural principles discovered by philosophers will be destroyed.[82]

Galileo followed Clavius's causal methodology in his own attempted proof for the Copernican system based on the tides, which was never viewed by him as merely "saving the appearances." He also used the canons of the *Posterior Analytics* in his final work, the *Two New Sciences*, wherein he proposed to found a new mixed science of local motion utilizing a method of demonstrating *ex suppositione*. This methodological aspect of Galileo's work has been examined at greater length by Crombie and by the writer in recent publications, to which the reader is referred for fuller details.[83] Here a few supplementary observations may suffice to indicate the main thrust of the argument.

Ockham had a theory of demonstration just as he had a theory of motion, but for him a demonstration is nothing more than a disguised hypothetical argumentation, thus not completely apodictic – a typical nominalist position.[84] Jean Buridan, here as in the case of the reality of motion, combatted Ockham's analysis and returned to an earlier position expounded by Thomas Aquinas, showing that in natural science, and likewise in ethics, truth and certitude can be attained through demonstration, but it must be done by reasoning *ex suppositione.*[85] Now Galileo, in this matter, turns out to be clearly in the tradition of Aquinas and Buridan, not in that of Ockham. He in fact mentions the procedure of reasoning *ex suppositione* at least six times in the writings published in the National Edition: once in 1615 when arguing against Bellarmine's interpretation that all *ex suppositione* reasoning must be merely hypothetical, once in the *Two Chief World Systems*, twice in the *Two New Sciences*, and twice in correspondence explaining the methodology employed in the last named work.[86] In practically every one of these uses Galileo gives the expression a demonstrative, as opposed to a merely dialectical, interpretation. Others may have equated suppositional reasoning with hypothetical reasoning, but Galileo consistently accorded it a more privileged status, seeing it as capable of generating true scientific knowledge.

Moreover, it seems less than coincidental that Galileo's earliest use of the expression *ex suppositione* occurs in his series of logical questions, the notes on Aristotle's *Posterior Analytics* referred to above, which otherwise are very similar to the physical questions that have been the focal point of this study. There, in answering the query whether all demonstration must be based on principles that are "immediate" (i.e., *principia immediata*), Galileo replies to the objection that mixed or "subalternated sciences have perfect demonstrations" even though not based on such principles, as follows:

I answer that a subalternated science, being imperfect, does not have perfect demonstrations, since it supposes first principles proved in a superior [science]; therefore it generates a science *ex suppositione* and *secundum quid.*[87]

Now compare this terminology with a similar sentence from Pererius, who, in a *reportatio* of his lectures on the *De caelo* given at the Collegio Romano around 1566, states:

The *Theorica planetarum*. . .either is a science *secundum quid* by way of *suppositio* or it is merely opinion, not indeed by reason of its consequent but by reason of its antecedent.[88]

Note that both Galileo and Pererius allow the possibility of science being generated from reasoning *ex suppositione*, and that Pererius explicitly distinguishes a science so generated from mere opinion. It thus seems far from unlikely that Galileo adopted the Jesuit ideal of a mixed or subalternated science (the paradigm being Clavius's mathematical physics) as his own, and later proceeded to develop his justification of the Copernican system, and ultimately his own 'new science' of motion, under its basic inspiration.[89] On this interpretation Galileo's logical methodology would turn out to be initially that of the Collegio Romano, just as would his natural philosophy – and this is not nominalism, but the moderate realism of the scholastic Aristotelian tradition.[90]

Viewed from the perspective of this study, therefore, nominalism and the *Doctores Parisienses* had little to do proximately with Galileo's natural philosophy or with his methodology.[91] This is not to say that either the movement or the men were unimportant, or that they had nothing to contribute to the rise of modern science. Indeed, they turn out to be an important initial component in the qualified continuity thesis here proposed, chiefly for their development of calculatory techniques that permitted the importation of mathematical analyses into studies of local motion, and for their promoting a "critical temper," to use John Murdoch's expression, that made these and other innovations possible within an otherwise conservative Aristotelianism.[92] But Galileo was not the immediate beneficiary of such innovations; they reached him through other hands, and incorporated into a different philosophy. What in fact probably happened is that the young Galileo made his own the basic philosophical stance of Clavius and his Jesuit colleagues at the Collegio Romano, who had imported nominalist and calculatory techniques into a scholastic Aristotelian synthesis based somewhat eclectically on Thomism, Scotism, and Averroism. To these, as is well known, Galileo himself added Archimedean and Platonic elements, but in doing so he remained committed to Clavius's realist ideal of a mathematical physics that demonstrates truth about the physical universe. And Jesuit influences aside, there can be little doubt that Galileo consistently sided with realism, as over against nominalism, in physics as in astronomy, from his earliest writings to his *Two New Sciences*. At no time, it would appear, did he subscribe to Duhem's ideal of science that at best attains only hypothetical results and at worst merely "saves the appearances."

What then is to be said of Favaro's critique of the Duhem thesis? An impressive piece of work, marred only by the fact that Favaro did not go far enough in his historical research, and thus lacked the materials on which a nuanced account of continuity could be based. As for Duhem's 'precursors,' they surely were there, yet not the precise ones Duhem had in mind, nor did they think in the context of a philosophy he personally would have endorsed. But these defects notwithstanding, Favaro and Duhem were still giants in the history of science. Without their efforts we would have little precise knowledge of either Galileo or the *Doctores Parisienses*, let alone the quite complex relationships that probably existed between them.

NOTES

* Research on which this paper is based has been supported by the National Science Foundation (Grant No. SOC 75-14615), whose assistance is gratefully acknowledged. Apart from the earlier version presented at the Blacksburg Workshop on Galileo on October 26, 1975, portions of this essay have been presented at the Folger Shakespeare Library in Washington, D.C., on October 20, 1975, and at meetings of the American Philosophical Association in New York on December 28, 1975, and of the History of Science Society in Atlanta on December 30, 1975. The author wishes particularly to acknowledge the comments, among many others, of Alistair Crombie, Stillman Drake, William F. Edwards, Paul O. Kristeller, Charles H. Lohr, Michael S. Mahoney, John E. Murdoch, Charles B. Schmitt, and Thomas B. Settle, which have aided him appreciably in successive revisions.

1 Galileo Galilei e i Doctores Parisienses', *Rendiconti della R. Accademia dei Lincei* 27 (1918), 3-14.

2 Léonard da Vinci a-t-il exercé une influence sur Galilée et son école?' *Scientia* 20

(1916), 257–265.

[3] The manuscript referred to here is contained in the Biblioteca Nazionale Centrale in Forence with the signature Manoscritti Galileiani 46 (= BNF/MS Gal 46). Its transcription is contained in Antonio Favaro (ed.), *Le Opere di Galileo Galilei*, 20 vols. in 21, G. Barbera Editore, Florence, 1890–1909, reprinted 1968, Vol. 1, pp. 15–177 (hereafter abbreviated as *Opere* 1:15–177). All manuscript material in this essay is reproduced with the kind permission of the Biblioteca Nazionale Centrale in Florence, Italy. This permission is here gratefully acknowledged.

[4] *Études sur Léonard de Vinci*, 3 vols., A. Hermann & Fils, Paris, 1913, Vol. 3, p. 583.

[5] "...che lo sviluppo scientifico e soggetto alla legge di continuità; che le grandi scoperte sono quasi sempre il frutto d'una preparazione lenta e complicatà, proseguita attraverso i secoli; che in fine le dottrine, le quali i più insigni pensatori giunsero a professare, risultano da una moltitudine di sforzi accumulati da una folla di oscuri lavoratori." – 'Galileo Galilei e i Doctores Parisienses,' p. 4 (translation here and hereafter by the writer).

[6] "Galileo Galilei e i Doctores Parisienses,' pp. 8–10.

[7] "Senza dubbio alcuno, dunque, ciò scriveva Galileo durante l'anno 1584." – Avvertimento, *Opere* 1:12.

[8] "Galileo Galilei e i Doctores Parisienses,' p. 10; cf. Avvertimento, *Opere* 1:12. The reference is to Francesco Buonamici, *De motu libri decem*. . .Apud Bartholomaeum Sermatellium, Florentiae, 1591.

[9] 'Galileo Galilei e i Doctores Parisienses,' p. 11; see also *Opere* 1:12–13 and 9:275–282.

[10] Favaro reports this inscription in his Avvertimento, *Opere* 1:9, as "L'esame dell'opera d'Aristotele 'De Caelo' fatto da Galileo circa l'anno 1590."

[11] The complete list of questions, as given by Galileo in MS Gal 46, is as follows:

Quaestio prima. Quid sit id de quo disputat Aristoteles in his libris De caelo.
Quaestio secunda. De ordine, connexione, et inscriptione horum librorum.

Tractatio prima. De mundo.
Quaestio prima. De opinionibus veterum philosophorum de mundo.
Quaestio secunda. Quid sentiendum sit de origine mundi secundum veritatem.
Quaestio tertia. De unitate mundi et perfectione.
Quaestio quarta. An mundus potuerit esse ab aeterno.

Tractatio de caelo.
Quaestio prima. An unum tantum sit caelum.
Quaestio secunda. De ordine orbium caelestium.
Quaestio tertia. An caeli sint unum ex corporibus simplicibus, vel ex simplicibus compositi.
Quaestio quarta. An caelum sit incorruptibile.
Quaestio quinta. An caelum sit compositum ex materia et forma.
Quaestio sexta. An caelum sit animatum.

[Tractatus de alteratione]
[Quaestio prima. De alteratione.]
Quaestio secunda. De intensione et remissione.
Quaestio ultima. De partibus sive gradibus qualitatis.

Tractatus de elementis
Prima pars. De quidditate et substantia elementorum.

Prima quaestio. De definitionibus elementi.
Quaestio secunda. De causa materiali, efficiente, et finali elementorum.
Quaestio tertia. Quae sint formae elementorum.
Quaestio quarta. An formae elementorum intendantur et remittantur.
Secunda disputatio. De primis qualitatibus.
Quaestio prima. De numero primarum qualitatum.
Quaestio secunda. An omnes hae quatuor qualitates sint positivae, an potius aliquae sint privativae.
Quaestio tertia. An omnes quatuor qualitates sint activae.
Quaestio quarta. Quomodo se habeant primae qualitates in activitate et resistentia.

Note in the above list that Galileo is not consistent in listing the divisions, i.e., that the first two tractates are given with the Latin title *Tractatio*, the last two with the title *Tractatus*, and that the last tractate is subdivided into a *Prima pars* and a *Secunda disputatio* (themselves inconsistent divisions), whereas the previous tractates are subdivided directly into *Quaestiones*. Such inconsistencies could be signs of copying from several different sources; see note 35 *infra*.

[12] Thus, in *Opere* 1, at p. 122, line 10 (hereafter abbreviated as 122.10), Galileo makes the reference to "[ea] quae dicta sunt a nobis 6° Physicorum. . .," an indication of a commentary on the eight books of the *Physics*, and at 137.14 he speaks of difficulties that will be solved "cum agam de elementis in particulari," a common way of designating the subject matter of the *Meteorology*, of which he apparently planned to treat. There is also an implicit reference to notes on logic that probably preceded the notes on natural philosophy; this occurs at 18.17, where Galileo writes ". . .de singularibus non potest esse scientia, ut alibi ostendimus," a possible indication of his having already explained this thesis from the *Posterior Analytics*. Other references to matter treated elsewhere occur at 77.18, 77.24, 113.13, 125.25–32, 127.30, 128.8, 129.24, 138.9, and 150.24–25.

[13] All of these references to Aquinas and the Thomists have been analyzed by the writer in W. A. Wallace, 'Galileo and the Thomists', *St. Thomas Aquinas Commemorative Studies 1274–1974* (ed. by A. Maurer), Pontifical Institute of Mediaeval Studies, Toronto, 1974, Vol. 2, pp. 293–330. At the time he wrote that article the writer accepted uncritically Favaro's judgment that the notes contained in MS Gal 46 were *Juvenilia*; now, as will become clear in what follows, he seriously questions such a characterization of the notes. See Essay 9, *supra*.

[14] For an account of the discoveries of Crombie and Carugo, see A. C. Crombie, 'Sources of Galileo's Early Natural Philosophy', in *Reason, Experiment, and Mysticism in the Scientific Revolution* (ed. by M. L. Righini Bonelli and W. R. Shea) Science History Publications, New York, 1975, pp. 157–175, 303–305. For the results of the author's researches, apart from those reported in note 13 above, see W. A. Wallace, 'Galileo and Reasoning *Ex Suppositione*: The Methodology of the *Two New Sciences*,' in Boston Studies in the Philosophy of Science, Vol. XXXII (Proceedings of the 1974 Biennial Meeting of the Philosophy of Science Association, 1974) (ed. by R. S. Cohen *et al.*), Reidel, Dordrecht and Boston, 1976, pp. 79–104; this paper, augmented with an Appendix, appears in the present volume as Essay 8, *supra*.

[15] The Latin text, also given below, is in *Opere* 1:38.4–9.

[16] The relevant folios of MS Gal 46 are fol. 16v to fol. 26r, which duplicate matter found on pp. 42–46, 55–71, and 135–143 of the 1581 edition of Clavius's *Sphaera*.

[17] This explains Galileo's apparent skipping of extensive passages in Clavius's text, viz., those on pp. 47–54 and 73–134.

[18] The ten folios of Galileo's notes contain about 5650 words, as compared with some twelve thousand words in the related passages of Clavius.

[19] Galileo himself alludes to this fact in his third letter on sunspots to Mark Welser, dated December 1, 1612; see *Opere* 5:190.2–7.

[20] Galileo could have used the 1570 edition for portions of his work, but he incorporates material not found in that edition, and for this he must have used (directly or indirectly) either the 1581 or the 1585 edition, both of which were printed from the same type. For a complete listing of Clavius's writings, see Carlos Sommervogel, S.J., *Bibliothèque de la Compagnie de Jésus*, Vol. II (Alphonse Picard, Paris, 1891) cols. 1212–1224; see also note 33 *infra*.

[21] It should be noted here, however, that other pieces of evidence count against this interpretation, and these will be discussed below in conjunction with the investigation of possible manuscript sources of MS Gal 46 (see text corresponding to notes 28 and 34, *infra*). Also noteworthy is Sommervogel's indication that Clavius's commentary was abridged in a printed edition published at Cologne in 1590 (see his Tom. II, col. 1213); the writer has not yet located this abridgment, which could of course be the source of Galileo's note-taking. If so, however, this would definitively date the composition of MS Gal 46 in 1590 or later.

[22] For the Latin text, see *Opere* 1:24.25–27. The reference to Pererius is apparently to his widely available *De communibus omnium rerum naturalium principiis et affectionibus libri quindecim*, printed at Rome in 1576 and often thereafter. It should be noted, however, that Sommervogel (Vol. VI, col. 499) lists an earlier work printed at Rome in 1562 with the title *Physicorum, sive de principiis rerum naturalium libri XV*; this is much rarer, and I have not yet been able to locate a copy. There are also manuscript versions of Pererius's lectures on the *Physics*, as noted *supra*, p. 212.

[23] Galileo's reply is on fol. 14v; Pererius's corresponding discussion is on p. 505 of the 1576 edition.

[24] For details, see R. G. Villoslada, *Storia del Collegio Romano dal suo inizio (1551) alla soppressione della Compagnia di Gesù (1773)*. Analecta Gregoriana Vol. LXVI, Gregorian University Press, Rome, 1954, especially pp. 89–91 and 329–335.

[25] The manuscript is in Rome in the Archivum Pontificiae Universitatis Gregorianae, Fondo Curia, MS 392 (= APUG/FC 392).

[26] Material enclosed in curly parentheses { } has been rearranged to show parallels in the texts.

[27] Note that for Pererius, as shown on p. 207 *supra*, the opinion of Philoponus and Henry of Ghent is his first, whereas for Galileo and Vitelleschi it is their third.

[28] This is one instance counting against the *prima facie* evidence for Galileo's copying from printed sources, alluded to above (note 21).

[29] For the Latin text, which is reproduced in part below, see *Opere* 1:138.3–9.

[30] The last folio of APUG/FC 392 concludes with the sentence: ". . .Et hec de elementis, nam quae spectant ad eorum figuram et quantitatem partim explicata sunt a nobis alibi, partim supponimus ex *Sfera*."

[31] In a private communication to the author, Stillman Drake has pointed out that Galileo could have had in mind the material he drafted for the *De motu antiquiora*, as in *Opere* 1:345–346, which gives such proofs and likewise has a marginal reference to "p°

Meteororum cap. 3º." This would establish a hitherto unnoticed connection between MSS 46 and 71.

[32] On the other hand, as Charles Lohr has indicated to me, the peculiar manner of citation, and the fact that Marsilius of Inghen is mentioned by Galileo whereas Buridan, Albert of Saxony, and Themo Judaeus are not (see Table I), suggests some dependence, possibly indirect, on a compilation such as Lokert's Paris edition of 1516, as Duhem originally speculated.

[33] The earliest edition recorded by Sommervogel (Vol. II, col. 1212) is a quarto volume published at Rome by Victorius Helianus in 1570. Some caution is necessary here, however, as the writer has seen reference to a quarto edition of this work published at Rome in 1565. This is in a handwritten inventory of books that at one time were in the personal library of Pope Clement XI, many of which are now in the Clementine Collection of the Catholic University of America, Washington, D.C.; unfortunately this particular Clavius volume is no longer in the collection.

[34] This is a second piece of evidence counting against Galileo's having copied directly from a printed source; see note 21 above.

[35] As remarked at the end of note 11, the inconsistencies of titling could be a sign of reliance on different sources. In this connection it is perhaps noteworthy that the primary division of Vitelleschi's work is the *Tractatio*, which is subdivided into the *quaestio*, as in Galileo's first set of notes; the primary division of Valla's work, on the other hand, is the *Tractatus*, and this is subdivided into the *disputatio* and the *pars*, and then finally into the *questio*, as in Galileo's second set of notes.

[36] See Favaro's Avvertimento in *Opere* 9:279–282.

[37] See Favaro's Avvertimento in *Opere* 9:275–276, together with his transcription of the text and an indication of the corrections, *ibid*., 9:283–284.

[38] For the Latin text, see *Opere* 1:27.

[39] See Favaro's Avvertimento in *Opere* 1:9 and 1:11–12; also 'Galileo Galilei e i Doctores Parisienses,' p. 8.

[40] This date was surely known in the time of Galileo; it is recorded, for example, by Joseph Scaliger in his widely used *De emendatione tempiris*, first printed in 1583, as taking place in A.D. 70.

[41] Rev. William Hales, D.D., *A New Analysis of Chronology and Geography, History and Prophecy*. . ., 4 vols., London: C. J. G. & F. Rivington, 1830, Vol. 1, p. 214.

[42] *Ibid*., pp. 211–214.

[43] In his researches the writer has uncovered only one book that does give Galileo's figure. This is Ignatius Hyacinthus Amat de Graveson, O.P., *Tractatus de vita, mysteriis, et annis Jesu Christi*. . ., Venetiis: Apud Joannem Baptistam Recurti, 1727, pp. 251–252: "Christus Dominus anno aerae vulgaris vigesimo sexto, imperii proconsularis Tiberii decimo sexto, anno urbis Romae conditae 779, anno a creatione mundi 4164. . ." The date of publication of this work obviously would rule it out as a source; but see note 50 *infra*.

[44] For details concerning Pererius's Averroism and internal controversies at the Collegio, see M. Scaduto, *Storia della Compagnia di Gesù in Italia*. L'Epoca di Giacomo Lainez, 2 vols., Gregorian University Press, Rome, 1964, Vol. 2, p. 284. Also R. G. Villoslada, *Storia del Collegio Romano*, pp. 52, 78ff, 329, and C. H. Lohr, 'Jesuit Aristotelianism and Suarez's *Disputationes Metaphysicae*,' to appear in *Paradosis: Studies in Memory of E. A. Quain* (New York 1976).

[45] Villoslada, *Storia del Collegio Romano*, p. 323.
[46] Benedictus Pererius, *Commentariorum in Danielem prophetar xdecim. . .* Romae: Apud Georgium Ferrarium, 1587; and *Prior tomus commentariorum et disputationum in Genesim. . .* Romae: Apud Georgium Ferrarium, 1589.
[47] We have used the Frankfurt 1593 edition, where the chronology is given on p. 377.
[48] For details, see the chronologies listed by Pererius on pp. 350–351 of his commentary on Daniel and on p. 130 of his commentary on Genesis.
[49] See Hales, *A New Analysis of Chronology*, p. 217.
[50] It is perhaps significant that the author cited in note 42, Amat de Graveson, taught at the Collegio Casanatense in Rome, which was adjacent to the Collegio Romano, and thus he could have had access to its archives.
[51] It should be noted, moreover, that the 5748-year total on fol. 10r was written with some hesitation: first Galileo wrote 50 in Arabic numerals, then crossed out these numbers, then wrote 'five thousand four and eighty' in longhand, then crossed out the 'four and eighty,' changed it to 'eight and forty,' and finally inserted a 'seventy' before the 'eight and forty,' all in longhand. Now none of the other figures on this folio were changed in any way; they all appear exactly as in the translation on p. 114. On fol. 15v, however, when recording what should have been the same number of years, Galileo did not give the same total but wrote instead 'six thousand seven hundred and 48 years,' without apparently noticing the difference of a thousand years. The successive revisions in the first sum could be an indication that this was Galileo's own calculation rather than something he copied from an existing source, the result of which calculation he had difficulty putting into Latin prose.
[52] See *Opere* 10:22; also Villoslada, *Storia del Collegio Romano*, pp. 194–199.
[53] For details, see Giuseppe Cosentino, 'L'Insegnamento delle Matematiche nei Collegi Gesuitici nell'Italia settentrionale. Nota Introduttiva', *Physis* 13 (1971), 205–217, and 'Le mathematiche nella "Ratio Studiorum" della Compagnia di Gesù', *Miscellanea Storica Ligure* (Istituto di Storia Moderna e Contemporanea, Università di Genova), II, 2 (1970), 171–213.
[54] See Charles B. Schmitt, 'The Faculty of Arts at Pisa at the Time of Galileo', *Physis* 14 (1972), 243–272, especially p. 260.
[55] Schmitt, 'The Faculty of Arts. . .,' p. 256; also Ludovico Geymonat, *Galileo Galilei* (transl. by Stillman Drake), McGraw-Hill Book Co., New York, 1965, pp. 10–11.
[56] Galileo's friend, Mario Guiducci, makes this point very well in his Letter to Father Tarquinio Galuzzi, where he states: "It seems to me that. . .it is wrong to call men copyists who, when treating a philosophical question, take an idea from one author or another and, as is not the case with those who merely copy the writings, make it their own by judiciously adapting it to their purposes so as to prove or disprove some or other statement. . . .To give an exceptional example, on these terms Father Christopher Clavius would have been a first-class copyist, for he was extremely diligent in extracting and compiling in his works of great erudition the opinions and demonstrations of the most distinguished geometers and astronomers up to his time – as seen in his compendious commentary on the *Sphere* of Sacrobosco and in so many other of his works." – *Opere* 6:189; see note 59 *infra.*
[57] *Opere* 10:44–45. This letter and its contents are discussed by Crombie in his paper cited in note 14 *supra*, pp. 167–68.
[58] This possibility, of course, reopens the question as to whether or not Galileo actually

did have recourse to primary sources, in some instances at least, when composing these notes; the mere fact that the notes are based on secondary sources, or bear close resemblances to existing manuscripts, does not preclude consultation of the originals cited therein.

59 Favaro speculates that Galileo might have had a hand in Guiducci's Letter, cited in note 56 *supra*; see *Opere* 6:6. For other collaboration between Galileo and Guiducci, see William R. Shea, *Galileo's Intellectual Revolution*. Middle Period, 1610–1632, Science History Publications, New York, 1972, pp. 75–76.

60 *Opere* 8:101.

61 See Schmitt, 'The Faculty of Arts...,' pp. 261–262.

62 Christophorus Clavius, *Euclidis Elementorum Libri XV*. . ., Romae: Apud Vincentium Accoltum, 1574; nunc iterum editi ac multarum rerum accessione locupletati, Romae: Apud Bartholomaeum Grassium, 1589. Galileo's own interpretation of Euclid, however, would still derive from Tartaglia's Italian translation, with which he was quite familiar, and which, as Stillman Drake has repeatedly argued, underlies his distinctive geometrical approach to the science of motion.

63 Schmitt, 'The Faculty of Arts...,' p. 262; also *Opere* 19:119-120; and Antonio Favaro, *Galileo Galilei a Padova*, Padua: Editrice Antenore, 1968, pp. 105–114, especially p. 108.

64 Four versions are listed by Favaro, *Opere* 2:206; Stillman Drake reports a fifth version, 'An Unrecorded Manuscript Copy of Galileo's *Cosmography*', *Physis* 1 (1959), 294–306.

65 See Schmitt, 'The Faculty of Arts...,' p. 206.

66 The table is reproduced in *Opere* 2:244–245; compare this with the table on pp. 429–430 of the 1581 edition of Clavius's *Sphaera*.

67 *Opere* 1:314.

68 Clavius, *Sphaera* (1581 edition), pp. 108–109.

69 Crombie has suggested a similar re-evaluation of the so-called *Juvenilia* in his 'Sources...' paper (note 14 above), pp. 162–170. In this connection it is not essential to the thesis here proposed that the notes actually have been written in 1590 – the date toward which the evidences adduced above appear to converge. If the dependence on the Collegio Romano materials be conceded, there are several possibilities for the transmission of these materials to Galileo. Galileo could have obtained them from Clavius as early as 1587, for use when tutoring in Florence and Siena or for securing the vacant teaching posts at Bologna or at Pisa. Again, he may have obtained them independently of Clavius, and thus at an even earlier date. For example, I have discovered that notes very similar to those of the Collegio were used by a Benedictine monk at the University of Perugia in 1590; it is quite possible that such notes were disseminated throughout other monasteries and religious orders in northern Italy. Now Galileo spent some time as a youth at the monastery of Vallombrosa and remained on friendly terms with the monks there, even teaching a course on the *Perspectiva* for them in 1588 (the latter information *via* a personal communication to the author from Thomas B. Settle). In such a setting he could have had access to *reportationes* of the type described and have made his own notes from them so as to be prepared for an eventual teaching assignment. The fact remains, however, that the materials discussed above are the only such *reportationes* that have been discovered thus far, and in defect of other evidences the 1590 dating alone has more than conjectural support.

70 For Duhem's own views see his *The Aim and Structure of Physical Theory* (transl. by P. P. Wiener), The Princeton University Press, Princeton, 1954; for their historical justification, apart from Duhem's monumental work on *Le Système du Monde*, see his *To Save the Phenomena*, 'An Essay on the Idea of Physical Theory from Plato to Galileo' (transl. by E. Doland and C. Maschler), University of Chicago Press, Chicago, 1969.

71 There are other flaws in Duhem's historical arguments, of course, and these have been well detailed by Annaliese Maier in her studies on the natural philosophy of the late scholastics. Other scholars have contributed substantial information, since Duhem's time, that support various aspects of his continuity thesis; among these should be mentioned Ernest A. Moody and Marshall Clagett and their disciples. In what follows, to the researches of these authors will be added a brief survey of sixteenth-century work that complements their findings but leads to slightly different philosophical conclusions than have heretofore been argued.

72 The last half of Duhem's third volume on Leonardo da Vinci is in fact entitled 'Dominique Soto et la Scolastique Parisienne,' pp. 263–581, of which pp. 555–562 are devoted to Soto's teachings. For a summary of Soto's life and works, with bibliography, see the article on him by the writer in the *Dictionary of Scientific Biography*, Vol. 12, Charles Scribner's Sons, New York, 1975, pp. 547–548.

73 *Storia del Collegio Romano*, pp. 60–61. Villoslada also calls attention to the professed Thomism of the theology faculty there, and to the tendency otherwise to imitate the academic styles then current at the Universities of Paris and Salamanca (p. 113).

74 Cited by Herman Shapiro, *Motion, Time and Place According to William Ockham*, Franciscan Institute Publications, St. Bonaventure, N.Y., 1957, p. 40.

75 *Ibid.*, p. 53; see also William of Ockham, *Philosophical Writings: A selection* (ed. by Philotheus Boehner), Thomas Nelson & Sons, Ltd., Edinburgh, 1957, p. 156.

76 *Subtilissime questiones super octo phisicorum libros Aristotelis*, Lib. 3, q. 7, Parisiis: In edibus Dionisii Roce, 1509, fols. 50r–51r.

77 W. A. Wallace, 'The Concept of Motion in the Sixteenth Century', *Proceedings of the American Catholic Philosophical Association* 41 (1967), 184–195; and 'The "Calculatores" in Early Sixteenth-Century Physics'. See Essays 4 and 5, *supra.*

78 W. A. Wallace, 'The Enigma of Domingo de Soto: *Uniformiter difformis* and Falling Bodies in Late Medieval Physics', *Isis* 59 (1968), 384–401. See Essay 6, *supra.*

79 The text of that paper, written around 1586, is cited by Cosentino, 'Le matematiche nella "Ratio Studiorum". . .,' p. 203, as follows: "Senza le matematichi 'physicam. . . recte percipi non potest, praesertim quod ad illam partem attinet, ubi agitur de numero et motu orbium coelestium, de multitudine intelligentiarum, de effectibus astrorum, qui pendent ex variis coniunctionibus, oppositionibus et reliquis distantiis inter sese, de divisione quantitatis continuae in infinitum, de fluxu et refluxu maris, de ventis, de cometis, iride, halone et aliis rebus meteorologicis, de proportione motuum, qualitatum, actionum, passionum et reactionum, etc., de quibus multa scribunt Calculatores.' "

80 Latin text cited above, note 30.

81 In this letter, dated April 12, 1615, Bellarmine commended Foscarini and Galileo for being prudent in contenting themselves to speak hypothetically and not absolutely when presenting the Copernican system, thus considering it merely as a mathematical hypothesis (*Opere* 12:171–172). Galileo, of course, quickly disavowed that such was his

intent (*Opere* 5:349–370, especially p. 360).

82 Latin text in Clavius, *Sphaera*, 1581 edition, p. 605. An English translation of this and surrounding passages is to be found in R. Harré, *The Philosophies of Science: An Introductory Survey*, Oxford University Press, Oxford, 1972, pp. 84–86.

83 Crombie, 'Sources. . .,' and W. A. Wallace, 'Galileo and Reasoning *Ex Suppositione*. . .,' both cited in note 14 above.

84 L. M. De Rijk, 'The Development of *Suppositio naturalis* in Medieval Logic', *Vivarium* 11 (1973), 43–79, especially p. 54. Other expositions of Ockham's theory of demonstration are E. A. Moody, *The Logic of William of Ockham*, Sheed and Ward, New York, 1935, and Damascene Webering, *Theory of Demonstration According to William Ockham*, Franciscan Institute Publications, St. Bonaventure, N.Y., 1953.

85 *In metaphysicen Aristotelis quaestiones*. . ., Lib. 2, q. 1, Parisiis: Venundantur Badio, 1518, fol. 9r; *Quaestiones super decem libros ethicorum Aristotelis*, Lib. 6, q. 6, Parisiis: Venundantur Ponceto le Preux, 1513, fols. 121v–123r. For a discussion of the first of these texts, and a critique of Ernest Moody's reading of it, see W. A. Wallace, 'Buridan, Ockham, Aquinas: Science in the Middle Ages,' *The Thomist* 40 (1976), pp. 475–483, revised as Essay 16, *infra*.

86 For the precise texts see *Opere* 5:357.22, 7:462.18, 8:197.9, 8:273.30, 17:90.74, and 18:12.52. The first of these uses the Italian equivalent (*supposizioni naturali*) but the remainder employ the Latin *ex suppositione* even when the surrounding text is in Italian.

87 Respondeo: scientiam subalternatam tamquam imperfectam non habere perfectas demonstrationes, cum prima principia supponat in superiori probata, ideoque gignat scientiam ex suppositione et secundum quid. . . – MS Gal 27, fol. 20v. This reading is quoted from the transcription of this manuscript kindly made available to me by Adriano Carugo; for a brief preliminary analysis of the place of this work in Galileo's thought, see Crombie, 'Sources. . .,' (note 14 above), pp. 171–174.

88 Theorica planetarum non est scientia, nam scientia est effectus demonstrationis, sed in illa nulla invenitur demonstratio, ergo. Est igitur scientia vel secundum quid ratione suppositionis vel opinio tantum, non quidem ratione subsequentis sed ratione antecedentis. – Österreichische Nationalbibliothek, Vienna, Cod. Vindobon. 10509, fol. 198r.

89 This theme is developed in full detail in Peter Machamer, 'Galileo and the Causes,' (1978), pp. 161–180.

90 For a good account of the Aristotelian revival in the late sixteenth century, which locates the work of the Collegio Romano in the larger context of European universities generally, see Charles B. Schmitt, 'Philosophy and Science in Sixteenth-Century Universities: Some Preliminary Comments', in *The Cultural Context of Medieval Learning* (ed. by J. E. Murdoch and E. D. Sylla), Boston Studies in the Philosophy of Science, Vol. XXVI, Reidel, Dordrecht and Boston, 1975, pp. 485–537.

91 Thus some of the claims made by Heiko A. Oberman, 'Reformation and Revolution: Copernicus's Discovery in an Era of Change', in *The Cultural Context*. . .(note 90), pp. 397–435, and by E. A. Moody in his *Studies in Medieval Philosophy, Science and Logic*, University of California Press, Berkeley, 1975, pp. 287–304 and *passim*, with regard to the role of nominalism in the Scientific Revolution, would seem to require revision in light of the findings reported in this paper.

92 See John E. Murdoch, 'The Development of a Critical Temper: New Approaches and

Modes of Analysis in Fourteenth-Century Philosophy, Science, and Theology,' Medieval and Renaissance Studies (Univ. of North Carolina), 7 (1978), 51–79; also his 'From Social into Intellectual Factors: An Aspect of the Unitary Character of Late Medieval Learning,' in *The Cultural Context* . . . (note 90), pp. 271–384, and 'Philosophy and the Enterprise of Science in the Later Middle Ages,' in *The Interaction Between Science and Philosophy* (ed. by Y. Elkana), Humanities Press, Atlantic Highlands, N.J., 1974, pp. 51–74.

APPENDIX

Since the original publication of this essay, the author has had the opportunity to investigate further the lecture notes of Mutius Vitelleschi and Paulus Valla, and in addition has studied *reportationes* of other lectures on natural philosophy given at the Collegio Romano by Antonius Menu in 1577–1579 and by Ludovicus Rugerius in 1590–1591. All of these Jesuits make references to the *Doctores Parisienses*, and thus reinforce the conclusion of the author that they are a likely source of Galileo's citation of the Parisian masters. To complete the account already given, in what follows we shall document the additional references to the *Parisienses* that have been uncovered, grouping them under the names of the lecturers making the citations.

Antonius Menu

Menu began teaching the course on the *Physics* more than a decade before Valla and Vitelleschi, and indeed only a year after the appearance of Pererius's *De communibus* (for details, see *Galileo's Early Notebooks*, pp. 13–24). This time span notwithstanding, his treatment of the question whether the world could have existed from eternity is similar in many respects to the replies of both Pererius and Vitelleschi, differing from the former mainly in that it references the *Doctores Parisienses*, and from the latter in the way in which it organizes the material. Whereas Vitelleschi lists basically only three opinions, including St. Thomas's view (which Galileo has made a fourth opinion) as a variation of his first, Menu classifies the alternative positions under six different headings. His first is that of St. Bonaventure, Henry of Ghent, Aureolus, Philoponus, and Burley, holding that no creature could have existed for all eternity; his second is just the opposite, allowing as probable the opinion that every type of creature, permanent and successive as well as corruptible and incorruptible, could have so existed – which he attributes to Ockham, Gregory of Rimini, Gabriel Biel, and John Canonicus. Menu's third opinion is similar to Vitelleschi's variation on his first, except that St. Thomas is not included. It maintains that both kinds of creatures, i.e., corruptible and incorruptible (see par. F18 in *Galileo's Early Notebooks*), probably could have existed from eternity, and that objections against either alternative can be answered, but that it is better to hold that the world's eternity is possible. Since the *Parisienses* are mentioned here, we provide the Latin text in our transcription:

Tertia opinio est Scoti in secundo [Sententiarum], distinctione prima, quaestione tertia, et *Doctorum Parisiensium*, octavo Physicorum, quaestione prima, qui dicunt utramque partem probabiliter posse defendi, quia rationes pro utraque parte possunt solvi probabiliter, licet magis accedant ad partem quod potuerit esse. – Cod. Ueberlingen 138. no foliation, De mundo, cap. 5, An potuerit mundus esse ab aeterno.

After that, Menu gives Durandus's view that incorruptibles could have existed from eternity, but not corruptibles, as a fourth opinion, and then two more opinions which he attributes to Aquinas's school. The fifth, he says, is that of certain Thomists, such as Domingo de Soto, who hold that successive entities (e.g., motion) could have existed from eternity, but not corruptibles, in view of the arguments based on "the pregnant woman" and infinites in act (. . . successiva potuisse ab aeterno; de corruptibilibus vero negat, propter argumenta de muliere pregnante et de infinito actu, *ibid*.). The sixth opinion is that of St. Thomas himself, followed by Capreolus, who holds that it does not imply a contradiction to hold either that any being could have been produced from eternity or that anything other than God could have existed from eternity, even though Thomas does teach in the *Summa theologiae* that men could not have existed from eternity, so as to avoid having an actually infinite number of human souls.

With regard to Valla's mention of the *Doctores Parisienses* in discussing the ratios between the sizes of the elements (see the essay above, as well as par. U7 in *Galileo's Early Notebooks*), Menu gives a similar citation, and also references Aristotle's text of the third *Meterology* in the same peculiar way. For comparative purposes, the portion of Menu's Latin text that parallels Galileo's U7 is given below:

Hactenus disputavimus de quantitate, raritate, densitate, loco, et figura, quae insunt elementis ratione quantitatis; reliquum videtur ut aliquid dicamus de proportione eorundam, quae est affectio quantitatis, explicantes an elementa habeant decuplam proportionem. Antiqui, quos sequuti *Doctores Parisienses*, primo Metheororum, quaestione tertia, et alii nonnulli, qui dicerunt elementa habere quidem secundum se tota proportiones aequales quoad extensionem, vero unum excedere aliud in decupla proportione. Et hoc idem videtur tenere Aristoteles in primo Metheororum, summa prima, capite tertio, ubi volans probare praeter tria elementa dari quartum prope concavum, id est, ignem, quia alias sequeretur inter elementa non dari proportiones; sed ista proportio est admittenda; ergo ... Sed contraria sententia est communis mathematicorum nostri temporis, qui negant talem proportionem. Probatur ... – Cod. Ueberlingen 138, De quantitate elementorum, cap. 10, De proportione elementorum.

Finally, it is noteworthy that in his concluding disputation on the elements (De qualitatibus motivis, cap. 6, De motu violento proiectorum gravium et levium), Menu does not mention the *Parisienses* as a group, but he does cite the individual Parisians, John Buridan and Albert of Saxony, as required reading for an understanding of the problem. His own conclusions are influenced by these authors and are quite different from those of Pererius, who had rejected impetus doctrine entirely. Menu holds the following: (1) it is probable that projectiles are moved by the medium, namely, air or water; (2) it is more probable that they are moved not only by air by also by some quality or *virtus impressa*; (3) this *virtus impressa* is reducible to a disposition in the first species of quality insofar as it is readily changeable, or to a passive quality like gravity or levity insofar as it causes motion; and (4) the *virtus impressa* has some of the attributes of a *qualitas spiritualis* and can be introduced into a body in an instant (*ibid.*). For more details, see Essay 15, *infra.*

Paulus Valla

As noted in *Galileo's Early Notebooks* (pp. 17–18), the materials surviving from Valla's lectures on natural philosophy are rather sparse, but they do include his treatise *De elementis*. In the latter parts of this treatise he makes further mention of the *Doctores Parisienses* when discussing projectile motion and problems relating to the gravity and levity of the elements. In so doing, he advances beyond the material contained in Menu's lectures and comes closer to the concerns that are manifested by Galileo in his early writings on motion. These additional references to the *Parisienses* will now be briefly described.

Valla's main citation of the Parisian masters occurs in the fifth part of his first disputation, devoted to *De elementis in genere*, which is entitled *De qualitatibus motivis*. The last question in this part, q. 6, bears the title, *A quo moveantur proiecta*? Valla has a lengthy discussion of this question, investigating thoroughly what is meant by a *virtus impressa*, to what species of quality it is reducible, and how it is produced and then corrupted. Throughout this treatment he pays special attention to the views and explanations of Marsilius of Inghen. He then raises the question how a projectile can be moved by the air or the surrounding medium, and details the various mechanisms that have been proposed by Aristotelian commentators. There are two main schools on this question, he continues, the first holding that the projectile is moved by the air in the ways just explained, and he labels this as Walter Burley's opinion, though he notes that it is commonly attributed to Aristotle himself, Themistius, Philoponus, Averroës, St. Thomas, and others, notwithstanding the fact that all of these authors also admit some kind of *virtus impressa*. The second opinion is that of those who say that the projectile is moved by impetus, and here Valla enumerates Albert of Saxony and John Buridan, although he also includes the *Doctores Parisienses* as a group, assimilating them to the teaching of St. Thomas Aquinas. The Latin text is worth transcribing in its entirety:

Secunda sententia est eorum qui dicunt motum proiectorum fieri a virtute impressa. Ita tenet Scaliger, exercitatione 28. Albertus de Saxonia, octavo Physicorum, quaestione 13, et tertio Caeli, quaestione 17; Buridanus, octavo Physicorum, quaestione 12; Paulus Venetus in Summa de caelo, capite 23; et Simplicius ibidem, videntur ponere hanc virtutem impressam in aere. Quod si ponatur in aere, nulla est ratio cur non possit dari in proiecto; et ideo ait [sic] posse fieri motum violentum in vacuo. D. Thomas etiam et *Doctores Parisienses*, octavo Physicorum et tertio Caeli, videntur admittere hanc virtutem in aere. Et communiter omnes fere antiqui, sicut etiam Aristoteles, indicant dari talem virtutem in medio, et tamen dicunt proiecta moveri a medio vel aere, ut ostendant in omni motu motorem distinctum a mobili. Potest autem haec sententia probari, primo . . . – Cod. APUG/FC 1710, no foliation, Tractatus quintus, Disputatio prima, Pars quinta, Quaestio sexta.

Valla's own view is that the medium alone is not sufficient to move the projectile, and that some other *virtus* is necessary for this, although it is probably helped by the motion of the medium along the lines of one or other of the mechanisms he has proposed. He concludes with the observation that those who wish more on this matter should consult the works of Vallesius, Albert, Soto, Gratiadei, Scaliger, Buridan, and others. (More details of his teaching on impetus are given in Essay 15, *infra*.)

Valla also mentions the *Doctores Parisienses* when discussing the question whether air, water, and earth gravitate in their own spheres (see Essay 7, *supra*). Here he lists three opinions, in the third of which he mentions the Parisians as answering the question in the negative. The Latin text of the reference follows:

Tertia sententia est nullum elementorum, sive medium sive extremum, levitare vel gravitare in sua sphaera. Ita tenet [sic] Ptolemeus in libro de sectionibus, ut refert Simplicius, quarto Caeli, commento 16; Themistius, referente Averroe, quarto Caeli, commentis 30 et 39; *Doctores Parisienses*, quarto Caeli, quaestione tertia; Albertus, quaestione etiam tertia; et Niphus, quarto Caeli, loco citato, qui dicunt quamvis elementa nec gravitent nec levitent, posse tamen dici potius gravitare quam levitare, quia magis apta sunt resistere trahendi sursum quam deorsum – *ibid.*, Disputatio secunda, Quaestio sexta et ultima (An aer, aqua, et terra gravitent in suis sphaeris).

Valla's own view is slightly more nuanced, and is otherwise similar to that adopted by Vitelleschi and explained in Essay 7, *supra*. Valla holds that all elements, wherever they are located, have their own "habitual motive powers" (*virtutes motivas habituales*), and so they possess gravity or levity even though they are completely at rest; this does not entail, however, that they actually gravitate or levitate when situated in their proper spheres.

Mutius Vitelleschi

Like Menu and Valla, Vitelleschi makes additional references to the *Parisienses* when discussing the motive powers of the elements, and this in the context of problems relating to gravity and levity. He also mentions them in a more general context, however, and this occurs at the outset of his exposition of the *Physics*, when he is discussing the proper subject matter of natural philosophy. On this question, he says, there are three opinions, two of which seem to differ more in terminology than in the substance of their positions. The first opinion is that the subject of natural philosophy is *ens mobile*, and this he attributes to St. Thomas, Ferrariensis, Soto, Javelli, Cajetan, Gratiadei, Aquarius, and other Thomists, noting that they are followed in it by John of Jandun, Albert of Saxony, and others (Cod. Bamberg. SB, 70, fol. 28r). To this question Vitelleschi also assimilates Albert the Great and Giles of Rome, saying that they agree with the position formally (*quoad formale*) but differ from it materially (*quoad materiale*); Toletus and others, he goes on, agree with it materially but differ from it formally (*ibid.*, fol. 28v). The second position is that of the Scotists, who say that the subject is finite natural substance, so as to exclude God and accidents from its ambit; the Parisian doctors hold a similar view, namely, that the subject is the natural body. The Latin for this citation reads as follows:

Secunda sententia est Scotistarum, qui dicunt obiectum esse substantiam naturalem finitam, ita ut ab obiecto physicae excludatur Deus et accidentia, ut caeterae substantiae etiam incorporeae comprehendantur. Ita Trombetta, quinta Metaphysicorum, quaestione secunda, Ioannes Canonicus, primi Physicorum, quaestio prima, articulus secundus, et *Parisienses*, quaestio quarta primi Physicae, qui tamen dicunt obiectum physices Aristotelis esse corpora naturalia, quia inquiunt Aristoteles existimavit intelligentias omnes esse immobiles motu physico. – *ibid.*, fol. 28v.

The third position, finally, and this seems to differ only verbally from the preceding, is that the subject of physics is the natural body precisely as natural (*corpus naturale ut naturale est* – fol. 29r); this view he attributes to Themistius, Avicenna, Averroës, Carpentarius, Buccaferreus, Paulus Venetus, and Aquilinus. Vitelleschi's own reply agrees with this last position: the adequate object of physics (*obiectum adaequatum physicae*) is the natural body precisely as natural.

In addition to this citation of the *Parisienses* and that discussed in the essay above (corresponding to *Galileo's Early Notebooks*, par. F18), Vitelleschi has two more references to them when discussing the motion of heavy and light bodies. The first occurs in a context similar to that of Valla's when treating the problem whether elements gravitate in their own spheres. Here Vitelleschi merely notes that the *Parisienses*, contrary to Borri and in agreement with Archimedes and Valesius, hold that the elements do not gravitate when in their proper place, and for this he cites the same locus as Valla, viz, *quarto Caeli, quaestione tertia* (Cod. Bamberg. SB, 70, fol. 370r). The second citation comes when Vitelleschi is listing opinions as to whether elements have only a passive principle of their natural motion within them, or whether they are moved actively by their own forms. The first alternative he lists as the teaching of St. Thomas, the Thomists, and Toletus, whereas the second he attributes to the Scotists, Walter Burley, the Parisian doctors, and others. The Latin text of the reference to those holding the second opinion is the following:

Secunda sententia affirmat elementa active moveri a propriis formis. Ita Avicenna, primi Sufficientiae, capite quinto; Thomas de Garbo, libro primo, quinto, in tractatu primo, capite tertio; Scotus, in secundo [Sententiarum], distinctione secunda, quaestione decima, et nono Metaphysicorum, quaestione quartodecima, ubi etiam Antonius Andreas, quaestione prima, et illud sequuntur Scotistae communiter; Gregorius, in secundo [Sententiarum], distinctione sexta, quaestione prima, articulo tertio; Achillinus, tertio de elementis, dubio secundo; Iandunus, octavo Physicorum, quaestionibus undecima et duodecima, et quarto Caeli, quaestione octodecima; Burlaeus, octavo Physicorum, in textu 33; Thiennensis, ibidem, in 28; Venetus, ibidem, in textu 82 et in Summa de caelo, capite 24 et 25; Albertus de Saxonia, octavo Physicorum, quaestionibus sexta et septima, et quarto Caeli, questionibus octava, nona, decima; *Parisienses*, octavo etiam Physicorum; Contarenus, primo de elementis; Philaleseus, quarto Caeli in textu 21; Buccaferreus, in textu 43, secundo De generatione; Zymara, tum theoremate 68 et in questione de movente re moto; Scaliger, exercitatione 28, numero secundo; Pererius, libro septimo, capite sextodecimo. – *ibid.*, fol. 373v.

Vitelleschi's own position is that the principle of motion within the elements is not merely passive but in some respects at least must be active, and that the substantial form is such an active principle, of which the motive quality within the element is merely the instrument. For fuller details on this and related teachings of Vitelleschi, see Essays 7, 14, and 15 in this volume.

Ludovicus Rugerius

Rugerius followed Vitelleschi in teaching natural philosophy at the Collegio Romano (see *Galileo's Early Notebooks*, pp. 18–20, for details), and does not have some of the references to the *Doctores Parisienses* that are found in his predecessor. He does make one general acknowledgment of them, however, and this is quite revealing, for it shows that the Parisian masters were among the authorities that were studied by the Jesuits of the Collegio when preparing their lectures in natural philosophy. Apparently they grouped their sources under three headings, viz, the Greeks, the Arabs, and the Latins, for Rugerius took account of the Parisians under the last category, along with Albert the Great, St. Thomas Aquinas, Albert of Saxony, John of Jandun, Nifo, and others. His acknowledgement of these sources is of interest because it occurs at the outset of his exposition of Aristotle's *De caelo*, where he is laying out the rationale of his treatment in general. For this reason the Latin text is here transcribed in its entirely:

Prima disputatio de universo seu de corpore simplici in universum, respondens fere primo libro De caelo:

EXORDIUM

Cum in superioribus disputationibus simul cum Aristotele egerimus de corpore naturali in universum, eiusque communissima principia et passiones investigaverimus; restat nunc ut progrediamur ad singulas species corporis naturalis. Dividitur autem corpus naturale in corpus simplex et corpus mixtum. De corpore autem simplici agitur in quatuor libris De caelo, ac primum quidem libro primo de corpore simplici in universum; secundo magis in particulari de corpore caelesti; tertio et quarto de gravi et levi, seu de quatuor elementis. Quem Aristotelis ordinem servabimus etiam nos in hac secunda disputationum physicarum parte, quae erit de corpore simplici. Agemus enim in hac disputatione de universo seu, quod idem est, de corpore simplici in universum, qua in disputatione non solum explicabimus multa quae pertinent ad librum De caelo, sed etiam plurima de quibus Aristoteles disputat etiam octavo Physicorum, quae a nobis in hunc locum reiecta sunt. Sed antequam ea aggrediamur, ne omnino ieiune hi quattuor Aristotelis libri praetereantur qui pro angustia temporis fusius explicari non possunt, praemittam quaestionem unam quae aliquam horum librorum explicationem utrumque continebit. Qui plura desiderat, eos consulat auctores qui in hos libros explicandos incubuerunt. Ii sunt quos viderim: ex Graecis, Simplicius qui scripsit commentarios, Philoponus qui scripsit 18 de mundi aeternitate adversus Proclum, licet hi maxime pertineant ad octavum

Physicorum. Ex Arabibus, Averroës scripsit commentarios et paraphrasim in libros De caelo. Ex Latinis, Albertus, Sanctus Thomas, Albertus de Saxonia, Iandunus, Niphus, *Doctores Parisienses* partim quaestiones partim commentarios in hos libros ediderunt. His accedunt Aquilinus, Zabarella; ac praeterea multi ex iis qui scripserunt in octavo Physicorum et in libros De generatione multa habent quae pertinent ad hunc locum. – Cod. Bamberg. SB, 62.4, pp. 1–2.

Note here that Rugerius refers the reader for fuller details to authors who have explained Aristotle, and makes this reference more explicit by listing "those whom I have seen," among whom he includes the *Doctores Parisienses*.

When treating the two matters discussed in the physical questions where Galileo cites the *Parisienses* (pars. E18 and U7 of *Galileo's Early Notebooks*), Rugerius has approximately the same coverage but does not reference them as authorities. A possible reason for this is that he organizes the matter in a way somewhat differently from Galileo's and his Jesuit predecessors' at the Collegio. Rather than treat the possibility of the eternity of the world in general, for example, he divides this question into parts, asking whether the eternity of motion is opposed to the natural light of reason (An aeternitas motus repugnet lumini naturae – *ibid.*, p. 56) and then whether the same is true of permanent things (An aeternitas rerum permanentium repugnet lumini naturae – *ibid.*, p. 61). Similarly, when discussing the ratios of the sizes of the elements, he contents himself with the statement that the peripatetics have great difficulty in this matter because of the objections the mathematicians bring against Aristotle, and then discourses at great length on the mathematical difficulties without referencing further the various authorities (*ibid.*, pp. 251–256).

Another citation of the Parisian doctors occurs in the question discussed by several of Rugerius's predecessors, namely, whether the elements gravitate and levitate in their proper places. Here he enumerates among those giving an affirmative answer Averroës, Nipho, the Conciliator, Aponensis, Gentili, Hieronymus Borrius in his treatise *De motu gravium et levium*, "and some others," and adds that this view is commonly attributed to Aristotle. The negative position he assigns to a number of ancients, including Ptolemy and Archimedes, and also to the *Parisienses*. The Latin text of this citation follows:

Secunda sententia est negantium elementa gravitare et levitare in propria sphaera. Est Themistii apud Averroëm, quarto Caeli, commentis 30 et 39; Simplicii, commento ibidem, qui citat pro eadem sententia Ptolemaeum in libro de erectionibus; et magnum Syrianum. Est Archimedis in libro de ponderibus; Philalesaei, quarto Caeli, 23; Hugonis, prima primi doctrinae, secundo, in fine; *Parisiensium*, quarto Caeli, quaestione tertia; Valesii in capite quarto quarti Physicorum, et est valde communis (*ibid.*, p. 179).

Rugerius himself holds the negative opinion on this matter, and gives arguments that show a sensitivity to mathematical details, thus indicating an awareness of the matter contained in the treatises of Archimedes and Ptolemy and the ability to apply it to the question at hand.

Rugerius's most significant mention of the *Parisienses*, however, occurs in a question where he is defining terms such as *uniformiter difformis*, in a passage that is very similar to Galileo's par. O4 (see *Galileo's Early Notebooks*, pp. 173–174, 280). It is this terminology in Galileo that led Duhem, as noted in the essay above, to claim the Parisian masters as the precursors of the famous physicist. Rugerius has a much fuller discussion of the calculatory terminology, and mentions the sources from which it is drawn, including the Calculator (Richard Swineshead), Burley, Albert of Saxony and the *Doctores Parisienses*, Francis of Meyronnes, James of Forli, John Canonicus, and various nominalists and Thomists (*ibid*., p. 280). A typical reference wherein the Parisians are mentioned explicitly by Rugerius is the following:

> Iam vero probo quod neque dici possit quod hi gradus qui supraadduntur sint homogenei; quia vel hoc explicari debet cum Scoto, primo [Sententiarum], distinctione 17, quaestione 4; Gregorio, ibidem; Occham, quaestione 6; Gabriele, ibidem; Francisco de Maior. [sic], quaestione 2, articulo 1; Forliviensi, De intensione formarum; Joanne Canonico, quinto Physicorum, quaestione 3; et *Doctoribus Parisiensibus*, in praedicamento qualitatis; quod hi omnes gradus differant materialiter et numero imperfecto, ratione principiorum individuantium et ratione entitatis, quia scilicet quilibet gradus habet suam propriam entitatem ab alterius entitate distinctam . . . – Cod. Bamberg. SB, 62.4b, p. 88.

Here, it should be noted, the reference to the *Parisienses* is to one of the logical treatises, which indicates that a rather extensive survey of their positions had been made, and that attention was not restricted merely to their natural philosophy.

Undoubtedly there are many more citations of the *Doctores Parisienses* in the *reportationes* of the lectures given at the Collegio Romano, but the ones just given amplify considerably the materials presented in the foregoing essay, and leave little doubt that the Jesuit professors to whom we have called attention could have been the proximate source of Galileo's knowledge of the Parisian tradition. Even more significant, perhaps, is the indication of Rugerius, to which we call attention in Essay 14 (at note 28), that the reader should consult the rules for calculating velocities given by Domingo de Soto and Franciscus Toletus (Soto's student) for a proper understanding of the mathematics of naturally accelerated motion. Here, as explained in Essay 6, the reader would have found the terminology of *uniformiter difformis*, etc., applied directly to the problem of free fall. Whether Galileo did read this far or not when working on the early treatises on motion is itself problematical. But there can be no doubt that the discussions therein, and in the Jesuit *reportationes*, are in essential continuity with the solution he was eventually to publish in the *Two New Sciences*, and which was destined to put the science of mechanics on its modern footing.

11. GALILEO AND THE SCOTISTS

Galileo's fame as the "Father of Modern Science" derives from his work, much studied by historians, while a professor of mathematics at the University of Padua (1592–1610). Previous to that he was associated over a period of ten years with the University of Pisa (1581–1591) both as a student and a lecturer, and these earlier years have recently interested historians also, especially for the light they shed on Galileo's knowledge of the medieval tradition. In notebooks that he wrote while at Pisa Galileo records the teachings of scores of prominent medieval and Renaissance thinkers on matters relating to the heavens, the elements, and scientific methodology.[1] Among these he mentions the views of John Duns Scotus in eight different contexts, and in elaborating on them he also cites Antonius Andreas, Joannes Canonicus, Franciscus Lychetus, and the *Scotistae*. An examination of these citations may help to shed light on Scotus's influence within the late sixteenth century, as well as provide background on Galileo's early philosophical orientation.

The references to Scotus or to Scotistic doctrine occur in four of Galileo's treatises contained in the notebooks, viz, *De mundo, De caelo, De elementis*, and *De praecognitionibus*, and may be enumerated and summarized as follows:

De mundo. (1) Concerning the universe as a whole, to the question whether God could add species to this universe or make other worlds having more perfect species essentially different from those of this world, Scotus replies in the negative (*In 3 Sent.*, d. 13, q. 1), maintaining that one must come ultimately to some finite creation than which nothing more perfect can be made (E8).[2] (2) To the query whether the world could have existed from eternity, (a) Joannes Canonicus (*In 8 Phys.*, q. 1) replies that this is possible for successive and permanent entities, for corruptibles as well as incorruptibles (F1), whereas (b) Scotus (*In 2 Sent.*, d. 1, q. 3) answers affirmatively for incorruptibles but admits that corruptibles present a problem (F18).

De caelo. (3) On the question whether the heavens are composed of matter and form, those holding that the heavens are a "simple body" as opposed to a "composed body" include Scotus "in the way of Aristotle" (*In 2 Sent.*, d. 14, q. 1), Joannes de Baccone and Lychetus (*ibid.*), and Antonius Andreas (*In 8 Meta.*, q. 4) (K9). (4) On the animation of the heavens, Scotus is listed as holding that the intelligences are not forms simply informing but rather forms merely assisting the motion of the heavens (L30).

De elementis. (5) Scotus is mentioned along with many others as having treated the intension and remission of qualities (N2). (6) On the question whether the forms of the elements undergo intension and remission, (a) the affirmative side includes Scotus (*In 8 Meta.*, q. 3), Antonius Andreas (*In 11 Meta.*, q. 1), Pavesius (*De accretione*), and Joannes Canonicus (*In 5 Phys.*, q. 1) (T2); further, (b) in an argument supporting the negative side, Scotus's interpretation of Aristotle's dictum, "substance does not admit of more or less," to mean substance according to quiddity, i.e., genus and difference, is rejected (T5). (7) On various problems relating to maxima and minima, (a) Scotus (*In 2 Sent.*, d. 2, q. 9) is cited as holding that all heterogeneous substances, such as living things and some compounds, have intrinsic termini of largeness and smallness, but that homogeneous substances, such as elements and some homogeneous compounds, have neither a maximum nor a minimum in any way (U32); (b) also attributed to Scotus (*In 3 Sent.*, d. 13) is the view that qualities such as grace cannot be increased intensively to infinity (U34); and finally, (c) an argument deriving from Scotus, viz, if quality is not intensively finite of its nature then there can be a quality that is infinite in intension and perfection, is rejected (U43–44).

De praecognitionibus. (8) On the question whether a science can prove the *esse existentiae* of its adequate subject, Scotus (*In 1 Sent.*, q. 3, and *In 1 Meta.*, q. 1) is cited as holding that a science can demonstrate *a posteriori* the existence of its total subject only, an opinion in which he is followed by Antonius Andreas and all Scotists (*Scotistae*).[3]

The foregoing eight points cover a considerable range of philosophical knowledge, and the citations are detailed and quite precise. The fullness of information contained in them and in the notebooks generally has in fact posed a problem for Galileo scholars, most of whom hold that the notebooks were Galileo's composition while only a student, and so are at a loss to explain the maturity and sophistication he then evidenced. The suggestion has been made that the notes were based on the lectures of a Pisan professor such as Francesco Buonamici, but such a conjecture has never been verified; indeed Buonamici's ponderous tome on motion, published in 1591, is almost totally different in style and content from Galileo's exposition.[4]

In a series of recent studies, Alistair Crombie, Adriano Carugo, and the author have been examining the Pisan notebooks with considerable care and have uncovered several printed works on which they might have been based. Among these, sources possibly used by Galileo were suspected by the author and independently identified by Carugo as books by Benedictus Pererius and Franciscus Toletus; subsequently Crombie added to these a book by Christopher Clavius.[5] The three writers thus identified – Pererius, Toletus, and Clavius – share a common characteristic: they were all Jesuits and professors at the Collegio Romano during Galileo's years at Pisa. Since their books contain passages similar to other sections of Galileo's notebooks, they could also be the source of Galileo's knowledge of the Scotistic tradition. Acting on this inspiration the author has checked the eight points listed above against the works by these writers known to contain matters similar to those treated by Galileo.[6] The results show the possibility of borrowing on only three of the eight points, and even for these three the detail is not sufficient to warrant the claim that the Jesuits' works are the unique source of Galileo's composition. To be more specific:

(2) Pererius maintains that (b) for Scotus this question is indeed problematic, citing *In 2 Sent.*, d. 1, q. 3, but he gives no details and (a) he makes no mention of Joannes Canonicus and his teaching.[7]

(3) Pererius cites Scotus (along with Durandus and Gabriel Biel) as holding that the heavens are a simple body in *In 2 Sent.*, d. 12 & d. 14, but has no reference to Joannes de Baccone, Lychetus, or Antonius Andreas.[8]

(7) Both Toletus and Pererius give Scotus's teaching (a) with regard to the maxima and minima of substances, but Toletus alone identifies this as Scotus's teaching and cites *In 2 Sent.*, d. 2, q. 9; moreover, neither (b) gives Scotus's position on qualities or (c) rejects his argument on qualities as does Galileo.[9]

This evidence, sparse though it is, for the possibility of Galileo's dependence on the printed works of these Jesuit professors led the author, in June of 1975, to search through manuscript sources at the Collegio Romano to see if there were any handwritten notes, or *reportationes* of lectures, that might supply more detail on these matters, and so might be an additional source of Galileo's information. This study is still in progress, but it has proved remarkably fruitful to date, and some results deriving from it can shed light on Galileo's knowledge of the Scotistic tradition. Little new information on Toletus or Pererius has turned up, but the lecture notes of four other Jesuit professors supply almost all of the information from which Galileo could have gleaned the eight points listed above, without having consulted any original sources. These professors are Antonius Menu, who taught logic at the Collegio in 1579–1580, natural philosophy in 1577–1578 and 1580–1581, and metaphysics from 1578–1582; Paulus Valla, who taught logic in 1587–1588 and natural philosophy in 1588–1589; Mutius Vitelleschi, who taught the same courses in 1588–1589 and 1589–1590 respectively; and Robertus

Jones, who also taught them in 1591–1592 and 1592–1593.[10] Not all of the notes of these men have survived, but from what has been located to date it is clear that the first of the points is covered by Menu, five others are covered by Vitelleschi, two by Valla, and one by Jones. These results will now be described, following the order of the professors as just named.

Menu's exposition of the *De caelo* of Aristotle contains a question relating to the perfection of the universe that is quite similar to Galileo's (1).[11] He asks whether God could make many worlds having species that are essentially different from, and more perfect than, those found in this world. The question itself, Menu continues, derives from another query, namely, whether there is a limit in specific perfections, or whether in this matter one ought to admit a regress to infinity. It is in answer to the latter query that Menu notes the teaching of Scotus and Durandus to the effect that there is such a limit, and therefore a creature might be so perfect that one could not be made more so. Like Galileo, Menu rejects this opinion and offers in its stead that of St. Thomas, which he proposes as common teaching among the scholastics. Not only is Menu's teaching the same as Galileo's, but his wording (as is that of the other *reportationes* on the remaining points) is strikingly similar, so much so that it seems impossible to attribute the agreement to coincidence, and therefore that some actual dependence was involved. To allow the reader to judge for himself, the following are transcriptions of the relevant passages:

ANTONIUS MENU	GALILEO GALILEI
[First Point]	
Quaeritur quarto an Deus possit facere plures mundos habentes species essentiales perfectiores essentialiter distinctas ab his quae sunt in hoc mundo. Ista quaestio pendet ex alia: An sit status in perfectionibus specificis vel potius sit admittendus processus in infinitum.	Quaeritur secundo an Deus potuerit addere aliquas species huic universo, vel efficere alios mundos habentes species perfectiores essentialiter ab his quae sunt in hoc mundo distinctas.
Scotus, 3° dist. 13, q. 3a, et Durandus, p° dist. 44, q. 2a, videntur defendere dari statum determinatum in perfectionibus specificis; unde notant esse deveniendum ad aliquam creaturam ita perfectam ut non possit dari perfectior.	Scotus, in 3° dist. 13, quo. prima, et Durandus, in p° dist. 44, q. 2a, negant, sentientes perveniendum tandem esse ad aliquam creaturam finitam, qua nulla perfectior effici potuerit.
Durandus vero ait esse valde probabile Deum de facto fecisse omnes species possibiles, quia non est maior ratio quam	Imo asserit Durandus, probabilissimum esse Deum procreasse in hoc mundo omnes species possibiles, atque ita,

alterius; et secundum hoc non possunt esse plures perfectiores, loquendo de perfectione essentiali. Verum haec sententia videtur absurda, et ideo dicendum est cum D. Thoma, prima parte, quaestione 25, art. 6, et cum aliis omnibus communiter, posse fieri a Deo plures mundos perfectiores, perfectiores, etc., in infinitum, loquendo etiam de perfectione essentiali, ac proinde non dari statum in perfectionibus. Probatur: tum quia Deus est infinitus et habet potentiam infinitam, tum quia est infinite participabilis a creaturis. Ex quo collige istum mundum potuisse fieri perfectiorem essentialiter semper et semper in infinitum.

consequenter, neque hunc mundum, neque alium, perfectiorem a Deo potuisse effici.

Melius tamen D. Thomas, in prima parte, q. 25, art. 6, et alii fere omnes, sentiunt Deum posse efficere perfectiores in infinitum propter suam vim infinitam: ex quo etiam patet, cum Deus possit efficere plures mundos in infinitum, posse etiam illos efficere perfectiores in infinitum.

In his lectures on Aristotle's *De caelo* Vitelleschi gives the answers of both Scotus (2a) and Joannes Canonicus (2b) to the question of the possibility of the world's eternity; he cites also Scotus's view (3) that the heavens are a simply body, but does not include the further references to the Scotists; and he explains Scotus's opinion (4) on the animation of the heavens.[12] Furthermore, in his lectures on Aristotle's *De generatione* Vitelleschi mentions that Scotus (5) treated the problem of intension and remission; and he explains various points relating to Scotus's teaching (7) on maxima and minima, namely (a) his doctrine on substances, (b) his treatment of qualities, and (c) the objection based on his position.[13] Thus, in all, the following five points relating to Scotistic doctrine that are found in Galileo's notebooks are contained also in Vitelleschi's lecture notes.

MUTIUS VITELLESCHI

[Second Point]

Prima sententia vult mundum eo plane modo quo nunc est et secundum omnia entia que continet tam permanentia quam successiva, tam corruptibilia quam incorruptibilia, potuisset esse ab aeterno. Ita Gregorius in 2° dist. p^a quest. 3^a art. p° et 2°; Occam ibidem quest. 8 et quodlibeto 2° quest. 5^a; Gabriel in 2° dist. p^a quest. 3^a. Hi tamen duo

GALILEO GALILEI

Prima opinio est Gregorii Ariminensis, in 2° dist. p^a q^e 3 art. p° et 2°, Gabrielis et Occam ibidem; quamvis Occam non ita mordicus tuetur hanc sententiam quin etiam asserat contrariam esse probabilem. Hos secuntur Ferrariensis, in 8 Phys. qu. 15, et Joannes Canonicus, in 8 Phys. q. p^a, et plerique recentorum, existimantium mundum potuisse esse ab aeterno, defendunt hanc sententiam tamquam probabiliorem, et volunt contrariam etiam probabiliter posse defendi; idem sentiunt Burlaeus et Venetus in t. 15 8[i] phy., Canonicus ibidem quest. 1[a], Gulielmus de Ralione in 2° dist. p[a] quest. 3[a] art. 2° par. 3[a]. Cum hac sententia convenit S. Thomas ... Scotus autem in 2° dist. p[a] quest. 3[a] neutram partem determinate sequitur sed putat utramque esse probabilem ....

tam secundum entia successiva quam secundum entia permanentia, tam secundum corruptibilia quam secundum incorruptibilia ....

Quarta opinio est D. Thomae ... Scoti in 2° dist. p[a] q. 3, ... Pererii in suo 15, et aliorum: qui putant mundum potuisse fieri ab aeterno secundum incorruptibilia ... at vero secundum corruptibilia problema esse ....

[Third Point]

Secunda sententia vult coelum esse corpus simplex expers omnis compositionis ex materia et forma; est Averrois multis in locis ... Eiusdem sententiae Scotus loco citato [in 2° dist. 14 quest. p[a]], Durandus in 2° dist. 12 quest. p[a], Occam loco citato, Aureolus apud Capreolum loco citato art. 2°, et demum Gabriel loco item citato.

Prima opinio est Averrois, sentientis caelum esse corpus simplex .... Hanc ... sententiam secuntur Durandus in 2° dist. 12, Scotus in via Aristotelis in 2° dist. 14 quo. p[a], Joannes de Baccone et Lychetus ibidem, Antonius Andreas 8 Met. quo. 4 ... et omnes Averroistae ...

[Fourth Point]

Tertia sententia dat caelo animam intelligentem, quidam enim volunt intelligentiam motricem esse formam informantem et dantem esse caelo ... Scotus in 2 dist. 4 q. 1 ait secundum Aristotelem intelligentiam solum assistere caelo ad motum, secundum veritatem tamen ... nihil est in [caelis] quod repugnet quominus sint animati ... Quarta sententia his omnibus contraria caelum facit inanimatum, ita ... Durandus ... Cirillus lib. 2 contra Iulianum.

Dico secundo quod, quamvis non videatur omnino improbabile secundum Aristotelem intelligentias esse formas informantes simpliciter, tamen et secundum Aristotelem et secundum veritatem, longe probabilius tantum esse assistentes ... Secunda parts probatur: primo, quia id sensisse Aristotelem docent D. Thomas locis citatis, D. Cyrillus lib. 2° contra Iulianum, Scotus, Durandus, et alii scholastici ....

[Fifth Point]

Quarta sententia dicit intensionem fieri per productionem novi gradus perfectionis qui virtute continet precedentes ... ; ita Burleus ... Gotfredus ... ; cuiusdem sententie meminit S. Thomas in p° dist. 17 quest. 2[a] art. p°, Scotus ibidem, Gregorius quest. 4 art. 2°, Capreolus ... Soncinas ....

Authores qui hac de re [i.e., de intensione et remissione] egerunt sunt: D. Thomas ... Capreolus ... Herveus ... Gandavensis ... Soncinas ... Aegidius ... Burleus ... Durandus in p° dist. 17 q. 7, Gregorius ibidem q. 4, art p°, Scotus ibidem q. 4, Occam et Gabriel ibidem q. 7.

[Seventh Point]

Tertia sententia. Convenit viventia habere ex sua natura maximum et minimum intrinsecum, negat de elementis et mixtis homogeneis, ita Scotus in 2° dist. 2^{a} quest. 9 et Scotiste, Occam in 2° quest. 8, Burleus in t. 18 p^{i} phy., Pererius lib. 10 cap. ult° et alii.

Quarta sententia est Scoti in 2° dist. 2^{a} q^{e} 9, Occam 2° q. 8, Pererii lib. 10 cap. 23, qui dicunt omnia heterogenea ut viventia et aliqua mixta habere terminos magnitudinis et parvitatis intrinsecos; homogenea vero, ut elementa et quaecunque mixta heterogenea neque habere maximum neque minimum ullo modo. . . .

Tertia propositio. Qualitates naturaliter determinate sunt quoad maximum, non tamen ita ut illis repugnet crescere in infinitum; non disputo autem de minimo qualitatis quia supra ostendi non dari. Hec conclusio est contra Durandum in p° dist. 17 quest. 9 et in 3° dist. 13 quest. p^{a}, Scotum ibidem quest. 4^{a} . . .

De qualitatibus vero Scotus et Durandus in 3° dist. 13 . . . negant qualitatem posse augeri in infinitum intensive; quia, cum aliae qualitates a gratia sint creatae et limitatae, proprietates essentiae debent habere certum terminum, licet intrinsecum, in intensione . . . .

The extant *reportationes* from Paulus Valla's lectures at the Collegio Romano are very incomplete, but bound at the end of his commentary on Aristotle's *Meteorologica* in one of the surviving codices is a treatise on the elements that supplements the materials contained in Vitelleschi's notes. In this Valla discusses Scotus's teaching (6) on the intension and remission of elemental forms, and also mentions the *Scotistae*, Antonius Andreas and Joannes Canonicus.[14] Moreover, although his logic notes are apparently lost, Valla published at Lyons a two-volume *Logica* in 1622; in its preface he states that the work contains materials he lectured on 34 years before at the Collegio Romano, i.e., in 1588. In treating *De praecognitione* in this work Valla holds (8) that it is not necessary that the *esse actualis existentiae* of the subject of a science be known previous to every demonstration in that science, but the actual existence of the total subject of the science must be foreknown; he does not, however, ascribe this teaching to Scotus, nor does he list Scotus's opinion among the seven opinions he cites.[15] Since the latter text did not appear until 1622, little would be served by reproducing it here, and thus only the transcriptions of the respective texts relating to the sixth point are given below:

PAULUS VALLA | GALILEO GALILEI

[Sixth Point]

Secunda sententia est Averrois . . . et

Prima opinio est Averrois [et Averrois-communis Averroistarum qui asserunt formas elementorum intendi et remitti et habere gradus sicut habent qualitaties; eadem sententia multo magis debent defendere Scotistae ut Antonius Andreas in met. quest^{e} unica, Joannes Canonicus 8° phys. q. ult° et Pavesius in tractatu de accretione, qui dicunt omnem formam quae habent esse a materia, per quod excluditur animal rationalis, posse intendi et remitti.

tarum] ... qui omnes dicunt formas substantiales elementorum intendi et remitti; quibus addi potest Scotus 8 Met. q. 3^{a}, quem sequitur Antonius Andreas 11 Met. q. p^{a}, Pavesius in lib. de accretione, Joannes Canonicus 5 Phys. q. p^{a}, qui idem affirmant de quacunque forma substantiali quae educatur de potentia materiae, ut excludatur anima rationalis.

Robertus Jones, a Jesuit from the British Isles, left a full set of notes from his logic course given at the Collegio and completed in 1592. In it he treated, as was customary, the foreknowledge required for demonstration, and there he teaches (8) that the existence of the subject of a science can be established by demonstration *quia*, i.e., *a posteriori*, but not by demonstration *propter quid*, and he ascribes this teaching to Scotus and to Antonius Andreas.[16] Effectively this is the same teaching as that contained in both Valla and Galileo, and the variant ways of expressing it can serve to tie all three expositions together. Again, for purposes of comparison, transcriptions of the relevant passages are reproduced below:

ROBERTUS JONES

GALILEO GALILEI

[Eighth Point]

Prima opinio est asserentium subiectum in aliqua scientia posse probari per demonstrationem quia, etiam si non possit probari per demonstratione propter quid ... Hanc sententiam videtur tenere Scotus Ioa. Andreas q^{e} p^{a} Meta., sed alii fere omnes in p° Posteriorum ....

Prima opinio est Scoti in primo Sent. quest^{e} 3^{a}, in p° Met^{es} quo^{e} p^{a}, quem sequitur Antonius Andreas ibidem et tunc omnes Scotiste. Haec opinio asserit scientiam a posteriori tantum posse demonstrare existentiam sui subiecti totalis.

These, then, are some preliminary results of comparative studies of Galileo's statements about Scotus and his school and those contained in the lecture notes of professors who taught at the Collegio Romano between the years 1578 and 1592.[17] At the time he originally wrote this essay the author had examined only the lecture notes of Valla, Vitelleschi, and Jones; since then he has studied additional *reportationes* deriving from Antonius Menu, Ludovicus Rugerius, and other professors listed in *Galileo's Early Notebooks* (pp. 12–21). The following Table, similar to that at the end of Essay 9, gives a fuller identification of the correspondences noted between Galileo's notebooks and these *reportationes*, for all of the references relating to Scotus and the Scotists reported in this essay. Here, as in the previous table, three degrees of correspondence are noted: *Excellent* signifies that the agreement is close enough to suggest almost verbatim copying; *Good* means that the expression and content are quite similar; and *Fair* means that the content is roughly the same, though the manner of expressing it differs. Fuller details relating to the composition of the notebooks, together with specific evidences of copying, are given in the commentary contained in *Galileo's Early Notebooks.*

TABLE

Paragraphs in Galileo's Text	Degree of Correspondence with Jesuit Lecturers at the Collegio Romano
E8	*Excellent*: Menu
F1	*Excellent*: Pererius (p); *Good*: Menu, Vitelleschi; *Fair*: Rugerius
18	*Excellent*: Pererius (p); *Good*: Menu, Vitelleschi, Rugerius
K1	*Good*: Pererius (p), Menu, Vitelleschi; *Fair*: Rugerius
9	*Good*: Pererius (p), Menu, Vitelleschi
L29	*Good*: Pererius, Vitelleschi
30	*Good*: Vitelleschi, Rugerius
N2	*Good*: Menu, Vitelleschi, Rugerius
T2	*Excellent*: Valla; *Good*: Menu, Rugerius
5	*Good*: Rugerius
U32	*Excellent*: Valla; *Good*: Menu, Vitelleschi, Rugerius, Pererius (p), Toletus (p)
34	*Good*: Valla, Vitelleschi
43	*Fair*: Vitelleschi
44	No recognizable agreement

Since on all these points the materials contained in Galileo's notebooks relating to the Scotistic tradition are so similar to those taught at the Collegio Romano, it is difficult to escape the inference that the lectures given at the Collegio are a likely source, either directly or indirectly, of Galileo's composition.

On the basis of these results two further conclusions may be suggested as of possible interest to Galileo scholars. (I) Since all the citations of Scotus and his school occur in Galileo's early notebooks, and no further references to Scotistic teaching are to be found in his letters or in his later writings, it seems unlikely that the knoweldge Galileo manifests of the Scotistic tradition exerted more than a background influence on his scientific work. In this respect the Thomistic tradition fared slightly better, for Galileo continued to cite Aquinas and Cajetan until at least 1615, and, as has been argued elsewhere, he made use of demonstration *ex suppositione*, one of Aquinas's methodological refinements, to the last years of his life.[18] (II) The lateness of the dates of the *reportationes* discussed in this paper – all except Menu's were written between 1588 and 1592 – suggests that Galileo did not compose the notebooks while only a student at Pisa, i.e., between 1581 and 1585. More probably he did so while preparing for a teaching career or while actually teaching at Pisa between 1589 and 1591. If this is true the propriety of referring to the Pisan notebooks as *Juvenilia*, using Favaro's initial caption for them, must be seriouslly questioned as a result of these studies.

NOTES

1 These notebooks are preserved in the Biblioteca Nazionale in Florence (BNF) in Manoscritti Galileiani 27 and 46; the first contains questions relating to logic and the second questions relating to natural philosophy. MS Gal. 46 has been transcribed by Antonio Favaro and is printed in Vol. I of *Le Opere di Galileo Galilei*, ed. A. Favaro, 20 vols. in 21, Florence: G. Barbèra Editore, 1890–1909, reprinted 1968, 15–177. Since preparing the original version of this essay, the author has translated these questions into English in a work entitled *Galileo's Early Notebooks: The Physical Questions*. A Translation from the Latin, with Historical and Paleographical Commentary. Notre Dame: University of Notre Dame Press, 1977. MS Gal. 27 is described and a list of its questions given in Vol. IX of the *Opere*; pp. 279–282; only a specimen question from it is transcribed, however, and this is given on pp. 291–292. Recently Adriano Carugo has transcribed this entire MS and is preparing it for publication. He has kindly given the author a copy of his transcription.

2 The abbreviation (E8) refers to the author's English translation of the physical questions referred to in the previous note. Here the letter E identifies the question, and the numeral 8 the paragraph number within the question, as they appear in the translation. The Latin texts for this and subsequent citations are given later in this essay.

3 BNF MS Gal. 27, fol. 8^{r}.

4 Francesco Buonamici, *De motu libri decem* . . . Florence: B. Sermatelli, 1591.

5 A. C. Crombie, 'Sources of Galileo's Early Natural Philosophy,' in M. L. Righini Bonelli & W. R. Shea, eds., *Reason, Experiment, and Mysticism in the Scientific Revolution*, New York: Science History Publications, 1975, pp. 157–175, 303–305; see also the original versions of the studies published as Essays 8 through 10 of this volume.

6 Actually only two of the books written by these three authors contain materials that relate to Galileo's citation of Scotus; these are Franciscus Toletus, *Commentaria una cum questionibus in octo libros Aristotelis de physica auscultatione*, nunc secundo in lucem edita, Venice: Apud Juntas, 1580, and Benedictus Pererius, *De communibus*

omnium rerum naturalium principiis et affectionibus libri quindecim, Rome: Apud F. Zanettum & B. Tosium, 1576.

7 *De communibus* . . . , p. 509C.

8 *Ibid.*, p. 182D.

9 *Ibid.*, p. 356D; Toletus, *Commentaria in libros de physica*, fol. 25va.

10 Further details concerning these *reportationes*, together with transcriptions of some passages from Vitelleschi and Valla, are given in the previous essays of this volume. See also *Galileo's Early Notebooks*, pp. 12–21.

11 For the excerpt from Menu, the author has used a manuscript now in the Leopold-Sophien-Bibliothek in Uerberlingen, West Germany, Cod. 138, entitled *In philosophiam naturalem, anno 1557*; for those from Vitelleschi and Valla he has used codices preserved in the Archivum Pontificiae Universitatis Gregorianae, Fondo Curia (APUG/FC); and for that from Jones, a codex in the Biblioteca Casanatense in Rome (BCR).

12 APUG/FC Cod. 392 (no foliation). Disputationes in libros de caelo. (2) Tractatio 1^{a}, disputatio 4^{a}; (3) Tractatio 2^{a}, disputatio 2^{a}; (4) Tractatio 2^{a}, disputatio 3^{a}. Although Vitelleschi does not refer to Lychetus, it is noteworthy that the latter is cited by another Jesuit, Ludovicus Rugerius, though in another context, in his lectures on the *De generatione*, Cod. Bamberg. SB 62.4, p. 223, line 11.

13 *Ibid.*, Disputationes in libros de generatione. (5) Tractatio de alteratione, disputatio 3^{a}; (7) Tractatio de augmentatione, disputatio 5^{a}.

14 APUG/FC Cod. 1710 (no foliation). Tractatus 5us, pars p^{a}, quaestio 4^{a}.

15 *Logica Pauli Vallii Romani*, duobus tomis . . . Lyons: Prost, 1622, Tom. II, pp. 163–167, especially p. 165.

16 BCR Cod. 3611, fol. 195^{v}.

17 For the lists of professors who taught courses at the Collegio Romano, year by year, see R. G. Villoslada, *Storia del Collegio Romano dal suo inizio (1551) alla soppressione della Compagnia di Gesù (1773)*, Analecta Gregoriana Vol. LXVI, Rome: Gregorian University Press, 1954, pp. 321–336; also *Galileo's Early Notebooks*, p. 23.

18 See Essays 8 through 10, and 14 through 16 in this volume.

12. GALILEO AND ALBERTUS MAGNUS

Albert the Great is justly regarded as one of the outstanding forerunners of modern science in the High Middle Ages. His contributions to all branches of learning earned for him the title of *Doctor universalis*, and he was heralded as "the Great" even in his own lifetime. Particularly noteworthy was his encyclopedic presentation, in Latin, of the scientific knowledge of the Greeks deriving especially from Aristotle and from his Greek and Arab commentators. To this corpus Albert himself added entire treatises based on personal observations of the heavens and of the mineral, plant, and animal kingdoms. From our vantage point in time we can therefore see him as a conserver and transmitter of the scientific knowledge of antiquity and of Islam, who also contributed to the advancement of science in his day, and who should, on both counts, be regarded as one of the key figures in the revival of learning within the thirteenth century (DSB 1: 99–103).

More difficult to assess is Albert's influence on later centuries, and the role he might have played in the scientific revolution of the seventeenth century. Like many great thinkers, Albert has been overshadowed by his students, and especially by his celebrated disciple, Thomas Aquinas. This fact, coupled with the change of mentality that is ascribed by intellectual historians to the Renaissance, which is usually seen as introducing a pronounced cleavage between medieval and early modern thought, may cause one to wonder whether there is any continuity whatever between the science cultivated by Albert and that associated with the names of, say, Galileo Galilei and William Harvey.[1] Even historians of science who specialize in the Middle Ages are prone to see the fourteenth-century development in mathematical physics as the main medieval contribution to the rise of modern science, and in this development it would appear that Albert the Great had but a small contribution to make.

It is the purpose of this essay to shed light on a possible connection between Galileo and medieval thinkers such as Albert by examining in some detail the knowledge that Galileo possessed of the German Dominican and of high medieval thought generally. Until quite recently Galileo was viewed as a sort of Melchizedek, an innovator without any intellectual forebears, who had little or no connection with the university tradition of his day, and who established his *nuova scienza* by dint of original investigation – working solely as a craftsman with a penchant for mathematical ways of thinking. Owing to the researches of the writer and others, however, this picture is gradually being revised.[2] The key to the revision is three sets of notes, or notebooks, written in Latin and in Galileo's hand that date from around 1590, when Galileo was beginning his career as a professor of mathematics at the University of Pisa. The first two notebooks are devoted to questions or problems arising within Aristotle's logic and physical science respectively, whereas the third contains Galileo's first attempts at constructing a science of local motion. Because of the affinity of its subject matter with that of the *Two New Sciences*, written at the end of Galileo's life, the third notebook has received some notice from scholars, but the first two have been neglected almost entirely. The editor of the National Edition of Galileo's works, Antonio Favaro, regarded them as *Juvenilia* or youthful writings, and dated the logical questions from his pre-university training at the Monastery of Vallombrosa and the physical questions from his student days at Pisa. As a consequence all three notebooks, and particularly the first two, were seen as trivial, having little or no bearing on the intellectual career of the Pisan physicist.

The later dating of the notebooks and their association with Galileo's career as a university professor, together with the uncovering in them of scholastic expressions that recur in Galileo's later writings,[3] have reopened the question of earlier influences on his thought. Albert the Great is mentioned four times in the notebook dealing with logical questions, and nineteen times in that dealing with physical questions. A study of these citations may help us ascertain the extent of Galileo's knowledge of Albert and the source, or sources, from which this derived. It may also enable us to date more precisely the time of Galileo's composition, and perhaps to determine the motivation behind the notebooks. To such a dual objective the present essay is directed.

I. THE PHYSICAL QUESTIONS AND THEIR SUBJECT

In view of the fact that the notebook dealing with the logical questions has yet to be edited definitively,[4] and the references therein to Albert are far fewer than in the notebook dealing with the physical questions, we shall restrict attention in what follows to the latter composition. This procedure benefits from the fact that we have published an English translation of the physical questions, together with a commentary that indicates some of the sources from which they derive, and thus there exists a work to which the reader can be referred for fuller particulars.[5]

Twenty-five questions are treated in the notebook under discussion, the first twelve (A through L in our system of reference) being concerned with the matter of Aristotle's *De caelo et mundo*, and the remaining thirteen (M through Y) with topics relating to his *De generatione et corruptione.*[6] Albert the Great is cited nineteen times in all, with twelve of these citations occurring in the treatises related to the *De caelo*, and the remaining seven in a treatise on the elements that makes up the major portion of Galileo's exposition of the *De generatione*. Our previous researches have shown that some 90% of the total number of paragraphs making up the physical questions have parallels in the lecture notes of professors teaching at the Collegio Romano, a Jesuit university in Rome, in the years between 1566 and 1597. The vast majority of these parallels can be found in *reportationes* of the lectures of four Jesuits, namely, Antonius Menu, who taught natural philosophy in the academic year 1577–1578; Paulus Valla, who taught the same in 1588–1589; Mutius Vitelleschi, the same in 1589–1590; and Ludovicus Rugerius, the same in 1590–1591. Of the nineteen references to Albert, ten have parallels in the lecture notes of these four Jesuits; several of the ten are found in the notes of one author alone, but most are found in two or three sets of notes. It is this circumstance that may shed light on the dating of Galileo's notebook, on the assumption that the closer the similarity of texts the shorter the temporal interval between their respective compositions.

The first treatise of the questions relating to the *De caelo* is made up of two questions, the first (A) inquiring about the subject matter of its various books and the second (B) about their order, connection, and title. Albert the Great is mentioned in two paragraphs of the first question (A4 and A6) and in two paragraphs of the second (B3 and B8). The first citation reads as follows:

> Albertus Magnus makes the subject bodies that are capable of movement to place. His reason is this: the subject of the whole of the *Physics* is bodies that are movable in general; therefore the subject of these four books, which are a part of the *Physics*, should be the first species of movable bodies, i.e., bodies movable to place. (A4)

No reference is given, but clearly Galileo has in mind Albert's exposition of the first book of *De caelo*, tract 1, chapter 1, where Albert makes the statements that the "mobile ad ubi est subiectum huius libri" and "suum subiectum . . . est simplex corpus mobile ad locum" (Borgnet 4: 1a–2b). Parallels for this paragraph are found in the lecture notes of both Vitelleschi and Rugerius, with the correspondences for the former being slightly more numerous than those for the latter. The Latin of Galileo's paragraph and of

Vitelleschi's parallel account are given below in facing columns, for purposes of comparison:

GALILEO	VITELLESCHI (1590)
Albertus Magnus point *obiectum corpus mobile ad ubi*. Ratio illius haec est: *quia obiectum totius Physicae est corpus mobile in* communi; horum ergo quatuor librorum, qui sunt una pars Physicae, obiectum debet esse prima species corporis mobilis, quod est *corpus mobile ad ubi*.	. . . *obiectum* esse *corpus mobile ad ubi*, est . . . *Alberti*, primo Physicae tr. p° cap. 4° et primo Caeli tr. p° cap. p°, et aliorum. Haec . . . sic explicatur: *obiectum totius Physicae est corpus mobile* . . . *in* universum [et] singulis speciebus . . . Primo modo consideratur in octo libris Physicae, secundo . . . *corpus* simplex *mobile ad ubi* . . . .

Rugerius's exposition is similar to Vitelleschi's, and is reproduced below in the same format:

	RUGERIUS (1591)
	. . . est *Alberti*, tr. 1 cap. 1, qui ait subiectum horum librorum esse *corpus mobile ad ubi*, . . . Fundamentum . . . est, *quia* in octo libris Physicae subiectum est . . . *corpus mobile* absolute sumptum, . . . ergo debuit post illam tractationem statim agi de *mobili ad ubi*.

Note in the citations above that 18 of the 43 words in Galileo's text have been italicized, and that 17 have correspondences in Vitelleschi's text, whereas only 11 are to be found in Rugerius's. It is on this basis that we state that Galileo's exposition is closer to Vitelleschi's than to Rugerius's, and possibly dates closer to the former in time of composition.

The remaining three citations of Albert in Galileo's introductory treatise do not have counterparts in the Jesuit *reportationes*, and thus are not helpful for dating purposes. They do have interest in another respect, however, and on this account are reproduced below:

Nifo, seeking to find agreement among these four opinions, holds with Alexander that the subject of aggregation is the universe; with Albert, that the subject of predication is bodies movable to place; with Simplicius, that the subject of attribution is simple bodies; and with Iamblicus and Syrianus, that the subject of principality is the heavens. (A6)

> Simplicius and Albertus Magnus, whom we and everyone else follow, hold that these books come after the eight books of the *Physics* and make up the second part of natural philosophy. (B3)
>
> Concerning the title, according to Alexander, Simplicius, and the Greeks, these books are entitled *De caelo* from the more noble portion; according to Albertus Magnus, St. Thomas, and the Latins, they are entitled *De caelo et mundo*. By the term *mundo* they understand the four elements, and this meaning was known also to Aristotle, in the first *Meteors*, chapter 1, saying that the lower world, i.e., the elements, should be contiguous with the movement of the higher. (B8)

With regard to the last two texts, B3 and B8, it is a simple matter to verify that their teaching is found in the place in Albert already referenced (Borgnet 4: 2b), and thus one might presume that Galileo himself had made the identification. With regard to the first text (A6), however, the matter is not so straightforward. When we check the teaching of Agostino Nifo contained in his *In quattuor libros de celo et mundo expositio*, printed at Naples in 1517, we find that the teaching ascribed to him by Galileo (fol. 1r) is not completely correct: the concordance of four opinions is there, as stated, but instead of the second opinion being identified as Albert's, in Nifo it is identified as Averroës'! Again, when we check Nifo to find his counterpart for Galileo's paragraph B3, we find that Nifo does not cite Simplicius and Albertus Magnus, as does Galileo, but rather Simplicius and Averroës. What are we to make of these substitutions? It is difficult to understand why Galileo himself would have made them, whereas it is comprehensible that one of the Jesuit professors at the Collegio Romano, who were being viewed with suspicion by their ecclesiastical superiors for being too partial to Averroist teachings, could have done so.[7] Thus we suspect, on the basis of this and other evidences, that Galileo probably did not have reference to original sources in composing his notebooks, but based them largely on the citations of others.[8]

II. THE UNIVERSE AND THE HEAVENS

The bulk of Galileo's questionary on the *De caelo* is made up of two additional treatises, one dealing with the universe as a whole and the other with the heavens. Of these the first is comparatively brief, taking up philosophical and theological queries regarding the origin, unity, perfection, and eternity of the universe. Only one reference is made to Albert the Great in this treatise, and that in the discussion of the unity of the universe. Galileo's citation reads as follows:

I say, first: there is only one universe. The proof: first, from Plato, there is only one exemplar of the universe; therefore [the universe is one]. Second, from Albertus, on the first *De caelo*, tract 3, chapters 5 and 6: because this is clearly gathered from the first mover, who is only one and cannot be multiplied, not being material, and from the places of the movable objects that are in the universe. Add to this: if there were many universes, a reason could not be given why these would be all and no more . . . . (E2)

The reference to Albert is correct, for in chapter 5 Albert gives the argument based on the immateriality of the prime mover (Borgnet 4: 79a) and in chapter 6 he discusses Plato and the problem of the exemplars of the universe (83a). He does not, however, give the argument that follows the words "Add to this." The latter, not unexpectedly, is to be found in the only parallel this text has among the Jesuit *reportationes*, that of Antonius Menu, which is given here along with the Latin of Galileo's text:

GALILEO	MENU (1578)
Dico, primo, *unum tantum esse* mundum. *Probatur*, primo, ex Platone: unum tantum est exemplar mundi; ergo . . . Secundo, ex *Alberto*, in *primo Caeli tr. 3, cap. 5* et *6*: quia ex primo motore, qui est tantum unus et non potest multiplicari, cum non sit materia, et ex locis mobilium quae sunt in mundo, id aperte colligitur.	Respondeo affirmative *unum tantum esse* . . . *Probatur* . . . Tertio . . . ad denotandum unitatem sui Creatoris . . . {De ista materia . . . *Albertus* Magnus libro *primo De caelo, tr. 3° cap. 6°* . . . }
Adde, quod *si* essent *plures mundi, non* posset assignari *ratio cur* essent tot et non plures	Quarto, *si* ponerentur *plures mundi* quam unus *non* est *ratio cur* constituuntur duo vel tres et cetera in infinitum. . . .

Here again 18 of Galileo's 71 words are to be found in Menu's notes, although not in the precise order; this is indicated by our enclosing some of Menu's expression in braces {}.

The treatise on the heavens shows a heavy dependence on the Jesuit writings, with the first two questions concerned with the number and order of the heavenly orbs being based almost verbatim on Christopher Clavius's commentary on the *Sphere* of Sacrobosco, and the remaining questions concerned with the composition, corruptibility, materiality, and animation of the heavens having many counterparts in lectures given at the Collegio Romano. It would be tedious to reproduce here all of these parallels, so we shall restrict ourselves in what follows to those with the larger number of coincidences or with readings of special significance.

In the latter category, actually a text for which we have not uncovered a parallel in the Jesuit notes, Galileo references Albert for a teaching that assumes considerable importance in his later treatises *De motu*, where he introduces the idea of a motion that is neither natural nor violent, but in some way intermediate between the two. Galileo's citation occurs in the context of his discussion whether the heavens are composed of fire or some other element, and reads as follows:

The first opinion was that of practically all ancient philosophers before Aristotle, who taught that the heavens are not different in nature from the elements; and this opinion originated with the Egyptians, as Albertus teaches in tract 1, chapter 4, of the first *De caelo*, the Egyptians thinking that the heavens were made of fire. For it is a property of fire to be carried upward, and then, when it can ascend no farther, to go around in a circle – as is seen in flame, which, when it arrives at the top of a furnace, circles around. From this it is apparent that the heavens are fiery, for they have the uppermost place and are moved circularly. (17)

This teaching is found in Albert pretty much as Galileo states it. To give the reader some idea of the relation of the two teachings we give the respective Latin texts in parallel column:

GALILEO	ALBERTUS (Borgnet 4:15b–16a)
Prima opinio fuit veterum fere omnium philosophorum ante Aristotelem, qui putarent caelum non esse naturae distinctae ab elementis: et haec sententia promanavit ab *Aegyptiis*, ut docet Albertus tr. p° cap. 4, primi De caelo, *qui* existimarunt *caelum* esse *igneum*. Nam ignis proprium est ut feratur sursum, deinde ut, cum non potest amplius ascendere, *volvatur in* girum; ut patet in *flamma*, quae, ubi pervenit ad summum *fornacis, circumvolvitur*; ex quo patet, cum caelum supremum locum obtineat et circulariter moveatur, esse igneum.	Scias autem quod omnia quae dicuntur, sunt dicta contra Platonem et philosophos *Aegypti, qui* dixerunt quod *coelum* est *igneum*, et non est motum circulariter nisi per accidens: dicebant enim quod ignis naturaliter ascendat, et quando non habet quo plus ascendat, tunc circum*volvitur in* seipso, sicut *flamma* ignis in fornace: praeter hoc solum quod concavum *fornacis circumvolvit* flamman, defectus autum ulterioris loci circumvolvit aetherem, ut dicebant. . . .

Words that are similar in the two texts have been italicized, and it would seem that there is sufficient resemblance here to maintain that either Galileo or the source used by him actually had an eye on Albert's exposition when composing this paragraph. The example discussed therein, viz, that of fire or flame having a circular motion when it rises to the top of a furnace, recurs in

Galileo's later exposition, for he goes on to inquire whether circular motion would be natural for fire under such circumstances, or, if not natural, then violent. Both possibilities he rejects, leaving only the alternative of a *tertium quid*, namely, a motion that is neither natural nor violent and therefore must be something intermediate between the two. Elsewhere we argue that this line of reasoning, already adumbrated in the lecture notes of Vitelleschi and Rugerius, could have led Galileo to the idea of circular inertia, which was seminal for his later treatment of local motion in the *Two New Sciences.*[9]

Other references to Albert that do have parallels in the Jesuit *reportationes* occur in paragraphs I10, K37–40, K170, and L11 of Galileo's notebook dealing with the physical questions. Of these I10 and L11 have counterparts in Menu alone, and K170 has a counterpart in Menu but without the explicit citation of Albert's teaching. The line of reasoning advanced in K37–40, however, has parallels in Menu, Vitelleschi, and Rugerius, and thus turns out to be an important text for the comparison of sources on which Galileo's composition could have been based. Galileo's words are the following:

The second opinion is that of those who think that the heavens are of an elementary nature, regarding the heavens as a composed body . . . (K37) Alexander was also of this opinion . . . Also all the Arabs, with the single exception of Averroës, attributed composition to the heavens; so Avicebron in the book *Fons vitae*, from Albertus and from St. Thomas in the First Part, question 66, article 2; Avempace, from the first *De caelo*, tract 1, chapter 3; Avicenna, in the first of the *Sufficientia*, chapter 3. So did a great number of the Latins, such as Albertus Magnus, in the first *Physics*, as above, the eighth *Physics*, tract 1, chapter 13, and in the book *De quatuor coaequevis*, question 4, article 3, where he teaches that Rabbi Moses was of the same opinion; St. Thomas . . . and likewise all Thomists, as Capreolus . . . Cajetan . . . Soncinas . . . Ferrariensis . . . [and] St. Bonaventure . . . (K38) However, the cited authors disagree among themselves. First, because some of them wish to define the matter of the heavens differently from the matter of inferior things. So Alexander . . . Simplicius . . . Albertus, in the first *Physics*, tract 3, chapter 11, and in *De quatuor coaequevis*, question 2, article 6, and St. Thomas in the places cited above. But others contend that heavenly matter is the same in kind as sublunary matter; so Philoponus, Avicenna, Avempace . . . , Avicebron, Giles, and Scaliger . . . (K39) Second, they disagree in this . . . , as Mirandulanus and Achillini . . . (K40).

The Latin text corresponding to this translation is given below on the left, and then, in parallel and under Galileo's wording, the corresponding passages in Menu, Vitelleschi, and Rugerius. Any expression in Galileo that has a coincidence in one of these three authors is italicized, and the corresponding expression in one or more of the three is also italicized. Since the ordering of the discussion varies in the different authors, it has been necessary to make some transpositions in the presentations of their texts, and these are indicated, as above, by the use of braces { }.

GALILEO

Secunda opinio est illorum omnium *qui putant caelum esse* naturae elementaris, sentientum caelum esse corpus *compositum* . . . Alexandrum etiam fuisse in hac sententia . . . Arabes etiam omnes, uno excepto Averroe, compositionem tribuerunt caelo; ut Avicembron in libro Fontis vitae, ex Alberto et ex *D. Thoma* in *Prima Parte, q. 66, art. 2.* Avempace, ex primo De caelo tr. p° cap. 3; Avicenna, in p° Sufficientiae cap. 3; et quamplurimi etiam Latinorum, ut *Albertus* Magnus in primo Physicorum, UBI SUPRA, octavo Physicorum, tr. p° cap. 13, et in libro *De quatuor coaequaevis, q. 4, art 3*, ubi etiam docet eandem sententiam fuisse Rabbi Moyses; D. Thomas . . . similiter omnes *Thomistae*, ut *Capreolus* . . . *Caietanus* . . . *Soncinas* . . . *Ferrariensis* . . . *S. Bonaventura* . . . Verum discrepant inter se citati authores. Primo quidem quia illorum nonnulli *volunt materiam caeli esse diversae rationis a materia horum inferiorum*, ut Alexander . . . Simplicius . . . *Albertus primo Physicorum tr. 3, cap. 11*, et in *De quatuor coaequaevis, q. 2, art. 6*, et D. Thomas, locis citatis supra; at vero alii contendunt esse *eiusdem rationis cum materia* sublunarium; ut Philoponus, *Avicenna, Avempace* . . . , *Avicembron, Aegidius*, et *Scaliger* . . . Discrepant secundo in hoc . . . ut *Mirandulanus* et *Achillinus* . . .

MENU (1578)

Secunda opinio est aliorum *qui putant caelum* tam secundum Aristotelem quam veritatem esse *compositum* ex materia et forma . . .

sic *D. Thomas Prima Parte, q. 66, art. 2*; octavo Physicorum, lect. 22; primo Caeli, lect. 6 et 8 . . .

{*Albertus* Magnus, *De quatuor coaevis, q. 2, art. 6*}

. . . *Capreolus*, ubi supra, *Ferrariensis* . . . , Dominicus de Flandria . . . *Soncinas* . . . Amadeus . . .

Tertia [opinion] est *Aegidii* Romani, qui ait caelum esse compositum ex materia et formam et materiam *eiusdem rationis cum materia* horum inferiorum. *Avicenna, Avempace, Avicembron* hoc idem senserunt . . . {*Mirandulanus* . . . *Achillinus* . . . }

VITELLESCHI (1590)

Prima Sententia vult *caelum compositum esse* ex materia et forma. . . .
{Eiusdem sententia est *S. Bonaventura* . . . *Capreolus* . . . *Caietanus* . . . *Ferrarinsis* . . . *Soncinas* . . . Flandria . . . Hervaeus . . . Amadeus . . . }

RUGERIUS (1591)

Unum caput est asserentium *caelum* ex materia et forma constare . . .
{quem sequitur *Capreolus* . . . Hervaeus . . . *Soncinas* . . . *Ferrariensis* . . . Flandria . . . *Caietanus* . . . et alii *Thomistae* }

Sed sunt in hac sententia duo modi dicendi . . . {Secundus . . . qui *volunt materiam caeli esse diversae rationis a materia horum inferiorum* . . . Est *S. Thomae Prima Parte*, loco citato, et alibi saepe, licet in 2° dist. 12, q., art. 1, indicatur sequi sententiam Averrois; *Albertus, primo Physicorum, tr. 3, cap. 11*, primo Caeli, tr. p°, cap. 8, et Prima Parte Summae *de quatuor coaevis, q. 4, art. 3*} Primus . . . qui dicunt materiam caeli *esse eiusdem rationis cum materia* horum inferiorum. Ita *Aegidius* in tractatu De materia caeli compositione, et octavo Metaphysicorum, q. 2 et 7, *Scaliger* . . . Idem sentiunt *Avicenna* . . . *Avicebron* . . . ut testis est *Albertus primo Physicorum*, tr. ultimo, cap. 11, et S. Thomas, Prima Parte, *q. 66, art. 2* . . .

Quo in capite sunt duo modi dicendi: . . . {Alter modus est ponentium *materiam diversae rationis a materia inferiorum*. . . . Ita *Albertus, primo Physicorum, tr. 3, cap. 11*, primo Caeli, tr. 1, cap. 8, et Prima Parte Summae quae est *de quatuor coaevis, q. 4, a. 3. D. Thomas, Prima Parte, q. 66, a. 2*, primo Physicorum, lect. 21 in text. 79, primo Caeli lect. 6 in text. 21 . . . }
Alter pontentium *materiam caeli eiusdem rationis cum materia* inferiorum . . . Ita censuit . . . *S. Bonaventura* . . . *Aegidius* in proprio tractatu De materia caeli compositione, et octavo Metaphysicorum, q. 2 et 3. Ex philosophis, *Avicenna* . . . *Avicembron* . . . Idem habet *Scaliger* . . .

According to our count, in these texts there are 185 Latin words of Galileo's composition that are relevant to our purposes, and of these, 64 words have counterparts in one or other of the Jesuit notes: 53 coincidences are found in Vitelleschi's lectures, 46 in those of Rugerius, and 33 in those of Menu. The patterns indicate that the content of the lectures remained fairly constant over the period of 13 years, but that the expression varied from year to year. Galileo makes reference to Albertus Magnus three times, and Vitelleschi twice, whereas Menu and Rugerius cite him only once; again, the order of treatment in Vitelleschi and Rugerius is the reverse of that in Galileo's exposition. In Galileo's text, however, it should be pointed out that Galileo's second citation of Albert is a blind reference: he cites Albert's exposition of the first book of the *Physics*, "as above" (UBI SUPRA, in capitals in the text), and actually he has no previous reference to that locus! There is a reference to it "below," on the other hand, and so Galileo should have written UBI INFRA. The mistake could be traceable to the fact that Galileo himself reversed the order of treatment from that found in the source from which he worked, and neglected to take account of the reversal in his first citation of Albert the Great. On the basis of this reasoning, Vitelleschi would be the closest to Galileo in time of composition.

Before concluding with Galileo's questions on the *De caelo*, we should note that all of his references to Albert in the treatises on the universe and the heavens can be verified in the Borgnet edition. In I10 Galileo states that "Albertus Magnus, in the first *De caelo*, tract 1, chapter 4 . . . maintain[s] that Plato . . . generally disagreed with Aristotle"; this is found in Borgnet 4: 15b. The references made in the text cited above, K37–40, are generally correct: the "ubi supra" citation of the first *Physics* should be to tract 3, chapter 11 (Borgnet 3: 68); that to the eighth *Physics* is found in Borgnet 3: 549–553; and the two citations of the *De quatuor coaevis* are in Borgnet 34: 335b and 34: 404a respectively – in Borgnet's division, however, the first would be referenced as tract 1, question 2, article 6, and the second, as tract 3, question 7, article 3. In K170 Galileo raises the question whether the matter of the heavenly spheres is one or many, and to this he replies: "We say that if the heavenly spheres differ specifically from each other, the matter of any one sphere is different in kind from the matter of another, as Albert holds in *De quatuor coaequaevis*, question 2, article 6 . . . "; this teaching is verifiable in a general way in Borgnet 34: 335, though not in the precise terms given in the text. Finally, in L11 Galileo observes that "some have thought that there ought to exist in the heavens, apart from intelligences, some kind of proper intellective souls; thus Alexander [of Aphrodisias] . . . " He goes on to recount that "Algazel, Rabbi Moses, and Isaac seem to feel the same, as Albert mentions in the seventh *Metaphysics*, tract 2, chapter 10" (Borgnet 6: 607a). The Latin of the last clause reads: "Idem videntur sentire Algazel, Rabbi Moyses, et Isaac, ut refert Albertus in 11 Metaphysicorum, tr. 2, cap. 10." The same reference is found in Menu, who writes: "Fuit Alexandri Aphrodisias . . . et Algazelli et Rabbi Moysi et Isaac Judaei, referente Alberto 11 Metaphysicorum, cap. 10, tr. 2."

III. THE ELEMENTS AND THEIR QUALITIES

This brings us to Galileo's last tractate of the physical questions, namely, that on the elements, which he divides into two parts or disputations, the first concerned with the elements in general and the second with their qualities. Albertus Magnus is mentioned twice in the first disputation and four times in the second, all four of the latter references being to Albert's teaching on how the primary qualities can be termed active and passive. The two citations in the part dealing with the elements are of some importance for the fact that they permit us to use the lecture notes of a Jesuit Professor not discussed thus far, Paulus Valla, the portion of whose lectures on the *De generatione* dealing with the elements has survived, though the remaining portions on that book and the *De caelo* are no longer extant.

Galileo's first mention of Albert occurs in a paragraph wherein he begins to enumerate the various meanings of the term element. He writes:

The first meaning, therefore, is that an element signifies the intrinsic causes composing a thing, i.e., matter and form. These seem especially apt to be called elements because from them a thing is first composed and into them it is ultimately resolved, and they are not in turn composed of others nor are they resolved into others; and this can in no way be said of the other meanings. Such causes, on the authority of Eudemus, based on Simplicius in his introduction to the *Physics*, were first called elements by Plato; the same usage was taken up by Simplicius, Philoponus, Averroës, and Albertus on the first chapter of the first *Physics*. For this reason Averroës, in the third *De caelo*, comment 31, says that Aristotle in the books of the *Physics* treated of the universal elements of all simple and composed bodies. Thus Philoponus, in the first text of the second *De generatione*, gives the reason why Aristotle, in the third *Physics*, 45, the second *De generatione*, first text, and the second *De partibus* [*animalium*], chapter 1, calls the four simple bodies elements. He gives the reason: because, he says, these bodies are not really elements themselves, since they are composed of other things that are prior, i.e., matter and form, which are most properly elements. For, although elements in this sense are said of any intrinsic cause, more commonly and more properly they are said of matter, as is apparent from Alexander, Eudemus, Simplicius, and St. Thomas on the first chapter of the first *Physics*. Since, however, matter is manifold, the first and most common matter of all, says Averroës, third *De caelo*, 31, second *De generatione*, text 6, first *Metaphysics*, text 4, and tenth *Metaphysics*, text 2, is primarily and most properly said to be an element, for elements are like the material parts of a thing; and this is the first meaning. (P8).

The Latin for this passage is reproduced below, and placed opposite it is the corresponding passage from the undated "Tractate on the Elements," composed by Valla probably in 1589:

GALILEO	VALLA (1589)
Prima igitur est, ut *elementum* significet *causas intrinsecas* rem componentes, idest *materiam et formam*: quae maxime videntur posse dici elementa, quia *ex* his *primo componitur* res et in haec ultimo resolvitur, et ipsa non amplius ex aliis componuntur neque in alia resolvuntur; quod non omnino caeteris significationibus aptari potest.	{tertio sumitur *elementum* pro *causis* tantum *intrinsecis, materia* scilicet *et forma* . . . }
Et haec, authore *Eudemo* apud *Simplicium in prooemio Physicorum, a Platone* primo fuerunt dicta *elementa*; et idem etiam usurpavit *Simplicius, Philoponus, Averroes*, et *Albertus, primo Physicorum, primo.*	*Eudemus*, ex *Simplicio*, primo *Physicorum in proemio* ait *Platonem* . . . ita appellasse elementum . . . Hoc tertio modo sumuntur elementum *Simplicio, Philopono, Alberto*, et aliis *primo*

Et hac ratione *Averroes*, tertio Caeli, com. 31, ait, Aristotelem in primis Physicorum egisse de elementis universalibus omnium corporum simplicium et compositorum: unde *Philoponus, in textum primum secundi De generatione*, reddit rationem quare Aristoteles, tertio Physicorum 45, *secundo De generatione t. p°*, et *secundo De partibus, capite primo, quatuor corpora* simplicia appellet *vocata elementa*: reddit rationem, *quia*, inquit, *non sunt ipsa vere elementa, siquidem ex aliis prioribus componuntur*, idest *materia et forma quae sunt propriisme elementa. Quamvis autem elementa* in hoc sensu *dicantur de utraque causa intrinseca*, communius tamen et *proprius* dicitur *de materia*; ut patet ex *Alexandro, Eudemo, Simplicio*, et D. Thoma *primo Physicorum, primo*. Cum autem materia sit multiplex, ideo, *primam* et *communissimam* omnium, inquit Averroes, tertio Caeli 31 et secundo De generatione t. 6, et *quinto Metaphysicorum t. 4*, et decimo Metaphysicorum t. 2, primo et *propriissime dici* elementum ipsam materiam; *elementa enim sunt quasi partes materiales rei*: et haec est prima acceptio.

Physicorum, textu *primo* . . . Avicenna, prima primi, elementa definit esse *corpora*, et videtur hoc colligi ex *Averroe, quinto Metaphysicorum t. 4* . . . ut videtur indicasse elementum dicere esse speciem aliquem, i.e., corpus aliquod completum, ex quo sequitur nomen elementum primo ici de quatuor elementis. Alii vero asserunt elementum primo dici *de materia*; ita *Eudemus* et *Alexander, primo Physicorum, t. p°*, referente *Simplicio*, et Averroes, tertio Metaphysicorum, com. 4, *Philoponus, secundo De generatione in t. p°*, et aliis communiter, primo Physicorum, t. p°, primo De generatione, t. p°, et primo Meteororum in principio, et hanc ob causam sequisse Averroem: quatuor elementa vocat *vocata elementa*, non autem simpliciter elementa. Ita habet tertio Physicorum, t. 45, *secundo De generatione, t. p°*, 4° et 6°, *secundo De partibus* animalium, *cap. p°*, et primo Physicorum, t. 22, quia quatuor elementa non sunt *communia* omnibus rebus, cuiusmodi est materia *prima*, et quia elementa debent esse prima et simplicissima, quod magis videtur convenire materiae primae.

There is no corresponding passage in the notes of Menu and Vitelleschi, but Rugerius discusses approximately the same matter, and in so doing has wording that is somewhat closer to Galileo's, as can be seen from the following text:

RUGERIUS (1591)

{Solet autem nomen hoc elementi variis significationibus usurpari . . . Aristoteles, *quinto Metaphysicorum*, quod quia *elementum* est id *ex* quo *primo componitur* aliquid inexistens indivisibile specie in aliam speciem . . . }
Primo advertendum quod *Eudemus*, referente *Simplicio* in prologo *Physicorum*,

elementaria principia, id est, intrinseca, materiam et formam ait primo *a Platone* appellata fuisse *elementa* . . .
Simplicius vero, *Philoponus*, *Albertus*, *Averroes*, et alii intelligunt duas *causas intrisecas, materiam et formam* . . . et *Philoponus* hic docet Aristotelem dixisse *vocata elementa, quia non sunt ipsa vere elementa, siquidem ex aliis prioribus componuntur, materia* scilicet *et forma, quae sunt propriisime elementa* . . .
Quamvis autem proprie *elementum dicatur de utraque causa intrinseca, magis tamen proprie dici* solet *de materia*, et *propriissime* quidem de prima; *elementa enim sunt quasi partes materiales rei*, et sic proprie videntur causae materiali convenire definitiones illae . . .
primum enim est materia prima, vel materia et forma ut modo dicebamus, immo haec propriissime sunt elementa . . .

Comparing all three passages, we note that Galileo's use of Albert is merely to list him among those who give the common interpretation to the use of the term element by Plato; this is verifiable in Albert's exposition of the first book of the *Physics*, tract 1, chapter 5 (Borgnet 3: 12a). Of the 233 Latin words in Galileo's paragraph, 96 have counterparts in either Valla or Rugerius, with 54 being coincident in Valla and 65 in Rugerius. Noteworthy perhaps is the fact that Valla has a fuller enumeration of texts and authorities, whereas Rugerius duplicates more of the content of Galileo's exposition; again, with Galileo and Rugerius this is the first meaning of element, whereas with Valla it is the third meaning. These indications perhaps favor Rugerius over Valla slightly; between the two, however, they can account for over 40% of the wordage actually used by Galileo.

The remaining citation of Albert in the general tractate on the elements is of special interest because it occurs in a passage where there are two *lacunae* in Galileo's exposition. As we have suggested in our commentary on this passage (1977a, pp. 286–287), these are probably explicable from the fact that Galileo used a secondary handwritten source in composing the passage and had difficulty deciphering either the words or their meaning. Fortunately similar passages occur in the lectures notes of Menu, Valla, and Rugerius, and they are all helpful for reconstructing the sense of Galileo's statement. As we reconstruct his meanin passage reads as follows in English translation:

The fourth opinion is that of those saying that the forms of the elements are substantial forms hidden from us [but knowable through their] qualities. This is the position of St. Thomas, Albert, and the Latins, in the second *De generatione*, 16, and the third *Metaphysics*, 27; likewise the Conciliator, [clarification of] difference 13; Giles, on the first *De generatione*, question 19; Jandun, *De sensu*, question 25, and the fifth *Physics*, question 4; Zimara in the *Table*; and Contarenus in the first and seventh *De elementis*. (S8)

Here the two passages enclosed in square brackets replace spaces left blank in Galileo's Latin composition. His Latin text, together with the corresponding texts of Menu, Valla, and Rugerius, are given below in parallel column:

GALILEO

Quarta opinio est dicentium *formas elementorum esse formas substantiales* [space for three words] *qualitates* nobis *occultas*. Est *D. Thomae, Alberti*, et Latinorum, *secundo De generatione* 16 et tertio Metaphysicorum 27; item *Conciliatoris* [space for one word] *differentiae 13*, Aegidii, primo De generatione, quaestione 19, *Ianduni De sensu, quaestione 25* et quinto Physicorum, quaestione 4, *Zimara in Tabula, Contareni, primo* et septimo *De elementis*.

MENU (1578)

Tertia opinio est communis aliorum qui asserunt *formas elementorum* esse quasdam substantias *occultas* qui explicantur per *qualitates* motivas et alterativas. Ita *Albertus* Magnus, *secundo De generatione*, tr. 2, chap. 7, et *D. Thomas*, primo De generatione super t. 18, *Conciliator, differentia 13, Iandunus*, libro *de sensu* et sensili, *quaestione 25*.

VALLA (1589)

Tertia sententia est communis omnium fere Perepateticorum, quia asserunt *formas substantiales elementorum* esse substantias quasdam *occultas*, quae interdum explicant per *qualitates* motivas, interdum per alterativas. Ita tenent . . . *D. Thomas, secundo De generatione* in t. 24 et primo De generatione, lect. 8, *Conciliator, differentia 13, Albertus* Magnus, secundo De generatione, tr.2, cap. 7, *Iandunus, De sensu* et sensili, *quaestione* 15 . . .

RUGERIUS (1591)

Altera sententia est communis, differentias essentiales elementorum *esse* veras et proprias *formas substantiales*, *qualitates* vero tam motivas quam alterativas fluere ex illis tanquam passiones proprias. Haec sententia . . . *Alberti* hic, tr. 2, cap. 7, *D. Thomae*, primo De generatione 18 et *secundo De generatione* 6. Legite *Conciliatorem, differentia 13, Iandunum De sensu* et sensili, quaestione 28 et quinto Physicorum, quaestione 4, *Zimaram in Tabula* . . .

Contarenum in *primo* libro *De elementis*, et alios . . .
Propositio: *formae elementorum* sunt vere formae substantiales . . . Illa ratio est quod quia rerum differentiae ignotae sunt, ideo maluerunt per qualitates tanquam notiores explicare naturas illorum . . .

With regard to Galileo's text, it should be noted first that his reference to the "third Metaphysics" is doubtful; the abbreviation he uses is "Met.," which he usually employs for *Metaphysics*, and not "Mete.," which is customary for the *Meteors*; since this reference does not occur in any of the Jesuit authors, however, it assumes no importance for our purposes. Again, Galileo's reference to the "seventh" book of Contarenus's *De elementis* is erroneous, since there are only five books in this work. The *lacunae*, on the other hand, are fairly important, since they offer a primary indication that the notebooks are derivative, but at the same time represent some thought and reconstruction on Galileo's part. It seems clear that the missing sense of the first *lacuna* in the passage is that conveyed in our English translation, with its substitution, but when we compare Galileo's Latin with the corresponding passages in the three Jesuit authors, we discover that it is almost impossible to find three Latin words that can be put in the empty space! This undoubtedly explains why Galileo left it blank. The second *lacuna* is also puzzling, but it does permit of solution. The Jesuit professors cite the Conciliator frequently, and in one of these citations (though not in the passage above), Rugerius writes out his citation more fully as "Conciliator in dilucidario ad differentiam 10." It is quite probable that Galileo saw the abbreviation for "in dilucidario" and was unable to decipher it (the expression being fairly unusual in scholastic texts); as a consequence he left a space, possibly to be filled in later. Note that Galileo's reference to Albert, moreover, is not specific, whereas it is clearly indicated in all three *reportationes* and can be readily verified in the Borgnet edition (4: 433a), where Albertus Magnus writes: " . . . et ideo [primae qualitates] sunt substantiales elementis alio modo quam formae substantiales, quia sunt substantiales sicut passiones quae fluunt a substantia." Rugerius incorporates Albert's expression into his own exposition, and it is noteworthy that his passage has the most coincidences with Galileo, namely 25 words, as compared to 18 each for Menu and Valla, and on this basis may be regarded as the closest to Galileo's.

The last four mentions of Albert the Great by Galileo all occur in the question wherein the Pisan professor is discussing the active and passive character of the primary qualities of the elements, and refers to Albert as providing the best explanation as to how heat assists dryness in producing the latter's proper effects. The specific references occur in paragraphs X7, X13, and twice in X15, and additional material is contained in paragraphs X8–10 and X16–17; all of them are further developments of a teaching of Albert in his exposition of the fourth book of the *Meteors*, tract 1, chapter 2 (Borgnet 4: 708b). Counterparts for these passages are found in Vitelleschi, and to a lesser extent in Rugerius, but not in Menu or Valla. For our purposes it may suffice to give only one paragraph in English translation, followed by the Latin texts of Galileo and Vitelleschi in parallel column:

On this account I assign the following reasons for these statements of Aristotle . . . The third reason is that given by Philoponus, second *De generatione* on text 8, St. Thomas, Averroës in the beginning of the fourth *Meteors*, and Pomponatius in the same place, third doubt, and Albert, same place, first treatise, chapter 2: namely, that Aristotle said this not absolutely but only with regard to compounds, wherein heat and cold act most effectively and wetness and dryness receive most effectively, though the former also receive a little and the latter act, as Albert and Buccaferrus have correctly noted. Yet they add that wetness and dryness act only with the aid of the other two, and particularly with heat; since wetness, for example, does not act in a compound *per se*, except insofar as previously, by virtue of the heat, a humid vapor is raised that can be mixed with the humidifying body; then, in virtue of the same heat, the body will be opened up so that the vapor can penetrate it and moisten it. Thus heat is said to aid the action of wetness, and in a similar way it aids the action of dryness also . . . . (X15)

GALILEO

Quare ego assigno has causas illorum dictorum Aristotelis . . . Tertia est quam reddit Philoponus, secundo De generatione *in t. 8, D. Thomas*, Averroes in *initio quarti Meteororum*, et *Pomponatius*, ibidem dubitatione 3, et *Albertus, ibidem tr. p° cap. 2°*: nimirum, *Aristotelem* illud *dixisse non simpliciter, sed* in ordine ad *mixtionem*, in qua calor et frigus potissimum agunt, humor et siccitas potissimum patiuntur; licet etiam illae aliquantulum patiantur et hae agant, ut recte notavit Albertus *et Buccaferrus*. Qui tamen *addunt*, humorem et siccitatem non agere nisi *iuvantibus* aliis duabus et praecipue *calore*: nam humor,

VITELLESCHI (1590)

S. Thomas, quarto Meteororum initio, Bannes hic, quaestione 4, *Pomponatius* loco citato, Turris primo Tegni, com. 18, et alii dicunt *Aristotelis dictum non* esse intelligendum *simpliciter* considerando eas qualitates secundum se quatenus per illas elementa agunt ad invicem *sed* respectu *mixtionis*, quia enim calor et frigus . . . dicuntur activae, reliquae passivae . . . *Albertus, quarto Meteororum, tr. p° cap. 2°, et Buccaferrus* ibidem, initio et hic *in t. 8*, explicant Aristotelem eodem modo quo praecedentes, sed *addunt* etiam *in mixtione* omnes qualitates agere et omnes pati; calorem tamen et frigus plurimum agere

verbi gratia, *in mixtione* per se non agit, nisi prius, vi caloris, elevetur vapor humidus, qui possit *admisceri corpori humefaciendo*; deinde, vi eiusdem caloris, aperiatur ipsum corpus, ita *ut* illud *penetrare possit* et *humefacere*. Et sic dicitur calor iuvare actionem humidi; et simili modo etiam iuvat actionem sicci . . . .

et ideo dici activas, alias plurimum pati et ideo dici passivas, cum praesertim actio humoris et siccitatis *iuvetur* actione *caloris*, qui et elevat vaporem humidum et exalationem siccam, quae debent *admisceri corpori humectando* et exiccando. Item corpus ipsum rarefacit et attenuat, et ita illud agit vapori et exaltationi, *ut possunt penetrare* ad illud *humectandum* et exiccandum.

Of the 140 relevant Latin words in Galileo's composition, 35 have coincidences with the terms employed by Vitelleschi. This in itself does not convey how important the notions deriving from Albert *via* this Jesuit and his confreres are for Galileo's treatment of primary qualities and their role in explaining the activities and passivities of the elements. The ideas presented by Vitelleschi (and also contained, in some instances in more detail, in the exposition of Rugerius) form the key to Galileo's understanding of the sublunary region, which he was to use to good effect in his later discussions of the comets and other heavenly appearances, and often in debate with other Jesuits whose intellectual formation (we now know) was not far different from his own.

IV. ALBERT'S IMPORTANCE FOR GALILEO STUDIES

This brief study of Albert the Great as seen through the use made of him by Galileo in his early notebooks is helpful on two counts: it furnishes a large amount of incidental information that can be used to date Galileo's writings, and it shows how seriously philosophers in the late sixteenth century took their thirteenth-century sources, and particularly the writings of Albert on physical science, when they were addressing the problems whose solutions were only to come in the early modern period.

With regard to the first point, it is possible to summarize all of the information on word coincidences in Galileo's Latin text and in the corresponding passages in the *reportationes* of lectures given at the Collegio Romano. A tabulation of the results of these word counts is given in Table I. Of the total number of words in passages relating in one way or another to Albert the Great, 37% have counterparts in lecture notes of Jesuits teaching at the Collegio Romano between 1578 and 1591. Only four sets of notes are available for this period, and of these only two sets are complete, viz, those

TABLE I

Number of Latin words coincident in Galileo's composition and in the notes of Jesuits teaching at the Collegio Romano between 1578 and 1591. The last column gives the number of words in Galileo's text that have counterparts in one or more of the four notebooks indicated.

Par. No.	Total No. Relevant	MENU 1578	VALLA 1589	VITELLESCHI 1590	RUGERIUS 1591	GALILEO c. 1590
A4	43	–	–	17	11	18
E2	71	18	–	–	–	18
I 10	137	22	–	–	–	22
K37–40	185	33	–	53	46	64
L10–11	82	30	–	–	–	30
P8	233	–	54	–	65	96
S8	55	18	18	–	25	27
X7	63	–	–	15	22	22
X8	60	–	–	9	20	20
X9	37	–	–	17	23	25
X10	214	–	–	63	27	73
X13	108	–	–	32	–	32
X15	140	–	–	35	–	35
X16	90	–	–	33	53	61
X17	146	–	–	70	38	79
Totals	1664	121	72	344	330	622
Percent of Total No.		7%	4%	21%	20%	37%

of Vitelleschi and Rugerius, who taught in 1590 and 1591 respectively. Of these two authors, Vitelleschi has 21% coincidences with Galileo's composition, and Rugerius has 20%; these are far in excess of the 7% and the 4% coincidences to be found in the notes of Menu and Valla respectively. Valla's low percentage is perhaps explicable by the fact that only a small portion of his notes have survived, but Menu's notes are fairly complete, and his coincidences are particularly sparse in the latter parts of Galileo's notebooks, even though he does treat the same subject matter as Galileo. The larger number of coincidences for the Jesuits writing in 1590 and 1591 would seem to confirm the dating of Galileo's composition as "around 1590," an inscription written on the notebook by one of the curators who bound together the folios as they are now found in the Galileo Archives.

It should be pointed out that this evidence is not being adduced to suggest that Galileo actually used the notes of either Vitelleschi or Rugerius when composing the physical questions. This would have been impossible if he composed them in 1589 or 1590, because Vitelleschi's lectures were just being given at that time in Rome, and we know that Galileo was not there then to hear them; *a fortiori* he would not have had Rugerius's notes, because they were not yet written. It is probable that Galileo had access to an earlier set of notes deriving from the Collegio Romano, possibly those of Valla, which we know existed at one time, or even more likely, those of Mutius de Angelis, who last taught in 1587 but whose notes also are no longer extant.[10] Our examination of notes coming from the Collegio in successive years show that there is very little evidence of verbatim copying from one year to another, but similar phrases continue to recur, and it is this type of repetition that could well explain the coincidences of words we have pointed out in this essay.

With regard to our second point, this relationship between the young Galileo and the Jesuits of the Collegio Romano helps explain a curious fact about the Pisan notebooks, namely, that they consistently pay as much attention to philosophers and theologians of the thirteenth century as they do to writers of the fourteenth and fifteenth centuries, including those of nominalist leanings. The standard account of medieval influences on the rise of modern science is that these all surfaced after the Condemnation of 1277 and particularly in the nominalist schools of Bradwardine at Oxford and Buridan at Paris. Galileo's citations of such authors is minimal, whereas his use of authors such as Albert the Great and Thomas Aquinas is quite substantial. The explanation of this fact is now clear: Galileo was influenced in his choice of philosophical authorities by Jesuit professors who, while taking a glance at these nominalist contributions, were much more influenced by the scholastic syntheses of the High Middle Ages. They were pronouncedly realist in their options, moreover, and this perhaps explains why Galileo himself turned out to be so doggedly realist in his ill-fated attempt to establish the truth of the Copernican system in his later years. The Jesuits, too, were not adverse to studying Averroës, though the effect of the Condemnation of 1277 was still felt in their day, and so occasionally they had to resort to someone such as Albert the Great as a cover for their explorations of Averroist teachings.

The substantive doctrine deriving from Galileo's citations of Albert is not extensive, but two points are worthy of mention, and with these we must conclude. The first is Albert's discussion of the circular motion of fire after it rises as far as it can through rectilinear motion. This phenomenon raised the question whether such circulation is natural for fire in these circumstances, or whether it is violent, and ultimately led to the proposal that it is neither natural nor violent but intermediate between the two. Such a proposal, explicit in Galileo's later writings but already adumbrated in Jesuit lecture notes, was seminal for the concept of circular inertia, from which it is a simple matter to trace the rise of modern mechanics from Galileo to Newton. The second area is the use made by Galileo of Albert's teaching on primary qualities in his attempt to unravel the activities and passivities of elemental bodies. Remotely, a study such as this undoubtedly influenced the new theories of primary and secondary qualities that would emerge as distinctive of modern philosophy from Galileo through Descartes to Locke and Berkeley. More proximately, it led to a development of scholastic teachings on qualities and powers, on ways of quantifying these, and ultimately to the employment of the force concept as more fruitful for analyzing elemental interactions. We have explored this point elsewhere,[11] but for purposes here it may suffice to note that Albert the Great was the author recognized by Galileo as having made the most significant contributions in these areas, and thus as influencing at least indirectly the *nuova scienza* he was soon to originate.

NOTES

[1] Albert and Harvey both studied at Padua, the latter while Galileo was a professor there, so one might suspect that there would be some continuity of thought among the three. For scholastic and Aristotelian influences on Galileo and Harvey, see Wallace (1974d), pp. 211–272; Albert is not explicitly mentioned, but the influences are typical of Albertine thought.

[2] See the four previous essays in this part.

[3] For example, the expression *ex suppositione*, discussed in Essay 8, *supra*.

[4] Adriano Carugo has transcribed the Latin text from Galileo's autograph but has not yet published it; some comments on its contents are given by Crombie (1975).

[5] W. A. Wallace, *Galileo's Early Notebooks: The Physical Questions. A Translation from the Latin, with Historical and Paleographical Commentary* (Notre Dame: University of Notre Dame Press, 1977). The introduction and notes to this volume give biographical information on the Jesuit professors who were the inspiration behind the notebooks. They also contain specific references to the manuscripts in which the lecture notes of these Jesuits are recorded, and also outlines of the contents of the manuscripts, which for the most part lack foliation and thus cannot be cited through folio number.

[6] In the translation (note 5, *supra*) we have used a capital letter to designate each question, and Arabic numerals to designate the paragraph numbers within each question. Thus "A4" designates the fourth paragraph of question A, which is the first question in the introductory treatise, translated on p. 26 of *Galileo's Early Notebooks.*

[7] For some of the tensions provoked by Averroism in the Collegio Romano, see M. Scaduto, *Storia della Compagnia di Gesù in Italia*, 2 vols. (Rome: Gregorian University Press, 1964) Vol. 2, p. 284.

[8] Evidences of Galileo's use of secondary sources in composing the physical questions

are described in detail in the author's commentary on the questions, *Galileo's Early Notebooks*, pp. 253–303.

9 See Essays 14 and 15, *infra.*

10 In the preface to his two-volume *Logica* published at Lyons in 1622 Valla indicates that he has commented on all the philosophical works of Aristotle and has these ready for publication; these are undoubtedly the lecture notes from his teaching at the Collegio Romano, which he probably reworked for later publication but which are not known to have survived. For information on Mutius de Angelis, see *Galileo's Early Notebooks*, pp. 22–23.

11 See Essay 7, *supra*; also Essay 14 and 15, *infra*, for the adumbration of the concept of circular inertia.

13. GALILEO AND THE CAUSALITY OF NATURE

The previous studies in Part III of this volume have related Galileo's early thought to that of the Thomists, the Scotists, and the nominalists, but little thus far has been said about the Averroists, or Galileo's relation to Arab thought generally. Yet, as was indicated in the opening essay of this volume, the Arabs had distinctive teachings on motion that strongly influenced the development of Aristotelian natural philosophy in both the medieval and the Renaissance periods. One such teaching was the concept of nature advanced by Avicenna and Averroës, and particularly their understanding of how nature itself is a cause of certain motions. As the author's translation of Galileo's early notebooks also makes clear, Galileo was quite aware of the positions of Averroës and other Arabs on the composition and animation of the heavenly bodies, and treated them with more than usual respect (Wallace, 1977a, pp. 103–158). Moody, in his essay on 'Galileo and Avempace' (1975, pp. 203–286), further speculated that Galileo's treatment of motion was strongly influenced by Avempace, whose analysis of falling motion was known to the Latin West only through the commentary of Averroës. Thus there are good reasons for examining Galileo's views on the causality of nature, particularly as this might reveal a preference on his part for the teachings of Averroës, Avicenna, or other Arabs on this matter, as opposed, for example, to those of Thomas Aquinas.

Apart from this historical question, Galileo scholars have long been concerned with the problem of causality and particularly how causes function within Galileo's concept of scientific explanation (Wallace, 1974a, pp. 176–184). As was noted in the Appendix to Essay 8 *supra*, Galileo's *a priori* justification of his definition of natural falling motion as uniformly accelerated was made on the basis that nature always employs the simplest and easiest means, one wherein the velocity of motion increases uniformly with time of fall (*Opere* 8:197). From this text it would appear that Galileo saw nature as the cause, the determining factor that serves to explain falling motion. However, in a passage much quoted by historians of science, when later Sagredo raises the question whether the new discoveries reported in the discourse reveal the cause of natural acceleration, Galileo has Salviati reply that seeking this cause is not really worthwhile; it is sufficient to investigate the properties of falling motion, whatever its cause might be (*Opere* 8:201-202). Some historians interpret this latter statement as Galileo's rejection of causal explanation – the beginning for them of the modern era, as opposed to that of medieval science – although in this interpretation they take no account of Galileo's earlier statement about nature (Drake, 1974, pp. xxvii–xxix, 159). Now it seems that we should accord Galileo some measure of consistency: if we attempt to reconcile his two statements, it then is of capital importance whether nature is a cause and, if so, in what way.

Before delving into this matter, it may be well to review briefly Aristotle's teaching on the causality of nature as this is laid out in Books 2 and 8 of his *Physics*. In Book 2, as already explained in Essay 7 *supra*, Aristotle defines nature as a principle and cause of motion and rest, without specifying which of the four types of causality is involved in this definition. He does maintain, however, that nature as a cause can be identified with both form and matter, though more properly with form, and so we may presume that for him nature exercises its influence as both a formal and a material cause. Aristotle makes no reference to nature's being an efficient cause, but since in his later works he identifies the soul of the living thing with its form, and regards the soul as the efficient cause of the self-motion of the living, we can further argue that, at least in some cases, the form as nature can function as an efficient cause. Here a special problem presents itself in the realm of the non-living, and Aristotle takes this up in Book 8 when defending his motor causality principle, axiomatized by the Latins as *Omne quod movetur ab alio movetur*. If everything in motion is moved by another, as the axiom states, what efficient mover is involved in the natural motion of a falling body? Were there no difference between the living and the non-living, and granted that the form could be the efficient mover in the case of the living, one would be tempted to say that the heavy body's substantial form is the efficient agent of the body's fall. Aristotle, however, does not make this identification in Book 8, but rather points to two other movers that are completely extrinsic to the falling body, viz, its generator as the *per se* initiator of its motion, and whatever removed the supports from under it as the motion's *per accidens* cause.

When touching on this matter briefly in Essay 1 *supra*, we noted that Averroës (supplying the missing identification in Aristotle) held that the form of the heavy body, meaning by this its substantial form, is the principal mover of the body as an active principle within it, and that its gravity, as an accidental form inhering in the body, is its secondary mover as an instrument of the substantial form. In this teaching Averroës was merely elaborating a doctrine he had already found in Avicenna, which has been analyzed by J. A.

Weisheipl and shown to involve strong Neoplatonic elements.[1] By Weisheipl's account, Avicenna taught that the form of the heavy body is its mover as an active principle, and indeed as the efficient cause of the body's fall. Aquinas, as opposed to both Avicenna and Averroës, denied this, for in his view such a teaching would blur the distinction between living and non-living, and would effectively raise the inorganic realm to the level of the animate. In Aquinas's understanding, therefore, the only efficient agents involved in falling motion are those explicitly identified by Aristotle as such, viz, the generator and the *removens prohibens*. The body's form and its gravity he saw as intrinsic principles of such motion, and these suffice in his estimation to characterize this motion as natural. But as principles they are merely passive, something "by which" (*quo*) the generator moves the body to its natural place in the universe, without themselves being active or exercising any efficiency in the process. John Duns Scotus, by contrast, sided with Avicenna and Averroës in this controversy, and described nature as an active principle that, in a sense, moves itself to activity. The answer he gives to the much-debated question, "What causes hot water to cool when left standing by itself?," is typical: the natural form of water moves the water to cool in the absence of heat.

To return now to Galileo, it is well known that he studied and taught at Pisa between 1581 and 1591, so it should not be difficult to ascertain the concepts of nature that were current at that time. In his notebooks dating from the Pisan period, Galileo in fact references two Spaniards, Domingo de Soto and Benedictus Pererius, both of whom discuss nature extensively in their commentaries on Aristotle's *Physics.*[2] Soto composed his commentary at Salamanca around 1545, and in it he explains the definition of nature in terms that are equivalent to Aquinas's teaching.[3] This is not surprising, for Soto was a Dominican and a Thomist. In light of the question that now concerns us, we may further ask whether Soto knew anything about Avicenna. The answer is yes, he did, for Soto quotes Avicenna and raises an objection to the definition of nature based on Avicenna's text. This objection raised difficulties that very much puzzled sixteenth-century thinkers.[4] The problem is that the heavens seem to be natural bodies, and their motion seems to be natural also, as coming from nature. But no active principle moves the heavens from within, since they are moved actively by intelligences, which are extrinsic principles; nor does their motion have an internal passive principle, for this could only be their matter, and matter naturally tends to rest, whereas the heavens appear to move eternally. Since a natural motion must derive from internal principles, either active or passive, and neither of these can be found in the heavens, their motion is not natural in Aristotle's sense. Soto goes on to mention also Avicenna's opinion that Aristotle is wrong in holding that nature is so obvious as not to require demonstration.[5] Neither of these positions is directly pertinent to our concerns, but they bear witness to the fact that Avicenna's opinions were known in the mid-sixteenth century, and they were discussed, though not the precise point that is the focus of this essay.

Pererius's textbook on natural philosophy was published at Rome in 1576, and, perhaps because Pererius was a Jesuit professor at the prestigious Collegio Romano, it exerted a great influence in Italy to the end of the century.[6] Pererius had read Soto, to be sure, and he likewise gives evidence of acquaintance with Avicenna's views.[7] He discusses nature under the title, *De forma ut est natura*, which seems to accent nature's role as a formal principle, and possibly even as an active principle, of natural motion.[8] Pererius is not as Thomistic as was Soto, for in many of his teachings he favors the interpretations of Averroës and of Scotus, both of whom were popular in northern Italy at the time of his writing. Indeed, when discussing the motion of the heavens, he aligns Scotus with Avicenna in holding that Aristotle's definition of nature does not fit the heavens, though he himself prefers to follow Aquinas in saying that it does.[9] But on the problem of interest here, namely, whether nature is an active principle of the motion of the elements, Pererius is in a quandary. He does not have a definite answer, but it seems probable to him that the position of Averroës and Scotus (he does not mention Avicenna) is correct: the elements are not moved by the generator, but they move themselves, and thus nature is an active principle within them.[10] Pererius notes, however, that he does not agree with Averroës that the medium is necessary for falling motion, nor does he think that Scotus explains himself very well; so he defers his own treatment of the problem to the time when he will write his own book on the elements, which unfortunately never did find its way into print.[11]

As remarked earlier, Galileo mentions Pererius in his notes written at Pisa, and there is evidence that Galileo may even have copied portions of those notes from Pererius's text.[12] Not all is there, however, and so it has been necessary to track down other sources of Galileo's notebooks. The most interesting results of this search, reported in Essay 10 *supra*, lie in parallels that have been discerned between Galileo's writings and *reportationes* of lectures given at the Collegio Romano around 1590 by other Jesuit professors.[13] One of these professors turns out to be significant for our purposes, by name, Muzio Vitelleschi.[14] In his *Physics* lectures Vitelleschi has an extensive treatment of nature and the causality it exercises in all forms of natural motion.[15] Since Galileo seems well acquainted with the positions Vitelleschi takes on other matters, it would be quite surprising if the latter's explanations did not match Galileo's at the time he was studying philosophy, and later teaching, at Pisa. Vitelleschi, it may be noted, is more consistently a Thomist than was Pererius, but there is some ambivalence in his exposition, which will now be detailed.

The passages of interest occur in a disputation on Aristotle's second book of the *Physics* that inquires "Whether any cause is included in the definition of nature, either universal, or accidental, or efficient?"[16] Vitelleschi replies with seven propositions, as follows: (1) the universal cause, as such, is not contained under the definition of nature; (2) no accidental principle is there contained; (3) nor is any efficient cause, precisely as it is efficient and acts on others; (4) nor is the final cause, as such; (5) nor is generation; (6) nor privation; (7) nor any composite. Of these seven points only the first three need concern us, as pertinent to our problem.[17]

The first proposition, that the universal cause as such is not included under the definition of nature, Vitelleschi says is directed against Philoponus [2]. Now this is an interesting observation, for if we consult Philoponus's commentary on the *Physics* we find that he is dissatisfied with Aristotle's definition of nature, and emends it considerably.[18] Philoponus's revised definition is the following: "Nature is a kind of life or force that is diffused through bodies, that is formative of them, and that governs them; it is the principle of motion and rest in things, and in such things alone, in which it inheres primarily and not incidentally."[19] As seen from this definition, Philoponus conceives nature as a type of *anima mundi*, an enlivening force operative throughout all of creation and accounting for its spontaneous movements. Whether Philoponous would equate this force with God is not clear from the text, but in his reply Vitelleschi understands him as doing so. The universal cause is not contained under nature's definition, he says, because the universal cause of all things is God and God is not a principle of motion in himself. Or perhaps you mean, he goes on, that the heavens are the universal cause? His answer to this is that the heavens do indeed have a nature, but they are not said to be natural because of the motions they effect in the sublunary region; rather this is because of motions that are properly their own [2].

Vitelleschi's second proposition, viz, that accidental principles are not included under the definition, is directed, he says, against Albert the Great, who taught that motive qualities are nature because they are primary, i.e., they are the immediate and proximate principles of natural motion [3]. Vitelleschi replies that when "primary" is used in the definition of nature it does not mean proximate; rather it means basic, as opposed to instrumental. Accidents such as motive qualities, however, are instrumental principles, and so they do not fit the definition [3]. At this point, however, he raises an objection that is somewhat illuminating. The objection is that if water vapor moves upward it would seem to do so naturally, since its motion is from an intrinsic principle; but such a motion is explainable only by the *levitas* of the vapor, and this is something accidental. The motion cannot come, moreover, from the substantial form of the vapor, for this is the form of water, and its natural motion is downward [4]. To resolve the difficulty Vitelleschi maintains that the upward motion of the vapor, in such a case, can be natural only *secundum quid*, because it comes from an instrumental principle, *levitas*, that has a natural inclination upward. Absolutely and *simpliciter*, on the other hand, the movement must be violent, for it is not consonant with the nature of water, which has a natural inclination downward [4].

This leads directly to Vitelleschi's third proposition, which is the key point of our inquiry: No efficient cause, precisely as efficient and as acting on another, is included in this definition [5]. Vitelleschi notes that this is the common teaching of peripatetics, Simplicius alone excluded. He explains its sense in this way: there are two kinds of natural motion, one that takes place in the thing itself, as when fire moves upward, another that takes place in something else, as when an animal moves by hand or by foot. Both can be said to result from the nature of the mover, and thus each is natural as proceeding from a principle within itself. In this way of speaking the essence of a natural thing is not referred to as "nature" with respect to motions it causes in something extrinsic to it, but only with respect to motions it causes within itself. Motions produced in another are not natural in this sense, i.e., they do not come from nature as it is a nature, but rather as it is an efficient cause [5].

There are two reasons, Vitelleschi goes on, why we speak in this way [6]. The first is that any motion that is manifestly produced in a thing from without is attributed by men to the cause that produces it; so the heating of the wood is attributed to the fire under it. Motions that occur without any apparent extrinsic cause, on the other hand, are attributed to the nature of the thing and are said to be "natural"; it is for this reason that men regard nature as the principle of motion within the thing itself [6]. The second reason is based on a tripartite division of all things into natural, artificial, and abstract [7]. Natural things are those we see possessing a certain force (*vim quandam*) that effects motions within themselves, that brings them to perfection, and that causes motion in other things. Abstract things (and here Vitelleschi seems to have in mind separated substances, or intelligences) have the power to produce (*vim efficiendi*) motion in other things but not in themselves, because they are unmoved movers, and so they differ from natural things. Artificial things are similar to abstract entities in that they cannot produce motion in themselves, but they differ in yet another way from natural things, for they do not so much produce motion in another as they modify the way in which motion results in more or less accidental fashion [7]. Of the three types of things, therefore, only the natural can be said to have its principle of motion within itself, and this in an essential as opposed to a merely accidental fashion, and that is why we define "nature" in precisely the way we do [8–9].

Here Vitelleschi raises an objection similar to that already noted with regard to the water vapor [10]. The objection proceeds as follows: Fire goes upward naturally, just as it heats naturally; but natural motions come from nature; so the form of fire must be nature with respect to both these motions. But in the case of the second motion, heating, nature acts as an efficient cause; therefore it must so act in the first case, that of upward motion [10].

Vitelleschi's reply is again based on a series of distinctions, this time focusing on the term "natural" rather than on "nature" itself[10]. The "natural," he notes, can be used in four different ways: first, as opposed to the "violent," since what is natural comes from some inclination or propensity within the thing, whereas the violent opposes such an inclination; second, as different from the "free," since what is natural comes from an inclination that necessitates in a particular way, whereas the free does not; third, as comprising everything that is realized when a natural agent does something in a natural way; and fourth, as opposed to the "artificial," in the sense that the natural has a principle of motion within itself, whereas the artificial does not. The response to the objection, therefore, is that the upward movement of fire is natural in the fourth way, whereas its heating of wood is natural only in the first three ways and not in the fourth. Fire's heating of wood is not violent, but is in accord with fire's natural inclination; it is not casual or free, but is necessarily determined; and it has all the features one would expect of a natural process. The only thing that prevents it from being completely natural is that the motion that results, the heating, is in the wood, and therefore its principle is not within the thing that is changed, but rather in the fire, which is the agent or the efficient cause of the wood's heating [10].

To sum up, in Vitelleschi's understanding there are different ways in which motions can be said to be natural, analogous to his previous reply wherein he qualified some motions as natural *simpliciter* and others as natural only *secundum quid*. Some motions are called natural, he now intimates, when they produce effects that are recognizably such as proceeding from an efficient cause; others are natural when they proceed from a principle within the thing itself. In the latter case, the principle need only be a passive instrumental principle, such as *gravitas* or *levitas*, and need not produce effects in another object, as does an efficient cause. So nature is truly a cause, and sometimes it produces effects in another, and thus can be called an efficient cause; but it is not always an efficient cause, and particularly not in the case of upward and downward motion, and thus efficiency is not a proper attribute and should not be included in nature's definition [8–10].

The foregoing clarification of the types of causality involved in natural motion may prove helpful for understanding Galileo's apparently inconsistent use of causal terminology in the passages referenced from the *Two New Sciences*. When referring to nature as the determining factor that explains how bodies fall Galileo clearly had in mind an internal cause, a fundamental principle from which their motion proceeds. In substantiation of this we would call attention to his explicit citation, in his notes on motion written around 1590, of a passage from Aristotle saying that the natural character of a motion requires an internal, not an external, cause (*Opere* 1:416–417). Like his contemporaries Galileo probably identified this internal cause with the substantial form of the falling body. Whether he saw this form, and the accidental form of *gravitas* associated with it, as an active principle or as a passive principle is problematical: he might have seen either of them in Platonic or Avicennian fashion, following Philoponus, as a kind of natural *vis*, or he might have viewed them, as did Aquinas, as merely passive principles. Regardless of how he decided this particular option, however, one can readily understand why Galileo would have been reluctant to label either form, as natural and internal, the proximate *efficient* cause of the body's fall. In the context of the question raised by Sagredo, on the other hand, the causes mentioned by Salviati as explaining the body's acceleration (viz, its nearness to the center of gravity and the actions of the medium on it) are all extrinsic to the body and are thought of affecting it in some efficient way (*Opere* 8:202). By Galileo's day much ink had already been spilled on identifying such agents that affect natural fall, all to no avail. Thus it was far from revolutionary for him to have Sagredo reply that further discussion of external causes was not really worthwhile. Suffice it to know that the motion proceeded from nature as from an internal cause, and then go on to investigate the properties of the resulting motion as *natural*, regardless of what its proximate efficient agents might be.[20]

APPENDIX

Lectiones R. P. Mutii Vitelleschi in Octo Libros Physicorum et Quatuor De Coelo, Romae, Annis 1589 et 1590, in Collegio Romano Societatis Jesu.

In secundum librum Physicorum. Disputatio secunda. An in definitione naturae contineatur causa aliqua universalis vel accidentalis vel efficiens?

[1] Explicatis breviter singulis partibus definitionis, easdem iterum sed accuratius discussamus. Primum igitur, circa illam particulam "principium et causa," videndum est quodnam principium quae causa sit natura. Sit ergo.

[2] Prima propositio. Sub definitione naturae non continetur causa universalis ut talis est; contrarium indicat Philoponus. Probatur: quia causa universalis omnium rerum, scilicet Deus, non est principium motus in se ipso. Confirmatur: quia causa efficiens non continetur in hac definitione; sed causa universalis est efficiens; ergo. Objicies: coelum est causa universalis eorum quae infra lunam sunt, et tamen coelum habet naturam, ut infra dicemus; ergo [sub nomine naturae continetur causa universalis]. Respondeo: in coelo non esse naturam ut est causa universalis, i.e., in coelo non dicitur esse natura ratione illius motus quem efficit in aliis, ratione cuius dicitur causa universalis, sed ratione illius motus quem habet in se, et ideo dixi non contineri causam universalem ut talis est. Dicitur tamen interdum causa universalis, scilicet Deus, autor naturae et etiam natura. Ita volunt aliqui accepisse Aristoteles nomen naturae, primo Coel. 20, cum ait "recte fecisse naturam coelum eximendo a contrariis," et quidem merito, quia ab ipso coelum et natura dependet, duodecimo Met. 38: primo enim, omnia efficit et dat omnibus naturam; secundo, singulis naturis perprior prestabit fines et eas ad eosdem dirigit; tertio, hunc tam admirabilem totius naturae seu universi ordinem et consensionem in tanta rerum diversitate conservat, sicut anima in corpore conservat omnem humorum temperiem; quarto, concurrit ad omnes operationes particularium naturarum, quae non solum ut sint sed etiam ut operentur indigent concursu primae et universalis causae, ut suo loco dicemus.

[3] Secunda propositio. Nullum principium accidentale continetur sub definitione naturae; est contra Albertum, trac. 1, c. 3, ubi ait qualitates motivas esse naturam, quia sunt primum, i.e., immediatum et proximum principium motus. Probatur: tum quia natura definitur principium primum, i.e. principale, motus, accidens autem est principium instrumentale; tum

quia omnes distinguunt naturam ab accidentibus, unde antiqui, quia solam materiam dicebant esse substantiam, eandem solum dicebant esse naturam; tum quia Aristoteles, quando docet formam esse naturam, aperte ostendit se loqui de forma substantiali, ut adnotavimus in textu; tum quia docet Aristoteles omnia quae natura constant esse substantias, t. 4, quomodo autem substantia constare potest ex non substantiis, t. 52 primi Phys.; tum quia si accidens esset natura, ipsius naturae et principii esset principium, nimirum substantia, a qua dependet accidens. Simili modo argumentatur Aristoteles, primo Phys. 52.

[4] Objicies: vapor naturaliter movetur sursum, movetur enim a principio intrinseco; sed hic motus non videtur posse referri nisi in levitatem, quae est quoddam accidens; ergo. Minor probatur: quia non potest provenire a forma substantiali vaporis; haec enim est forma aquae cui naturaliter debetur motus deorsum; ergo [a forma accidentali levitatis]. Respondeo: si vapor distinguitur ab aqua essentialiter, dicendum est motum illum esse naturalem et fieri a forma substantiali vaporis tanquam a principali causa; si vero non distinguitur, ut argumentum supponit, dicendum est motum illum esse violentum simpliciter loquendo, non quia non fiat a principio aliquo inhaerente, sed quia huiusmodi principium non est connaturale ipsi vapori, quod necessarium est ad hoc ut motus aliquis absolute et simplicter dicatur naturalis; potest tamen aliquo modo et secundum quid dici naturalis ratione principii instrumentalis, scilicet levitatis, cui naturale est movere sursum.

[5] Tertia propositio. Causa efficiens ut efficiens est et agit in aliud non continetur in hac definitione; est communis omnium sententia Peripateticorum excepto Simplicio, com. suo 12. Explico: duplex est motus, quidam in seipso, ut cum ignis v.g. fertur sursum, quidam in alio, ut cum manu vel pede moveo; non est autem negandum utrumque motum provenire ex natura moventis, ac proinde utrumque motum esse suo principio naturalem, si per naturale intelligamus omne id quod est debitum et conveniens naturae sui principii. Recte tamen dicimus essentiam rerum naturalium non dici naturam respectu motus quem efficit in aliquo extrinseco, sed solum respectu motus quem efficit in iis rebus in quibus est, ac proinde motum in aliud non esse naturalem in hoc sensu, i.e., non fieri a natura ut natura est, sed ut est causa efficiens.

[6] Huius autem rei duplex potissimum assignari potest causa: quia motus qui ab extrinseco fiunt in aliqua re, cum fere habeant evidentem causam, attribuuntur ab hominibus propriae causae, v.g. calefactio quae fit in ligno igni calefacienti; at vero motus qui fiunt in ipsamet re, cum non appareat causa aliqua extrinseca, attribuuntur ipsi naturae rei, et maxime dicimus

naturales, ac proinde homines per naturam intelligunt maxime principium motus in eo in quo est.

[7] Secunda causa: res omnes vel sunt naturales vel artificiales vel abstractae. Res naturales videmus habere vim quandam qua et in se ipsis motum efficiunt, quo se ad propriam perfectionem provehunt, et simul causant motum in aliis rebus. Res abstractae habent quidem vim efficiendi in aliquo motum aliquem, in quo conveniunt cum naturalibus, non tamen habent vim efficiendi motum physicum in seipsis, sed immotae movent, hic t. 71, in quo differunt a naturalibus. Demum res artificiales non habent principium efficiendi motum in se ipsis, in quo conveniunt aliquo modo cum rebus abstractis, differunt vero a naturalibus. In alio vero convenit aliquo modo ars cum natura, nam licet fortasse ars non tam efficiat motum quam illum modificet, tamen dicitur esse productivum motus in alio. Et quidem merito, quia est valde difficile distinguere quo modo ars efficiat illam modificationem et non motum; et vere homines communiter existimant et dicunt artem efficere opus artificiosum, quare videtur ars convenire cum natura in hoc quod efficiat motum in alio.

[8] Cum enim natura ut constituit res naturales debeat easdem distinguere a rebus non naturalibus, ut bene hic philosophatur Aristoteles, merito natura ita appellatur ex eo quod est principium motus in eo in quo est, non in alio. Licet enim fortasse res naturales distingui possunt ab artificiatis etiam in eo quod illae efficiunt motum in alio, hae vero non efficiant motum sed modificant, ut dictum est, tamen hoc non distinguit res naturales ab abstractis, ut dixi. Nec etiam recte distinguit ab artificiatis. Nam est valde occulta haec differentia, est et contra communem omnium sensum et sermonem. Unde Aristoteles, qui a posteriori et sensu duce res naturales vult distinguere ab artificiatis, eius differentiae omnino non meminit. Universalis ergo clarissima et optima differentia inter res naturales et alias omnes ea est, quod illae in se ipsis habent principium quo moventur motu physico et naturali, hae vero non habent, et ideo natura recte definitur principium motus et quietis in eo in quo est et non in alio.

[9] Hoc posito probatur nostra propositio, tum quia Aristoteles definit naturam principium motus in eo in quo est, quod etiam docet hic, t. 84, et primo Coeli t. 5, quinto Met. 5, sexto Met. initio, nono Met. 13, undecimo Met. c. 1° et summ. 3, c. item 1°, et quidem ita clare ut nullum relinquat dubitationi locum. Idem colligitur ex tex. 86 huius libri, ubi docet naturam esse similem medico sibi ipsi medenti, quia videlicet est principium motus in eo in quo est, et aliis multis locis. Unde idem Aristoteles initio huius libri distinguit naturam ab arte, quia haec non est principium motus in quo est nisi

per accidens; tum quia aliter definit Aristoteles causam efficientem, t. 29, nec potest illa definitio convenire cum hac definitione naturae; tum quia nomine naturae in presentia intelligimus id quo naturalia differunt ab aliis, sed differunt eo quo habent principium motus in se ipsis, ut dixi. Confirmatur: quia natura non solum distinguit naturalia ab artificiatis et abstractis, sed etiam ipsa distingui debent a violentia; sed motus in alium potest esse violentus, motus in se non potest; ergo natura ut natura est tantum principium motus in eo in quo est.

[10] Objicies primo: sicut naturaliter ignis ascendit sursum, ita naturaliter calefacit; sed motus naturalis est a natura; ergo forma ignis respectu utriusque motus est natura; sed in secundo motu habet rationem causae efficientis; ergo [etiam in primo]. Respondeo, sicut natura multis modis accipitur, ita etiam naturale: primo quidem sumitur ita ut opponatur violento, quo modo naturale est id quod fit secundum aliquam inclinationem et propensionem ipsius rei, violentum vero quod fit repugnante ipsa re, ut cum lapis sursum fertur; secundo, naturale opponitur libero, quo modo naturale est id quod necessario fit et ab aliqua inclinatione naturali et determinata; tertio, naturale complectitur omne id quod fit ab agente naturali modo naturali; quarto, accipitur naturale ut opponitur artificiali, et ita naturale dicitur id quod in se habet sui motus principium. Ad argumentum igitur respondeo motum ignis sursum esse illi naturalem hoc quarto modo, calefactionem vero respectu ignis calefacientis esse naturalem aliis modis, ea videlicet ratione qua omnes operationes debitae et convenientes alicui principio naturali et qui fiunt modo naturali dicuntur esse illi naturales. Et quia necessario ignis calefacit et non est contra inclinationem quo modo etiam dicimus naturale esse intelligentiis intelligere, non est autem naturalis illa calefactio igni quarto modo, qua ratione eae tantum operationes dicuntur naturales quibus res naturales ab aliis omnibus distinguuntur, quo modo nos accipimus naturale in presentia.

[11] Objicies secundo: Aristoteles, t. 13, probat formam esse naturam quia ex homine generatur homo; ergo forma dicitur natura quia est principium motus in alio. Respondeo: Aristotelem ibi convincere formam esse naturam eodem argumento quo antiqui probabant esse naturam materiam, ut ipse Aristoteles docet, t. 7. Et haec de iis quae in titulo proposuimus. . . .

NOTES

[1] Aristotle's Concept of Nature: Avicenna and Aquinas,' a paper read at the Twelfth Annual Conference, Center for Medieval and Early Renaissance Studies, State University of New York at Binghamton, 1976; devoted to the theme "Nature in the Middle Ages."

2 *Opere* 1:24, 35, 144–145; English translation in Wallace (1977a), pp. 38, 54, 208, and 210.
3 Dominicus Sotus, *Super octo libros physicorum Aristotelis quaestiones*, Salamanca: Andrea a Portonariis, 1555, fols. 29v–33v. This is the second edition, which is more widely diffused than the first; there are no significant changes in this particular question, viz, Lib. 2, q. 1, Utrum diffinitio naturae sit bona.
4 *Ibid.*, fol. 31r; see also fols. 29v and 32v.
5 *Ibid.*, fol. 33v.
6 Benedictus Pererius, *De communibus omnium rerum naturalium principiis et affectionibus*, Rome: Franciscus Zanettus, 1576.
7 Pererius makes frequent reference to Soto's commentary throughout his work; in his treatment of nature he mentions Avicenna on pp. 249 and 253.
8 Lib. 7, De forma ut est natura, pp. 242–270.
9 Pp. 249D, 251A, and 253B.
10 P. 262B.
11 P. 263B–C; see also p. 260B. Pererius probably treated the mover of the elements in his lectures on the *De caelo*, but no section entitled *De elementis* survives in the *reportationes* of these lectures preserved in the Oesterreichische Nationalbibliothek, Cod. Vindobon, 10509. However, he does have a disputation *De elementis* in his course on the *De generatione*, Cod. Vindobon. 10470, fols. 167–173, and this includes a discussion of their motive qualities, but not how these influence the local motion of bodies. Other manuscripts in which such a teaching might be contained are listed in C. H. Lohr, 'Renaissance Latin Aristotle Commentaries, Authors N-Ph,' *Renaissance Quarterly* **31** (1979), 532–603.
12 Galileo's answer to this question, *An mundus potuerit esse ab aeterno*, in Opere 1: 32–37, parallels very closely Pererius's treatment of the same question in *De communibus*, pp. 505–512, and could have been copied from it; see Essay 10 *supra*. Consult also *Galileo's Early Notebooks*, pp. 49–57, and the commentary on pp. 260–262. Galileo apparently extracted also a portion of Pererius's treatment of the cause of acceleration in free fall; see *Opere* 1:318, 411.
13 Additional details are given in Essays 9, 11, 12, 14, and 15.
14 Muzio Vitelleschi taught logic at the Collegio Romani in 1588–1589, natural philosophy in 1589–1590, and metaphysics in 1590–1591; he was also prefect of studies in 1605–1606, and later served as general of the entire Jesuit order. See R. G. Villoslada, *Storia del Collegio Romano dal suo inizio (1551) alla soppressione della Compagnia di Gesù (1773)*, Analecta Gregoriana Vol. LXVI, Rome: Gregorian University Press, 1954.
15 These are preserved in the Staatsbibliothek Bamberg, Cod. 70 (H.J.VI.21), which also contains Vitelleschi's lectures on the *De caelo*, both given at Rome in 1589–1590.
16 This is but one of many disputations included in the lectures, which expose the text of Aristotle and then raise questions relating to its understanding; see Essay 7 *supra*. The titles of disputations relating to the subject of nature include the following:

De definitione naturae

An in definitione naturae contineatur causa aliqua universalis vel accidentalis vel efficiens

An secundum Aristotelem natura sit solum principium passivum, an solum activum, an utrumque

Quinam motus et respectu cuius principii sint a natura
An possit demonstrari naturam esse
An materia vel forma vel utraque sit natura
Cuinam formae conveniat ratio naturae

[17] For the Latin text, see the Appendix to this article which contains a transcription of most of the second disputation listed in the previous note, made by the author from the Bamberg manuscript, fols. 112r–114r. Paragraph numbers have been inserted in square brackets for purposes of reference; these will be cited in what follows.

[18] A Latin translation of Philoponus's commentary on the *Physics* was published at Venice in 1539, and another translation in 1554. For a detailed analysis of Philoponus's analysis of falling motion, based on the Greek text, see Michael Wolff, *Fallgesetz und Massebegriff*: Zwei wissenschaftshistorische Untersuchungen zur Kosmologie des Johannes Philoponus. Quellen und Studien zur Philosophie, 2. Berlin: Walter de Gruyter & Co., 1971.

[19] Natura est quaedam vita sive vis quae per corpora diffunditur, eorum formatrix et gubernatrix, principium motus et quietis in eo cui inest per se primo et non secundum accidens – Aristoteles, *Physicorum libri quatuor*, cum Ioannis Grammatici cognomento Philoponi commentariis, quos . . . restituit Ioannes Baptista Rosarius, Venice: Hieronymus Scotus, 1558, p. 67, col. b.

[20] In the broader context of the role of causes in scientific explanation generally, as the author has stressed in his works on that subject (1972, 1974a), much harm has been done by philosophers who focus on efficient causation to the exclusion of other types of causal explanation. The history of science is replete with explanations made through final, formal, and material causes, most of which have proved easier to discover than those through efficient agents, particularly when the latter are conceptualized as forces, occult or otherwise. It is to Galileo's credit that he did not show such a narrow preoccupation, but based his *nuova scienza* on principles broad enough to include even those of Aristotelian natural philosophy.

PART FOUR

FROM MEDIEVAL TO EARLY MODERN SCIENCE

14. PIERRE DUHEM: GALILEO AND THE SCIENCE OF MOTION

The pioneers of the history of science movement, with an important exception, were not much interested in the Middle Ages. One can read George Sarton, William Dampier, and Alexandre Koyré, to name but three, and not generate any great excitement for the medieval period.[1] Such historians may not have referred to these as the "Dark Ages," but one gets the impression from them that they were pretty murky times – an uninteresting interlude between the close of classical antiquity and the rebirth of learning at the end of the Renaissance,[2] which would lead directly to the seventeenth century and its "Scientific Revolution" so graphically portrayed by Herbert Butterfield.[3] The important exception was Pierre Duhem, who first wrote his fascinating essay on '*Sozein ta phainomena*: An Essay on the Notion of Physical Theory from Plato to Galileo,' then produced his three-volume study on Leonardo da Vinci, subtitled 'Those Whom He Read, and Those Who Read Him,' and finally his monumental ten volumes on *Le Système du monde*.[4] In all these works the Middle Ages got their fair share of attention. Moreover, in the course of their writing, Duhem, the scientist and the philosopher, became Duhem the medievalist, a man passionately in love with the Middle Ages. And out of this research Duhem was emboldened to propose a daring two-part thesis: (1) that the condemnations of 1277 marked the origin of modern science, the decisive break with Aristotle and the beginning of new, imaginative cosmologies to replace his;[5] and (2) that the fourteenth-century development following the condemnations gave birth to important new concepts, such as impetus and uniformly difform motion, whose proponents, the *Doctores Parisienses*, were the precursors of Galileo.[6] This is sometimes referred to as "Duhem's continuity thesis"; it advocates a bond of continuity between medieval and modern science, but in such a way as to invite critical appraisal and revision in what follows.

To put the matter somewhat differently, historians of science who, following Duhem, have addressed the continuity of thought between the Middle Ages and the modern era usually speak of the transition from *late* medieval to early modern science.[7] Here the expression "late medieval science" designates that following 1277 and elaborated throughout the fourteenth and fifteenth centuries – the science of the Oxford Mertonians, the Paris terminists, and the Paduan Averroists, all regarded, initially through the efforts of Duhem and then through the emendations of Anneliese Maier, Ernest Moody, John Herman Randall, Jr., Marshall Clagett, and others, as the "medieval precursors of Galileo."[8] The revised thesis to be defended in this essay is that the *late* medieval period was not the only one that contributed to the rise of modern science; more important, perhaps, was what we shall refer to as "*high* medieval science," the science developed mainly by thirteenth-century thinkers, such as Robert Grosseteste at Oxford and Albertus Magnus, Thomas Aquinas, and Giles of Rome at Paris, all of whom did their work before the condemnations of 1277, or in essential independence of them.[9] Our contention will be that this earlier group was just as influential in the birth of Galileo's *nuova scienza* as was the later group; indeed, it was only when the ideas of both were put in juxtaposition that the elements were at hand for Galileo's genius to become operative. Our procedure will be to start with the corpus of Galileo's writings, with a special emphasis on his early notebooks, and to show how the ideas that were seminal in these and later guided Galileo's mature works derive in a fairly direct line from both "*high* medieval science" and "*late* medieval science." Thus we too are advancing the continuity thesis first proposed by Pierre Duhem, only enhancing it now to show a fuller dependence on medieval thought than has hitherto been proposed by historians of science.

That much said, there will be three parts to the presentation: (1) a brief topology of medieval science, explaining "high" and "late" science, and indicating somewhat schematically the role of the condemnations of 1277 in their differentiation; (2) a description of Galileo's early notebooks, their contents, the authorities cited in them, and their sources, and (3) an analysis of the ideas that were seminal for Galileo's more mature thought, to ascertain his debt to his predecessors, first to those of the sixteenth century, and then, working back through the three previous centuries, to those of the high medieval period.

I. TOPOLOGY OF MEDIEVAL SCIENCE

For our purposes medieval science may be divided into three chronological periods, corresponding to the generally accepted division of scholasticism into early, high, and late:[10]

(1) "*Early*" *medieval science*. This is the science of the twelfth century and before. During this period there was very little by way of accomplishment; it saw the recovery of some nature studies and of some Aristotle, but little else. From a methodological point of view the analysis of this material

requires a historiography unlike that used for later periods. Brian Stock's *Myth and Science in the Twelfth Century* (1972) is about the best anyone has done with it thus far, apart from encyclopedic accounts of individuals and their writings. We mention this period only to exclude it; it will not be of interest or of relevance to our general thesis.

(2) *"High" medieval science*. The span envisaged here covers the entire thirteenth century, extending perhaps into the first decade of the fourteenth, so as to include the researches of Theodoric of Freiberg on the rainbow. The distinctive characteristic of this period was the complete recovery of Aristotle, including the *Posterior Analytics*, the *Physics*, the *De caelo*, etc., with commentaries such as those of Simplicius and Averroës. Central to it was Grosseteste's rediscovery of the distinction between knowledge "of the fact" (*quia*) and knowledge "of the reasoned fact" (*propter quid*), and his recognition that mathematics can supply "reasoned" or *propter quid* explanations for many physical phenomena. The contributions were mainly in optics, but some thought was given to astronomy also. This period was pronouncedly realist in orientation; "science" was understood in the hard sense of *scientia*, true and certain knowledge of the physical universe that could not be otherwise, though requiring special methodological techniques for its attainment, such as Aquinas's refinement of the procedures for *ex suppositione* reasoning, already implicit in Aristotle. The most severe blow this period received was that of the condemnations of 1277. These were directed by the bishops of Paris and Oxford against the Averroists, and, incidentally, against Aquinas and Giles of Rome, but they had the general effect of weakening knowledge claims, especially in the area of Aristotelian natural philosophy. Their occasion was a growing rationalism in the arts faculties that seemed to menace traditional teachings of Catholic theology. Faced with this threat, the bishops in some ways anticipated Immanuel Kant, who, as is well known, later found it necessary to deny knowledge to make room for faith. The knowledge the bishops denied was effectively Aristotle's science of nature; they struck at its "true and certain" character, emphasized its fallibility, and so opened the door to the possibility of alternative explanations of the universe God had made.

(3) *"Late" medieval science*. This is the science of almost all the fourteenth century and continuing through the fifteenth, centered, as already indicated, at Oxford, Paris, and Padua. Throughout this period the characteristic movement was nominalism, itself a type of skepticism compatible with the condemnations, and its chief proponent was William of Ockham. The emphasis was still mathematical, but the focus shifted toward the logical,

toward *dubitabilia* and *sophismata*, especially at Oxford. The language of quantification grew, and calculatory techniques were devised for discussing all types of imaginary motions, involving infinites and even foreshadowing infinitesimals. Thomas Bradwardine, William Heytesbury, and Richard Swineshead laid the foundations for a new kinematics, but it all concerned points and imaginary bodies moving in empty space, with little or no concern for the world of nature. At Paris, Jean Buridan, Albert of Saxony, and Nicole Oresme, the so-called *Doctores Parisienses*, are usually known as nominalists, but they were actually more realist in their interests, at least in the sense that they applied the calculatory techniques of the Mertonians to some motions found in nature. They developed the concept of impetus, and sought examples of motions in the universe that were *uniformiter difformis* (one would now say "uniformly accelerated"), but they failed to identify this as the characteristic motion of freely falling bodies. Still intimidated by the condemnation of the bishop of Paris, they too shied away from strong knowledge claims; their science was not as hard-headed as that of Grosseteste, Albert, and Aquinas. Apparently they were surer of their logic than of their physics, and on this account are aligned – in standard histories of philosophy – more with Ockham than with the realists of the thirteenth century (Gilson, 1955, pp. 511–520). Finally, at Padua, in the early fifteenth century Paul of Venice brought the *calculationes* back from Oxford and Paris, and trained his students in their use. Generally the northern Italians combined the realist interests of the high scholastic period with nominalist methods, although the nominalism was inhibited somewhat by the strong attraction of Averroës. A representative thinker was Gaetano da Thiene, who studied under Paul of Venice. There were many others, however, who developed techniques proposed by Aristotle in the *Posterior Analytics* to put the science of nature on firmer ground. Isolated from, and unaffected by, the condemnations that had been in effect at Oxford and Paris, they could still maintain the ideal of *scientia*, and in this sense, at least, were in essential continuity with the high medieval period.

So much, then, for a quick topology of medieval science.

II. GALILEO'S EARLY NOTEBOOKS

Let us turn now to the questions of how much Galileo knew of these contributions and how they may have influenced his own scientific development. If we subscribe to William Shea's characterization of Galileo's intellectual life, we may divide this into an early, a middle, and a late period (1972, pp. vii–viii). The first extends from about 1580 to 1610, and embraces Galileo's career as a student and later a teacher at Pisa, plus his professorship at Padua. The middle period extends from 1610 to 1632, and was centered mainly in Florence and Rome, where Galileo was embroiled in his long disputes over the truth of the Copernican system, leading to his trial and condemnation after the publication in 1632 of the *Dialogue on the Two Chief World Systems*. The third period, finally, comprises the last ten years of his life, from the condemnation to 1642, spent largely in Arcetri, outside Florence, where he concentrated on mechanics, wrote his *Discourses on the Two New Sciences*, and thereby laid the foundation for what is now known as "modern science." Recent research on Galileo has shown that the long middle period devoted principally to astronomy represents something of a detour in the genesis of the "new science."[11] The early and late periods, on the other hand, are seen now as in essential continuity; in fact, evidence is available to show that practically all of the experimental and theoretical work necessary for the writing of the *Two New Sciences* had already been done by 1609, at which time Galileo perfected the telescope and set out to prove the truth of the heliocentric system.[12] In light of this new information Galileo's early period, which has hitherto been somewhat neglected, is now being studied with care as of seminal importance for his later thought. The researches reported in this essay have concentrated on the first part of that period, the time Galileo spent in Pisa preparatory to his professorship at Padua, when it apparently all began.

While at Pisa, and probably in conjunction with his teaching there or in immediate preparation for it, Galileo wrote, in Latin, three sets of notes, or, to simplify somewhat, three notebooks.[13] These notes are scholastic in inspiration, and they treat questions or difficulties raised by the text of Aristotle's *Posterior Analytics, Physics, De caelo*, and *De generatione*. The first set, now contained in MS 27 of the Galileo Collection in Florence, may be referred to as the logical questions; it contains treatises on demonstration and on the foreknowledge required for its attainment.[14] The second set, that of MS 46, includes the physical questions; it has lengthy disputations on the universe, the heavens, qualitative changes, and the elements.[15] Finally, the third set, that of MS 71, contains drafts of several essays, including a dialogue, on the subject of motion.[16] Only the latter notebook has received any attention from scholars, and this because of its obvious relation to the tract on motion in the *Two New Sciences*; all agree that it was composed "around 1590," when Galileo was already teaching mathematics at Pisa.[17] The logical and physical questions, on the other hand, have not been taken seriously, possibly because of their more pronounced scholastic style. Antonio Favaro, the editor of the National Edition of Galileo's works, labeled them *Juvenilia*, or youthful writings, and saw the physical questions as dating from 1584, when Galileo was still a medical student at Pisa. Favaro further accorded to Galileo only the role of an *amanuensis* in their composition, suggesting that they were copied from the lectures of one of his professors at Pisa, Francesco Buonamici (*Opere* 1: 12). Now, in the essays of this volume contained in Part III and in *Galileo's Early Notebooks*, we question Favaro's use of the label *Juvenilia* and the dating on which it is based. New evidence has been offered, moreover, to show that all of these notes derive from roughly the same sources and that they probably date in their entirety from the same period. It is also noteworthy that the present numbering of the manuscripts, i.e., 27, 46, and 71, follows Favaro's system of cataloguing, which is by subject matter rather than by chronology. In the original collection of Galileo's manuscripts, made by his student Vincenzio Viviani, all three notebooks were put in successive volumes;[18] again, the curator had clearly written on the second notebook that this was Galileo's examination of Aristotle's *De caelo* made "around 1590" (*Opere* 1: 9). If such be true, then all the notes would represent the work of Galileo's head as well as his hand, either while he was a young professor at Pisa, 1589–1591, or at the earliest while he was aspiring to a teaching position in the year or so immediately preceding.

Galileo's citations of authorities in these notebooks, concerning which further details are given in Essay 10 *supra*, reveal his rather extensive debt to authors of classical antiquity, the patristic period, the Middle Ages, and the Renaissance. Among classical Greek authors, in the three notebooks Galileo has over 200 references to Aristotle, many favorable but a goodly number critical of his thought, and scores to Aristotle's Greek commentators, including 31 citations of Simplicius, 20 of John Philoponus, and 19 each of Alexander of Aphrodisias and Themistius. He further cites Plato 23 times, Ptolemy 20 times, Galen 15 times, and Archimedes and Hipparchus 6 times each. There are 31 citations of the Church Fathers, with St. Augustine being referenced most frequently, 8 times, St. Basil 4 times, Gregory of Nyssa 3 times, and Ambrose, Bede, Chrysostom, Damascene, and Jerome twice each. Among medieval authors there are 65 references to Averroës, 14 to Avicenna, and 3 to Avempace, to name but three Arabs; Robert Grosseteste is cited only once, from his commentary on the *Posterior Analytics*; but Aquinas is cited 54 times, Albert the Great 21 times, Giles of Rome 14 times, Scotus 11 times, Ockham 6, and so on. Finally, there is a very extensive citation of Renaissance authors, including many Aristotelians teaching at the universities of northern Italy, some Neoplatonists, and a goodly number of scholastic commentators in the traditions of the various Schools.

A study of such citations first alerted us to the fact that Galileo, in his use of medieval authors, relied much more heavily on those of the high scholastic period than on those of the late period, itself a reason for suspecting the accuracy of the Duhem thesis. In Essay 9 we have called attention to Galileo's particularly good knowledge of the Thomistic tradition; he made intelligent reference, for example, to the writings of Capreolus, Cajetan, Soncinas, Ferrariensis, Hervaeus Natalis, Domingo de Soto, and Chrysostom Javelli. When one totals all of these, there are about 80 references to Aquinas and the Thomistic school, which would make these authors the most frequently cited after Aristotle himself. Now that is a surprising and somewhat unexpected result, for if these notebooks have any bearing on Galileo's intellectual development, and particularly if they contain materials that are in continuity with the writings of Galileo's later life, one would have to admit that the high medieval period was by no means a factor that can be neglected when assessing Galileo's debt to ages past.

One may well wonder about the source of all these citations, and whether Galileo himself had a first-hand knowledge of the authors he cites – over 150 in all. A partial answer to this question is suggested by the circumstance that the earlier of the notebooks, those containing the logical and the physical questions, show considerable evidence of copying, or at least of being based on other sources (*Opere* 1: 9–13; 9: 279–282). This indication notwithstanding, it has proved remarkably difficult to identify the books or manuscripts Galileo might have used in their composition. Owing to the researches of the author and others, however, we now have the answer to the puzzle. Practically all of this material, with its very erudite citation of authors, derives from textbooks and lecture notes that were being used at the Collegio Romano, the prestigious Jesuit university in Rome, from the 1570's through the 1590's. A full documentation of this statement with respect to the physical questions will be found in *Galileo's Early Notebooks*. In its light, of course, it becomes a simple matter to understand the broad citation of writers, the humanist knowledge of classical antiquity, and especially the detailed scholarship relating to the patristic, medieval, and Renaissance periods. Galileo may have studied at Pisa, but in no sense do his notebooks derive from the meager philosphical instruction available at that university. No, when writing his notes, he went to the most scholarly sources he could find, to some of the most learned professors of his day, and from them derived his understanding of Aristotle's *Organon* and his natural philosophy.

III. GALILEO'S DEBT TO HIS PREDECESSORS

Our previous essays discussing Galileo's use of these Jesuit sources have concentrated on topics in logic and in Aristotle's *De caelo* and *De generatione* – the matter of the first two notebooks – to show their continuity with the high medieval tradition. The focus of interest in this essay will be topics in the third notebook, that relating to motion, to make essentially the same point. Now the standard account of the origin of this third set of notes, sometimes called the *De motu antiquiora* to distinguish it from the treatise on motion in the *Two New Sciences*, is that they were prompted by Galileo's interest in mathematics, and were not associated in any way with his pursuit of natural philosophy (Geymonat, 1965, pp. 5–16). Archimedes has been thought to be their inspiration more than Aristotle, and Galileo is said to have become interested in Archimedes' work through the mathematician Ostilio Ricci, who had studied under Niccolò Tartaglia and is known to have privately tutored Galileo in mathematics. The concepts that characterize the third set of notes, moreover, are not those to be found in the third book of Aristotle's *Physics*, where a philosophical account of motion is given in terms of potency and act. Galileo never discusses the Aristotelian definition, but prefers rather to discourse on the nature of heavy and light; the role of the medium in falling motion; the way in which rate of fall is determined by the buoyancy of the medium, and so is better related to the specific gravity of the falling object than to its absolute weight; the possibility of there being "neutral" motions, i.e., neither toward nor away from a center of gravity; what it is that causes the motion of projectiles after they have left the hand of the thrower, with a detailed explanation of a theory of impetus; the way in which impetus itself can be used to explain changes in the speed of fall of bodies; and so on. Galileo's discussion of all these topics is unlike that found in books then available in northern Italy, and since his citation of sources in the third notebook is much sparser than in the previous two, scholars have rightly been puzzled as to the sources from which this composition derives. Girolamo Borro, who taught at Pisa while Galileo was a student there, and who is cited in this notebook, and also Francesco Buonamici and Giovanni Battista Benedetti, who are not cited but who touch on some of these matters in their treatises, have been mentioned as possible sources, but the details are vague and the problem is generally regarded as unsolved.[19]

It is our contention that the key concepts to be found in the third notebook derive, like those in the first two, from *reportationes* of lectures given by Jesuit professors at the Collegio Romano. These professors, although teaching the course in natural philosophy, were more sympathetic to mathematical approaches to physical problems than were the Aristotelians at Padua and elsewhere. Their openness in this regard is probably traceable to the influence of Christopher Clavius, also a Jesuit at the Collegio and the pre-eminent mathematician of the sixteenth century, who had stressed the importance of training in mathematics for anyone who would be proficient in the physical sciences.[20] Most of this mathematics, it is true, focused on Euclid's *Elements* and on the geometrical astronomy contained in Clavius's own commentary on the *Sphere* of Sacrobosco. But the newer types of mathematical reasoning introduced by the fourteenth-century *calculatores* for dealing with problems of motion – and here we have in mind the Mertonians, the Parisian *doctores*, and the fifteenth-century Paduans – were not unknown to Clavius and his fellow Jesuits. Rather than treat these problems in the commentary on the third book of the *Physics*, however, as was the practice in the early part of the sixteenth century at Paris, they reserved such treatises for the latter part of the course in natural philosophy, when they had been through the *Physics* and had already covered the doctrine of the elements in the *De caelo* and the *De generatione*. The motion of heavy and light bodies, *De motu gravium et levium*, was thus delayed until the students had studied the motions of the heavens and had a general qualitative knowledge of the elements and their properties. When one searches in the latter part of the course in natural philosophy given at the Collegio, therefore, one finds precisely the topics discussed by Galileo. To date no evidence of direct copying on Galileo's part has been found,[21] but the similarity is there, and it is no difficult matter to trace the few changes Galileo could have made to produce the *De motu antiquiora*, and then proceed from this on to the more systematic account he was to give in the *Two New Sciences*.

In what follows we shall discuss several of the topics treated in Galileo's third notebook, as these are found in *reportationes* given by three Jesuits, by name Paulus Valla, Mutius Vitelleschi, and Ludovicus Rugerius. All of these men taught "around 1590," and upon their courses much of the material contained in Galileo's first two notebooks is quite clearly based.[22]

1. The key concept of the *De motu antiquiora* is what later generations would refer to as "specific gravity." Galileo does not use this expression, but speaks rather of *gravitas propria* (*Opere* 1: 251), which has been translated into English as "essential heaviness."[23] It is this concept that ties Galileo's analysis to Archimedes, and that permits him to introduce hydrostatic considerations into the discussion of bodies falling through a medium, where their effective weight will be determined by the medium's buoyancy. Now in

a related text, when treating of weight or *gravitas*, Valla, Vitelleschi, and Rugerius all make explicit reference to Archimedes.[24] Vitelleschi has the fullest discussion, cites Archimedes' treatment of both weights (*De ponderibus*) and bodies that float in water (*De iis quae aqua invehuntur*), and makes a distinction similar to Galileo's and on which the latter's could have been based. Vitelleschi notes that different weights will produce different velocities of fall, but that one must be careful as to how he reckons gravity in order to compute this. In his mind, one gravity can be greater than another in two ways, either intensively or extensively. Gravity considered *intensive* respects the intensity or degree of weight proper to the material, whereas gravity considered *extensive* takes into account the size of the body and the number of its parts. To calculate velocity, he says, one must consider gravity both *intensive* and *extensive*, because though the intensity is the primary determiner of the motion, the extent of the body also affects the resistance it encounters from the medium and so limits its rate of fall.[25] In this context Vitelleschi mentions Bradwardine's treatise *De proportione motuum* and also Jean Taisnier's work on the same subject, which gives all the mathematics necessary for calculation.[26] And Taisnier's book, as is well known, was plagiarized from Benedetti's treatise on the ratios of motion, which, as has already been remarked, bears a strong similarity to Galileo's own account (Drake & Drabkin, 1969, p. 402).

2. A related topic on which Galileo discourses at length is the cause of speed in natural motions, i.e., what determines how fast a body falls. Consistent with his discussion of specific gravities, he assigns the cause to the weight of the medium as well as to the weight of the falling body, and exemplifies this with the classical case of the bladder filled with air, whose motion varies in different media (*Opere* 1: 260–273). The same cases are discussed by Vitelleschi, and analogous conclusions drawn. Paralleling the criticisms contained in Taisnier's (actually Benedetti's) treatise, which rejects the rules of speed formulated by Aristotle in the fourth and seventh books of his *Physics*, Vitelleschi likewise questions their validity.[27] Galileo's rejection of them has frequently been seen as an innovation on his part, whereas these were far from being accepted dogma, save among the more conservative Averroist Paduans, who, unlike the Jesuits and other scholastics, searched for their philosophy in the text of Aristotle and there alone. Rugerius also rejects Aristotle's rules of speed and refers the reader to the elaborate treatise on this subject written by Domingo de Soto and further expanded by Franciscus Toletus, who had studied under Soto at Salamanca.[28] And Soto, as we have documented in Essay 6 *supra*, was the author who thoroughly understood the

uniformiter difformis doctrine of the *Calculatores*, and who was the first to apply it to the case of free fall.

3. A third point Galileo makes, which again puts him in opposition to Aristotle, is his contention that air cannot have weight in its natural place, or, put otherwise, that air has no weight when itself surrounded by air (*Opere* 1: 285–289). Such a conclusion clearly follows from the application of Archimedian principles, but not from the texts of Aristotle found in the *De caelo*. It is noteworthy that Valla, Vitelleschi, and Rugerius all discuss these texts, as well as the many opinions of commentators, the experiments that have been adduced for and against Aristotle's position, and the arguments based on Archimedes.[29] Their conclusions, again, are quite consistent with the teachings contained in Galileo's third notebook. Both Vitelleschi and Rugerius, moreover, cite the account of Girolamo Borro in his *De motu gravium et levium*, which describes experiments with falling bodies similar to those apparently performed by Galileo, and thus could be the source of the latter's mention of Borro in the same notebook.

4. Another topic taken up by Galileo that has proved particularly difficult to locate in previous writers is his mention of a third or intermediate type of local motion that is neither natural nor violent – the only two types hitherto allowed by Aristotle and the peripatetics (*Opere* 1: 304–307). This is also commonly thought to have been Galileo's innovation, found first in his early treatises on motion, and then later developed by him into the concept of circular inertia. Now it may be of interest that in the second notebook, the one containing the physical questions, Galileo has already adumbrated that concept, a fact completely overlooked by scholars. The topic comes up in his treatise on the heavens, when he is inquiring whether the heavens are made of a separate element and whether they are really incorruptible.[30] Fire, according to the accepted doctrine of the time, was the lightest earthly element, and thus the one whose natural place would be closest to the lowest of the heavens, i.e., the orb of the moon. Galileo observes that when flame rises to the top of a furnace, its motion there becomes circular, and on this basis the motion of the fire in contact with the orb of the moon would also appear to be circular, that is, neither up nor down, but remaining always at the same distance from the center of the earth. The question he proposes is whether such circular motion would be natural for fire in these circumstances, and his answer is in the negative. Then later, returning to the same phenomenon, he inquires whether this kind of motion could be called violent, and that too he answers in the negative.[31] Now if a particular motion occurring in the universe is neither natural nor violent, then in some sense it has

to be "neutral," or intermediate between the two. As might be expected, these portions of Galileo's treatise on the heavens have counterparts in lectures given at the Collegio Romano.[32] Moreover, when one searches through the *reportationes* deriving from Vitelleschi and Rugerius, one finds in them explicit discussions of motions that are intermediate between the natural and the violent.[33] Vitelleschi explains this in a way that is quite consonant with Galileo's later development. A motion is violent, he says, if it is imposed from without and the object acted upon contributes no force at all. But for a motion to be violent in the strict sense, he continues, the action from without must be opposed to the natural inclination of the object. Should the body be acted upon in a way that does not oppose the body's natural inclination, then the resulting motion is actually intermediate between the natural and the violent, and so may be regarded as neutral, i.e., as beyond nature but neither according to it nor contrary to it. Vitelleschi goes on to observe that the generally accepted statement that no unnatural motion can be perpetual must be understood of the violent only in the strict sense, for it is this type that takes away the force of nature and so depletes the body's source of motion.[34] Implicit in this statement, of course, is the admission that a motion such as the circular motion of fire around the earth's center, not being opposed to fire's natural inclination, could go on forever – itself an adumbration of circular inertia.

5. Another characteristic of Galileo's third notebook is the use there of the medieval concept of impetus; he regards impetus as the agent in projectile motion and maintains that it continually weakens after the projectile motion has begun (*Opere* 1: 307–315). Such a theory of "self-expending impetus," as it is called, is usually traced to Buonamici, who is thought to have been the source of Galileo's teaching (Koyré, 1939, pp. 18–41). A careful check of Buonamici's *De motu*, however, shows that, although he gives an account of impetus theory, he himself rejects it;[35] thus it is difficult to see how Buonamici could be the source of its adoption by Galileo. Vitelleschi, on the other hand, is more open to impetus theory, and Valla, his immediate predecessor in the chair of natural philosophy at the Collegio, has a thorough explanation and justification of precisely the theory Galileo adopts.[36] Thus again the Jesuits appear as a likely key to the doctrines contained in the *De motu antiquiora*.

6. As a final point, we could cite Galileo's celebrated explanation there as to why a body accelerates during the course of falling motion (*Opere* 1: 315–323). Some years ago, in a much-cited article entitled 'Galileo and Avempace,' Ernest Moody traced Galileo's explanation to the teaching of the

Arab philosopher Avempace, who held that the velocity of any motion results from an excess of the motive force over the resistance it encounters, and so the velocity of fall increases as the resistance grows less.[37] Galileo's explanation employs this principle, but the mechanism he adopts is actually that of the Greek thinker Hipparchus, who conceived of the resistance that must be overcome as a residual lightness or impetus previously impressed on the body, which is gradually dissipated during the fall (*Opere* 1: 319–320). Now, as already remarked in Essay 7, Vitelleschi formulates the same principle also, mentions the teachings of both Avempace and Hipparchus, and gives a more sophisticated explanation than Galileo for the mechanisms involved in acceleration. Far from this being a novel contribution, this too is a matter about which the Jesuits were speculating "around 1590," which stands in rather clear relation to the central theses of Galileo's third notebook.

IV. "HIGH" VS. "LATE" MEDIEVAL SCIENCE

It is time now to take account of the significance of these new findings and how they relate to Duhem's two-pronged thesis about the continuity between medieval and early modern thought. Obviously it is not our intention to negate Duhem's conclusions entirely, for it should be clear from what has been said that Galileo remains in considerable debt to late medieval thought as well as to that of the High Middle Ages. Our point is that even late medieval thought was transmitted to Galileo by sixteenth-century writers, who themselves modified the distinctive theses of their immediate predecessors, incorporated elements from high scholasticism as well as from a mathematical tradition that was beginning to emerge during that period, and so contributed substantially to the new synthesis Galileo himself was to elaborate in his later writings. A fuller account would thus see Galileo's ideas originating in a progressive, somewhat eclectic, scholastic Aristotelianism, otherwise quite Thomistic, which, unlike the Aristotelianism in the Italian universities under Averroist influences, was sufficiently open-ended to incorporate the techniques of the *Calculatores* and the Archimedian ideal of physico-mathematical reasoning applied to the world of nature.

To spell out more clearly how the Duhem continuity thesis might be revised on this basis, we must refer back to its two parts as set out at the beginning of this study. The first part saw the condemnations of 1277 as marking the birth-date of modern science. This has been roundly criticized by intellectual historians, mainly because condemnations, being themselves a repressive tactic, were hardly compatible with the spirit of free inquiry that should characterize scientific investigation, and resulted in no recognizable spurt in scientific activity after they were pronounced (Grant, 1971, pp. 34–35, 83–90). Such criticism notwithstanding, however, Duhem still had a valuable point to make. This point is that Aristotelianism, particularly when conceived as defending the text of Aristotle and that alone, did constitute an impediment to scientific inquiry. Aristotle had to be approached with a "critical temper," to use John Murdoch's expression,[38] if progress was to be made in the true understanding of nature. But such a critical temper did not have to await the edict of Etienne Tempier for its genesis. It was already present in the Christian philosophers of the High Middle Ages, in Grosseteste and Albert and Aquinas and Giles of Rome, who were far from being disposed to seeing Aristotle as a god.[39] The same mentality, quite obviously, is apparent in the *reportationes* of the philosophers of the Collegio Romano. They were teaching natural philosophy in its entirely, but they were also preparing students to be theologians, and they could not afford to be uncritical in evaluating Aristotelian doctrine. The same unfortunately cannot be said of the Averroists at Padua and elsewhere, so much extolled by John Herman Randall, Jr. (1961), for their role in the genesis of Galileo's methodology. No, it was these secular professors, not the clerics, who turned out to be the slaves to Aristotle. They were the ones who refused to look through Galileo's telescope, against whom he could direct the barbed criticism that *he* was the true heir of Aristotle, and not those who called themselves by his name (Geymonat, 1965, pp. 192–197).

The second part of Duhem's thesis we would also revise, and this because of its excessive commitment to nominalism as the only philosophy consistent with the scientific enterprise. As is well known, Duhem was a positivist.[40] His ideal of scientific explanation was merely "to save the appearances," and this is all he felt his nominalist heroes were concerned to do. Whatever one wishes to say about that evaluation of fourteenth-century physicists, and this subject would bear much fuller investigation, there can be no doubt that such was never the mentality of Galileo in his early, his middle, or his late periods. As a philosopher Galileo was a realist through and through. He knew that the Ptolemaic system could "save the appearances," and yet he fought to go beyond this to ascertain the actual structure of the universe. He knew that Aristotelian "rules of speed" could give a rough phenomenological account of falling motion, but he wanted a "new science" of motion, one that could demonstrate properties of motions that are found in nature. Whenever Galileo spoke of science and demonstration, as we have indicated in the previous essays, his was the hard-headed notion of *scientia* found in thirteenth-century writers, not the weakened account that is attributed to those of the fourteenth and fifteenth centuries.

If we wish, therefore, to do justice to the orientation Galileo received at the outset of his intellectual life, we do well to credit properly the scholastic Aristotelians on whose thought his three notebooks are obviously based. When we do so, we can readily discern a continuity between early modern science and that of the High Middle Ages. Galileo stands in debt not only to the fourteenth and fifteenth centuries, but to the thirteenth century as well.[41] He took his beginnings from sixteenth-century writings that were progressive and open-ended, that were knowledgable in the extreme, and that incorporated the best to be found in the three preceding centuries. To neglect this fact is to fail to understand what Galileo's writings, both early and late, are all about, and how he could make the contributions that merit for him the title, "Father of Modern Science."

NOTES

1 Sarton's evaluation of St. Thomas Aquinas, for example, is revealing in this regard. He states of Aquinas: "Though interested in science, he utterly failed to understand its true spirit and methods, and no scientific contribution can be credited to him. Indeed his mind was far too dogmatic to be capable of disinterested scientific curiosity." – *Introduction to the History of Science* (Baltimore 1931) II/2, p. 914. Such a statement obviously says more about Sarton that it does about Aquinas. For Dampier's view see the following note. Koyré's attitude is sketched accurately by Edward Grant in his *Physical Science in the Middle Ages* (New York 1971), p. 114, where he gives references that substantiate Koyré's belief in an "essential discontinuity between medieval physical science and the achievements of Galileo and the scientific revolution of the seventeenth century."

2 As Dampier puts it, "To us [historians of science], then, the Middle Ages have their old significance – the thousand years that passed between the fall of the ancient learning and the rise of that of the Renaissance: the dark Valley across which mankind, after descending from the heights of Greek thought and Roman dominion, had to struggle towards the upward slopes of modern knowledge. In religion, and in social and political structure, we are still akin to the Middle Ages from which we have so recently emerged; but in science we are nearer to the ancient world. As we look back across the mist-filled hollow, we see the hills behind more clearly than the nearer intervening ground." – *A History of Science* (4th ed. Cambridge 1968), pp. 60–61.

3 *The Origins of Modern Science* (rev. ed. New York 1965).

4 The essay appeared in French in 1908 in the *Annales de philosophie chrétienne*; it has been translated into English by E. Doland and C. Maschler with the title, *To Save the Phenomena* (Chicago 1969). The other volumes remain in their original language: *Etudes sur Léonard de Vinci* (3 vols. Paris 1906–1913) and *Le Système du monde* (10 vols. Paris 1913–1959).

[5] See Vol. 6 of *Le Système du monde*, which bears the title, 'Le Reflux de l'Aristotélisme: Les condemnations de 1277.'
[6] See Vol. 3 of *Études sur Léonard de Vinci*, titled 'Les Précurseurs parisiens de Galilée.'
[7] Anneliese Maier, for example, entitled her five-volume series pursuing and revising Duhem's theme *Studien zur Naturphilosophie der Spätscholastik* (Rome 1949–1958).
[8] For Maier, see the previous note; Ernest Moody, *Studies in Medieval Philosophy, Science, and Logic*: Collected Papers 1933–1969 (Berkeley 1975); J. H. Randall, Jr., *The School of Padua and the Emergence of Modern Science* (Padua 1961); Marshall Clagett, *The Science of Mechanics in the Middle Ages* (Madison 1959); also the writings of Clagett's students, John Murdoch and Edward Grant.
[9] A. C. Crombie has already supplied considerable background evidence in support of this theme; see his *Medieval and Early Modern Science* (2 vols. New York 1959) and *Robert Grosseteste and the Origins of Experimental Science* (Oxford 1953).
[10] This general parallelism is indicated in the opening essay of this volume, to which the reader should refer for fuller details.
[11] For a full documentation, see the monograph by W. L. Wisan, 'The New Science of Motion: A Study of Galileo's *De motu locali*,' *Archive for History of Exact Sciences* **13** (1974), pp. 103–306.
[12] Some of this evidence is set out in the author's 'Three Classics of Science: Galileo, *Two New Sciences*, etc.,' *The Great Ideas Today* (Chicago 1974), pp. 211–272.
[13] These are not notebooks in the sense of *quaderni*, but more like looseleaf notes relating to a distinct subject matter that were subsequently bound together by others; thus the ordering of the folios is not necessarily chronological.
[14] A description of these is given by A. C. Crombie, 'Sources of Galileo's Early Natural Philosophy,' in M. L. Righini-Bonelli and W. R. Shea, eds., *Reason, Experiment, and Mysticism in the Scientific Revolution* (New York 1975), pp. 157–175.
[15] These have been translated into English and commented on by the author in his *Galileo's Early Notebooks: The Physical Questions* (Notre Dame 1977).
[16] These materials are described by Raymond Fredette, 'Galileo's *De motu antiquiora*,' *Physis* **14** (1972), pp. 321–348.
[17] Substantial portions of this notebook have been translated into English in Galileo Galilei, *On Motion* and *On Mechanics*, trs. & eds. I. E. Drabkin and Stillman Drake (Madison 1960); here the 1590 date is assigned to its composition. Additional portions are to be found in *Mechanics in Sixteenth-Century Italy*, trs. Stillman Drake and I. E. Drabkin (Madison 1969).
[18] We are indebted to Dr. Fredette for his information; more details are contained in his unpublished paper, 'Bringing to Light the Order of Composition of Galileo Galilei's *De motu antiquiora*,' as well as in the paper cited in note 16.
[19] The treatises referred to are Girolamo Borro, *De motu gravium et levium* (Florence 1576); Francesco Buonamici, *De motu libri decem* (Florence 1591); and Giovanni Battista Benedetti, *Demonstratio proportionum motuum localium* (Venice 1554).
[20] Some details are given in Essay 10 *supra*; for a fuller account see Giuseppe Cosentino, 'L'Insegnamento delle Matematiche nei Collegi Gesuitici nell'Italia settentrionale,' *Physis* **13** (1971), pp. 205–217, and 'Le mathematiche nella 'Ratio Studiorum' della Compagnia di Gesù,' *Miscellanea Storica Ligure* **2** (1970), pp. 171–213.
[21] Many of the topics touched on by Galileo in his Memoranda on Motion (*Opere* 1: 409–417), translated in *Mechanics in Sixteenth-Century Italy*, pp. 378–387, are very

much like those in the Jesuit *reportationes*, and could have been culled from these or similar sources.

22 For particulars on these three authors see *Galileo's Early Notebooks*, pp. 17–20.

23 By Drabkin, Galileo's *On Motion*, p. 13.

24 Valla, *Tractatus quintus de elementis*, disputatio secunda, quaestio ultima, conclusio secunda, Archivum Pontificiae Universitatis Gregorianae, Fondo Curia (henceforth APUG/FC), Cod. 1710, no foliation; Vitelleschi, *Lectiones in octo libros Physicorum et quatuor De caelo*, Staatsbibliothek Bamberg, Cod. 70. H. J. VI. 21, fol. 370r–371r; Rugerius, *Quaestiones in quatuor libros Aristotelis De caelo*, SB Cod. 62. H. J. VI. 9, p. 179. Some additional details are given in the Appendix to Essay 10, *supra*.

25 Cod. Bamberg. SB 70 fol. 363v–366r. See also Essay 7, *supra*.

26 *Ibid*. fol. 365r. Vitelleschi simply states: "Legendum est Thomas Bradwardinus in sua tractatione de proportione motuum, et Ioannes Thaisnerus in tractatione de eadem rei . . . " The latter treatise, according to Drake and Drabkin (1969, p. 402), is contained in Taisnier's *Opusculum perpetua memoria dignissimum* (Cologne 1562).

27 fol. 365v.

28 Cod. Bamberg. SB 62, pp. 215–216.

29 See note 24: Valla's discussion is in his last question (no foliation), whereas Vitelleschi's is on folios 369r–373r and Rugerius's on pp. 178–183.

30 *Galileo's Early Notebooks*, pp. 81–102, especially pars. I2, I7, and I37–38.

31 *Ibid*., p. 97 par. J16.

32 See the commentary on questions I and J, *ibid*., pp. 266–270.

33 See note 24: Vitelleschi's discussion is on fols. 257–259v, Rugerius's on pp. 209–211.

34 fol. 259r; see Essay 15, *infra*.

35 *De motu libri decem*, pp. 503–512.

36 APUG/FC Cod. 1710, *Tractatus quintus*, disputatio prima, pars quinta, quaestio sexta: "A quo moveantur proiecta?" See Essay 15, *infra*.

37 The article is reprinted in Moody's *Studies*, pp. 203–286; see Essay 16, *infra*.

38 In his essay, 'The Development of a Critical Temper: New Approaches and Modes of Analysis in Fourteenth-Century Philosophy, Science, and Theology,' *Medieval and Renaissance Studies* (Univ. of North North Carolina) 7 (1978), 51–79.

39 So Albert the Great could write: "Whoever believes that Aristotle was a god, must also believe that he never erred. But if one thinks that he was human, then doubtless he was liable to error just as we are." – Liber VIII Physicorum, tractatus primus, cap. 14, *Omnia opera*, ed. A. Borgnet (Paris 1890–1899) 3: 553.

40 This is apparent in his *The Aim and Structure of Physical Theory*, tr. P. P. Wiener (Princeton 1954).

41 Moody himself signalled this conclusion when he noted, in the 'Galileo and Avempace' article, that "Galileo's Pisan dialogue seems to move wholly within the framework of the thirteenth century formulations of the problem of motion," and that its sources "are to be sought elsewhere than in the tradition of fourteenth century Oxford or Paris, or than in the tradition of fifteenth century Padua." – *Studies*, p. 274.

15. ANNELIESE MAIER: GALILEO AND THEORIES OF IMPETUS

In one of her late essays Anneliese Maier took up a theme that had interested her throughout a good part of her life and attempted to clarify, once again, the extent to which Galileo was indebted to fourteenth-century theories of impetus for the elaboration of his "new science" of motion (1967, pp. 465–490).[1] In the course of her exposition she pointed out how Galileo's concept of impetus changed from his earliest writings on motion to his more mature works, but that even when such changes are taken into account, the way in which he thought of impetus was quite different from the view of the fourteenth-century Scholastics. She also raised a question about his reading of Aristotle's texts bearing on this subject, suggesting that he might have gotten his knowledge of Aristotle "at second hand," and so implying that the ideas about impetus being entertained in the latter part of the sixteenth century were different from those that were current in the fourteenth century (*ibid.*, p. 468). Regardless of how Galileo came by his understanding of impetus, however, Maier contended that even his earlier view was closer than the medievals' to the modern concept of inertia, and thus that it was easier for him to make the transition to the "new science" than it would have been had he well understood and subscribed to the fourteenth-century concept (*ibid.*, pp. 468–469).

It is the purpose of this essay to show that Maier's intuition was basically correct: that there had been an evolution of the concept of impetus from the fourteenth to the sixteenth centuries, and that Galileo derived his early ideas from contemporary sources that incorporated changes facilitating the transition to the inertial concept. The identification of these contemporary sources has not been easy; only in the very recent past has progress been made in uncovering them, and the work is not yet completed. In what follows only a few elements of the teaching contained in them will be touched on, as relevant to the points made by Maier in her essay.

I. CONCEPTS OF IMPETUS

Before discussing these proximate sources of Galileo's early writings it will be well to review the differences between the concept of impetus contained in his composition and that proposed by fourteenth-century writers, as noted in the essay under discussion.

The origin of impetus theory is closely associated with the Scholastic axiom, *Omne quod movetur ab alio movetur*, which leads one to search for the mover that is involved in every instance of motion. In the normal case the mover will be in contact with the thing moved, and when both mover and moved are so identified, the axiom presents no problem. When visible contact is lacking, however, then it becomes a matter of either abandoning the axiom or of identifying a mover that enables it to be retained. Usually the choice was made for the latter alternative, and then two troublesome cases where contact was not visibly in evidence had to be explained. These were the case of falling bodies and the case of projectiles, the first an example of a natural motion and the second that of a violent or forced motion. If one sought a *motor coniunctus*, i.e., a mover that would be conjoined with, or inherent within, the falling body and the projectile, the further option was for *gravitas* as a motive power or force in the falling object and for *impetus* as an impressed force in the projectile. Both of these in turn were regarded as instruments of some primary agent, usually the generator in the case of the falling body and the projector in the case of the projectile. Thus, in the latter instance, the projector would truly be the agent responsible for the movement of the projectile, but he would effect its movement by means of the *impetus* that he impressed on it and that continued along with it in its movement.

In medieval mechanics, moreover, motion was thought of as a continuous reduction of the object moved from potency to act, which could only occur so long as the mover, as the agent in act, was present to effect the reduction. One important consequence of this was that motion would only occur so long as the *vis motrix* actually inhered in the moving object. Both *gravitas* and *impetus* satisfied the definition of a *vis motrix* in this sense, but there was a further difference between them. The *gravitas* of a body was a natural quality or property that was always present in the body and could not be dissipated or used up; on this account it was thought of as *infatigabilis* or indefatigable. *Impetus*, on the other hand, could be used up, for projectiles seem to slow down and eventually stop, and so it was seen as *fatigabilis* or dissipatable. In both cases, moreover, the motion would terminate with rest: the falling object would come to rest in its natural place, while still retaining its gravity; and the projectile would come to rest when its *impetus* was exhausted, being then without *impetus* to carry it farther in its movement (*ibid*., pp. 470–475).

Finally, it should be noted that the division of motions into two types, natural and violent, the first illustrated by the falling object and the second by the projectile, was thought to be dichotomous and exhaustive. Some complex motions, such as those of animals, might combine elements of both, but no one motion, at one and the same time, could be both natural and violent. There was no thought, moreover, of a *tertium quid* or third type of motion that would be neither natural nor violent but intermediate between the two. Thus all cases to which the principle *Omne quod movetur ab alio movetur* applied could be accounted for, and its universal validity therefore assured.

Galileo apparently understood some elements of this account, but the context in which he worked out his first theory of *impetus*, in the *De motu antiquiora* written around 1590, was a polemic against Aristotle and his followers (presumably those of the sixteenth century, which could well explain why he was not attuned to the nuances of fourteenth-century thought). His earliest use and defense of the concept occurs in a draft of an essay on motion wherein he is attempting to explain why falling bodies accelerate, i.e., move faster at the end of their fall than at the beginning. Here he considers the case of the heavy body that is thrown upward and inquires into its velocity of movement at various stages of its rise and fall. The case is interesting because it combines elements of the violent and the natural, since the first part of the motion, when the body is projected upward, is violent, whereas the second part, its downward movement, is natural. Here Galileo conceives of *impetus* as a *privatio gravitatis*, i.e., as a temporary removal of the gravity that is normally found in the heavy object (*ibid.*, pp. 476–479). Like that of the fourteenth-century thinkers, his *impetus* is *fatigabilis* and expendable, and it begins to diminish as soon as the heavy object has left the hand of the thrower. The various stages of the motion are then the following: the upward movement, during which the *impetus* decreases toward the point where its upward force will just balance the downward force of the body's *gravitas*; the turning point at the top of the body's trajectory, at which both forces exactly balance; and the downward movement, during which the force of gravity more and more overcomes the residual *impetus* and causes the body to accelerate in its fall. It is in this explanation that Maier saw a radical departure from fourteenth-century theory: for Galileo, as she understood him, the *vis impressa* is conserved in the body even when it comes to rest at the top of the trajectory, whereas for the fourteenth-century Scholastics this would be unthinkable, since the *impetus* would be dissipated completely with the body's coming to rest and thus could no longer be present to act against the force of the body's gravity (*ibid*., pp. 480–485).[2]

Another context in which Galileo employs impetus theory is likewise present in the *De motu antiquiora*, though it is also prominent in his later writings and developed there to become an important component of the inertial concept that is characteristic of modern science. This is in the discussion of motion along planes inclined at various angles to the horizontal, which involves an examination of the hypothetical case of motion along a horizontal plane from which all friction and impeding forces have been removed. Here Galileo makes the claim that the latter motion would be neither natural nor violent, and that a body could be moved along such a plane by a *minima vi*, indeed by a force smaller than any given force. He further raises the question whether a motion that is neither natural nor violent would be perpetual, but postpones his answer until he has considered the agent that would cause such a motion. Unfortunately he never returns to this question in the early *De motu* drafts, so one can only speculate as to what his answer was going to be. In his later writings, however, an affirmative answer becomes more and more explicit. Horizontal motion, or, more correctly, motion along a frictionless plane that remains always at the same distance from the center of gravity, when once initiated, will continue *ad infinitum*. Thus the *impetus* that accounts for its motion is no longer self-expending, as it was when explaining why falling bodies accelerate, but remains constant throughout the indefinite duration of the body's motion. Here again Maier saw a departure from fourteenth-century theory, for the *vis motiva* associated with *impetus* was there inherently *fatigabilis* and self-expending, diminishing of its own accord even in the absence of external resistance (*ibid*., pp. 486–490).

The main difference, therefore, between Galileo's understanding of *impetus* and that of fourteenth-century Scholastics, as Maier saw them, was that he did not see its existence in a body as incompatible with the body's being at rest and thus could postulate its being conserved, though continuing to diminish in intensity, when a heavy object thrown upward reached the turning point of its trajectory. Later, he apparently accorded *impetus* even more of a self-subsistent character, conceiving it in some special cases as able to perdure indefinitely in undiminished form and thus able to move a body along a frictionless horizontal plane perpetually. The latter example would furthermore be an instance of a motion that was neither natural nor violent, but somehow intermediate between the two, and thus different from any terrestrial motion discussed by the earlier Scholastics.

II. GALILEO'S EARLY 'DE MOTU'

The ambience in which Galileo worked while composing the *De motu antiquiora*, as also while writing his other Latin compositions on the center of gravity of solids and on Aristotle's logic and natural philosophy, has proved difficult to ascertain. There are very few clues that serve to date these compositions or to identify their sources, and thus scholars have had to rely mainly on internal evidence, such as the works cited by Galileo and the concepts he discusses in them, when attempting to reconstruct the path that led him to their composition.[3] So long as investigation centered around books printed in the sixteenth century the results were sparse and inconclusive. The most important discovery came during the past decade, when attention was shifted from printed books to manuscript sources that Galileo could have used.[4] An exhaustive study of Galileo's questions on Aristotle's *De caelo* and *De generatione*, as it turned out, revealed numerous evidences of copying, with some signs of his inability to decipher the handwriting or the abbreviations used, and thus suggesting that he worked from a handwritten source. Further research showed that the contents of Galileo's "questionary" agree very well with *reportationes* of lectures given on the *De caelo* and *De generatione* by Jesuit professors at the Collegio Romano. Of these, the lectures of four professors who taught between the years 1577 and 1592, and whose notes are the only ones extant for that interval, show striking parallels with Galileo's text. The obvious inference is that Galileo used the lecture notes of some professor at the Collegio Romano as the basis for his own notes, which he probably intended to use for lectures himself. The precise set of notes he used has thus far eluded discovery, and indeed may be lost completely, but one can reasonably estimate that they date from 1587 or thereabouts. It seems plausible, moreover, that Galileo used them around 1589 or 1590, in conjunction with his own teaching of astronomy at the University of Pisa at that time.

If one admits that Galileo's *De motu antiquiora* dates from about 1590, or was composed between 1590 and 1592, as is generally agreed among scholars, then this discovery of the provenance of his questions on Aristotle's natural philosophy takes on additional significance. One need not search elsewhere, for example, for the source of Galileo's knowledge of medieval theories of *impetus*, or of his awareness of motions intermediate between the natural and the violent, if such materials were already present in lecture notes deriving from the Collegio Romano. A further circumstance heightens the import of such an observation: Galileo's questions on Aristotle are patently incomplete, since some folios are missing, and since there is internal evidence to indicate that they are but part of an entire course of natural philosophy, comprising questions not only on the *De caelo* and *De generatione* but also on the *Physics* and the *Meteorology*, that he himself had written out. When one searches, therefore, for the counterparts of these notes in the surviving lectures given at the Collegio Romano, one has access to a first-hand indication of Galileo's understanding of Aristotle and his natural philosophy in a sixteenth-century context. Such an understanding was indeed different from that of the fourteenth-century Scholastics, as Maier suspected, and can shed considerable light on the origin and development of Galileo's "new science" of motion.

The four Jesuit professors whose lecture notes are most similar to Galileo's text are Antonius Menu, who taught matter pertaining to natural philosophy at the Collegio between 1577 and 1582; Paulus Valla, who taught the same between 1585 and 1590; then Mutius Vitelleschi, between 1589 and 1591; and finally Ludovicus Rugerius, between 1590 and 1592.[5] Of these, Valla has the most extensive teaching on *impetus*, followed by Menu and Rugerius in that order; Vitelleschi has no formal treatment of the concept, though he does discuss matters germane to its development. In what follows a summary exposition of each of the four will be given, in chronological rather than in topical order, after which a brief assessment will be undertaken of the use Galileo made of them in developing his own ideas on the subject.

III. ANTONIUS MENU

Antonius Menu began his course on natural philosophy at the Collegio some years after the more famous Franciscus Toletus and Benedictus Pererius had already taught it, and only one year after the publication of Pererius's *De communibus omnium rerum naturalium principiis et affectionibus* (Rome 1576), which was to become a standard textbook on natural philosophy until the end of the century. Pererius's book is important because of its adherence to a strict Aristotelian teaching on motion, more along Averroist than along Scholastic lines, which entailed a rejection of *impetus* as offering a correct explanation of what moves the projectile after it has left the hand of the thrower. In light of this rejection it is surprising that Menu reopens the question and departs from Pererius's teaching, favoring the Scholastic over the Averroist doctrine, and thus suggesting the possibility that a *virtus impressa* might offer a better explanation of projectile motion than that of the traditional Aristotelians.

Menu's treatment of impetus occurs in the context of his discussion of motive qualities, which is the fifth disputation in his treatise on the elements, itself a part of his lectures on the *De generatione.*[6] Earlier disputations are concerned with the nature of the elements, their number and distinction, their quantity and other attributes, and their active or alterative qualities. (It is noteworthy that Galileo's notes terminate with the last topic mentioned, and thus that his treatment of motive qualities either was not completed or has since been lost.) Among the questions argued in Menu's treatment of motive qualities are five concerned with *gravitas* and *levitas* as these function in the natural motion of the elements, after which comes his question "On the violent motion of heavy and light projectiles."[7] At the outset of the latter he refers the reader to the works of Themistius, Simplicius, Philoponus, Albertus Magnus, Buridan, Albert of Saxony, Gratiadei (John of Ascoli), Paul of Venice, Scaliger, Domingo de Soto, and Benedictus Pererius – all with references to appropriate passages – for fuller treatments of the subject.[8] The two competing views he goes on to juxtapose are that the projectile is moved by the medium, i.e., by the air through which it is thrown, which is the traditional Aristotelian view, and that the projectile is moved by a *virtus impressa*, the view of the later Scholastics. Menu decides that both are correct, and so comes to two conclusions: it is probable that projectiles are moved by the media through which they pass; but it is also probable, and indeed more probable, that the projectiles are moved not only by the medium but also by some quality such as a *virtus impressa* that inheres in them.[9]

The structure of Menu's argument is that of a scholastic disputation, wherein he first lists the arguments against his first conclusion, then those against his second conclusion, after which he states each conclusion and follows each with the replies to the arguments that have been urged against it. For the sake of continuous presentation, in what follows the respective conclusions and the reasons in support of them will be given first, then the arguments that have been brought against them, together with their responses, so as to enable the reader to follow the thought more readily.

That the projectile is moved by the air, says Menu, can be seen to be true provided one understands the mechanism that brings this about. The thrower impels the air by his hand and communicates to it a *vis impellendi* that spreads through its various parts; this results, as explained by Giles of Rome and Walter Burley, from alternate compressions and expansions, and in the process the *vis* loses its force because it spreads out like a wave, which serves to explain why the violent motion ultimately ceases.

Arguments that are brought against this type of explanation, as Menu sees it, are eight in number: (1) effectively the problem has been transferred from the projectile to the air, and then the question remains as to what moves the air; (2) violent motion in a void would be ruled out because of the absence there of air; (3) on this explanation, a strong contrary wind should stop the projectile; (4) the nature of air works against the explanation, since air diffuses and does not push directly; (5) again, air gives way and does not impel; (6) a feather would be projected farther than a heavy object; (7) many experiments cannot be explained in this way, for example, (i) why a ball thrown to the ground bounces back; (ii) why a sling throws a stone faster and farther than the hand; (iii) why a thrown ball hurts less when it impacts closer rather than farther away; (iv) why a taut string continues to vibrate after being struck; (v) why a top, wheel, or hoop continues to rotate so rapidly; and (vi) why air close to a stone, when moved, does not move the stone; and finally, (8) it is not clear what air it is that moves the projectile, viz, that preceding it, or following it, or travelling along by its side.

For all of these Menu attempts an answer, usually brief and yet consistent, more or less, with both of his conclusions. In particular: (1) this is true, since a *vis* is impressed on the air; (2) also true, since in this view violent motion in a void is impossible; (3) not necessarily, since the projectile receives a greater *vis* from the impelling air than it does from a contrary wind; (4) again not so, since the nature of air is what permits the impelling mechanism to work; (5) in this case, the swiftness of the motion makes the air press harder and so impel the projectile; (6) not so, since feathers are hard to move anyway; (7) the experiments are capable of different interpretation, viz, (i) the ball bounces because the air too is reflected back and pushes upward; (ii) the sling moves the stone faster because it has more leverage, according to the principles of the *Quaestiones mechanicae*; (iii) the ball's motion is faster at the beginning, but later the motion is more level and uniform and thus produces the greater impact; (iv) the taut string continues to vibrate because it moves the air, which continues to move it; (v) air can assist the motion of the wheel, and it can also explain why any rotating object comes to rest; and (vi) air pushed from behind cannot be impelled as well as a stone pushed directly; and (8) the reply is clear from the mechanism invoked in explaining the conclusion.

Menu's second conclusion, as already noted, is that it is also probable, and indeed more probable, that the projectile is moved by the air but also by some quality, such as a *virtus impressa*. For this he offers eight proofs, some of which are obviously based on the arguments he previously listed against his first conclusion. These are all brief, and may be summarized as follows: (A) a lance with a sharp cone at its rear cannot be impelled by air alone; (B) if a cloth or fan of some kind is used to shield the air from a rotating wheel, the wheel still continues to rotate; (C) air is easily divisible and gives way without effort, and so it is incapable of supporting weights of three pounds or more; (D) a stone suspended by a thread is not moved by air agitated in front of it; (E) the experiments discussed above are more readily explained by a *virtus impressa*; (F) a man whom one hits with his fist is not moved by the force of the air; (G) even air cannot be compressed without some type of quality being impressed upon it; and (H) if a circular segment is removed from a tablet, and then inserted back into the hole it previously occupied, it can be made to rotate and will continue to do so, even though there is no space for the air to act upon its circumference. The last proof, it may be noted, is taken from Scaliger, for it uses the same words and the same Latin constructions as are found in the latter's text,[10] which are quite different from those normally employed in the Collegio lectures.

Against this conclusion Menu had earlier listed seven different objections, which now must be taken into account: (a) such a doctrine was not taught by Aristotle and the Peripatetics; (b) on the part of the thing moved: for, when a *virtus impressa* is employed, either the projectile would be moving from an internal principle and thus naturally, or it would be moved violently and yet by itself, which implies a contradiction; (c) from the nature of qualities: since there is no species of quality in which such a *virtus* can be placed, for its existence would posit two contraries opposed to only one natural quality (e.g., *levitas* and the upward *virtus*, both opposed to *gravitas*); or the *virtus* would be natural and thus a property of the body; or it would offer resistance and thus be produced in time with greater or less intensity, and so could not come to be instantaneously; or it would actually be *gravitas* or *levitas*, and there would be no way of distinguishing it, for example, from the *levitas* of fire, since it would exhibit the same properties; (d) from its generation: for no quality is acquired through local motion alone; or it would be acquired by squeezing a stone in one's hand, which does not in fact happen; or because an infinite number of entities would be continuously generated; (e) from its manner of generation: for such a quality would have to be produced throughout the entire projectile, and this is opposed to its being indivisible; or it would exist only in part of the projectile, and this is opposed to the *virtus motiva*'s moving the whole; (f) from the motion's being violent: for then the projectile's motion would be one and continuous, contrary to Aristotle; or there would be no reason why its motion would be faster [in the middle] than at the beginning, again contrary to Aristotle; and (g) motion in a void would be possible, yet again contrary to Aristotle.

For each of these arguments Menu has a reply, and the responses are most helpful for clarifying his understanding of the *impetus* concept. (a) With regard to the Aristotelian tradition in this matter, for example, he passes over the complex problem of ascertaining what Aristotle himself actually held, noting that both Themistius and Simplicius placed a *virtus impressa* in the air, and that since putting it in the air is little different from putting it in the stone, Philoponus expressly located it in the projectile – also to allow the possibility of motion in a void.[11] Menu here attributes Philoponus's teaching to St. Thomas, the *Doctores Parisienses*, Albert of Saxony, Paul of Venice, Buridan, Scaliger, and others. (b) With regard to the naturalness of the motion, he explains that local motion is not always natural, unless it comes from a form that is natural to the object moved and exists in it; also the projectile does not move itself, since the projector moves it by means of the *virtus impressa*; again, the *vis* is not a property of the stone but rather is the instrument of the projector, much as is the magnetism impressed on iron by a magnet. (c) On the nature of the quality involved, Menu claims that the *virtus impressa* is reducible to a disposition of the first species of quality, since it is *facile mobilis*; also, there is no incompatibility in the *gravitas* of the stone having two contraries, provided one is natural and the other violent; again, the *virtus* is an imperfect being and as such does not have a perfect *esse*, thus does not have to be natural to the stone; moreover, this particular *virtus* is like a spiritual quality, and so can be induced instantaneously; yet again, such a quality, properly speaking, is not contrary to the *gravitas* of the stone, especially in act, for two reasons: (i) contraries have to be of the same kind, but the impressed quality is an intentional quality whereas the *gravitas* of the stone is a material quality; and (ii) if it were a contrary, it could not be impressed on the stone without diminishing its *gravitas*, and this is contrary to experience; finally, such a quality is itself neither heavy nor light, but rather a type of force that is able to move up or down or back or forth, just as the magnetic quality induced in iron or the force that moves an element to fill a vacuum. (d) With regard to the generation of the quality, Menu answers that it is possible for local motion to produce an imperfect quality, such as occurs when sound results from rubbing or friction; the impressed force, moreover, is produced by the *virtus motiva* of the projector, which is lacking in the pressure of squeezing; and there is nothing to prevent imperfect entities from being generated a great number of times. The remaining difficulties are similarly answered: (e) the quality is produced throughout the entire projectile, and this is possible for a quality that has a spiritual and intentional character, similar to magnetism; (f) the inference is correct, and it is rather the contrary that is absurd; again, the greater velocity comes from the motion's being more level and more uniform, as already explained, and not from the force that moves the projectile; and (g) the conclusion is correct, and is to be conceded. Menu then concludes his exposition by appending two remarks, one clarifying how the *virtus impressa* can explain the bounce of a ball, not by having it corrupt at the point of impact but by having it reflected, a phenomenon not uncommon with intentional species; and the other explaining that the *virtus* does not corrupt instantaneously, even though it comes into being in this way, since it is an imperfect quality that gradually weakens once it has come into existence.

IV. PAULUS VALLA

Paulus Valla, as already noted, has the fullest treatment of *impetus* of any of the four professors being discussed. Like Menu's analysis, Valla's occurs in his tractate on the elements, in the fifth part of that tractate dealing with motive qualities, and in the sixth question of the part which inquires *A quo moveantur proiecta*?[12] There are other similarities between the lectures of Valla and Menu, and these give good indication that the former probably used the latter's lecture notes when preparing his own lectures. Menu's ideas are thus clearly present in Valla's *reportationes*, but Valla's explanations are fuller and better thought out, and there is a more pronounced acceptance of the *impetus* concept in the way he states his results than can be found in Menu. It is also noteworthy that the portions of Galileo's questions on the *De generatione* that are closest to this locus in both Menu's and Valla's lectures show more marked similarities with Valla's notes than with those of Menu; thus it seems plausible to maintain that Galileo's own views would be closer to those adopted by Valla than to those expressed by Menu.[13] So as to avoid duplication, however, in what follows only parts of Valla's exposition that express more nuanced conclusions than Menu's, or that give fuller explanations, will be dealt with.

Valla begins his question on projectiles with the notation that the term projectile means anything moved from without in such a way that the mover does not accompany the object moved, but first pushes it and afterwards no longer touches it. The problem then is this: granted that the projector is the mover of the projectile, what is the instrument through which he moves it after he no longer touches it? Is it some *virtus impressa* inhering in the projectile, or the action of the medium through which the projectile moves, or something else entirely? In order to appreciate the problem, Valla notes that there are a number of difficulties that have to be clarified relating to the concept of *virtus impressa* and to the way in which the medium might move the projectile, and he discourses on these before giving his conclusions and the arguments in their support.

The observations on the nature of *impetus* and its mode of generation and corruption are particularly significant. The *virtus impressa*, for Valla, is an imperfect quality after the fashion of an intensional or spiritual form, somewhat like light and color, and as such has no contrary in first act although it has a contrary in second act. Because it has no contrary in first act it can be introduced into a stone instantaneously. When the stone is thrown upward, the *virtus* is not contrary to the stone's *gravitas* in first act, though it is contrary to it in second act; as a consequence the stone's *gravitas* remains in the stone, even though it is impeded by the *virtus* in second act, i.e., in its motive effects. That the two are not contraries in first act is obvious from the fact that *gravitas* is a natural quality whereas the *virtus impressa* exists after the fashion of an intensional or spiritual quality. One need not fear, therefore, that if a massive *impetus* were imparted to the stone sufficient to impel it all the way to the inner concavity of the orb of the moon the stone would lose its natural gravity; it would still retain the latter in first act, though its second act would be impeded by the action of the *virtus*.

The problem of locating the *virtus impressa* in a species of quality is solved by Valla in two ways. According to Marsilius of Inghen, he says, it is reducible to the first species of quality, to a disposition that is easily impressed and just as easily removed. According to others, and Valla prefers their view, it is a *qualitas passibilis* (therefore in the third species of quality) that is impressed on an object the way in which magnetism is impressed on iron. One could call the magnetic quality *levitas* when it moves the iron upward and *gravitas* when it moves the iron downward, but properly speaking it is neither *levitas* not *gravitas*; rather it is a kind of quality that is capable of moving the object moved in any way the one impressing it desires. Similarly the *virtus impressa* can move the projectile in any direction whatever, and, being like a *species intensionalis*, it can do so without being the contrary of any of the projectile's natural qualities.

With regard to the *virtus*'s mode of production and corruption, Valla holds that it is produced by local motion, just as is heat, from the motive power of the thrower; thus it is generated either in an instant or during the time the thrower is touching it. Since the *virtus* is, like light, a diminished and imperfect entity (*diminuta et imperfecta entitas*), it is easily corrupted. It is distributed throughout the entire volume of the projectile, just as the magnetic quality is distributed throughout the magnetized iron, and this presents no difficulty for intensional species of this type. Though produced instantaneously, it corrupts successively and in time, diminishing both by itself and from the fact that it is resisted in second act. Valla's statement regarding the mode of corruption is cryptic, and is worth quoting in his own words:

I say that this quality corrupts successively and in time, since it corrupts by itself (*a seipsa*). For the impressed quality has degree-like parts (*partes graduales*) and lacks a contrary, and it corrupts only because it gradually loses existence, becoming weaker the more it is removed from its principle, since it has a latitude which causes such succession. For this quality is corrupted from the fact that it is resisted in second act; its second act is contrary to the second act of the *virtus motiva* of the stone, and therefore there can be succession in this corruption. Nor is there any absurdity in something being produced in an instant and corrupted in time, because in [its] production it has no contrary and is produced after the motion [of the mover], whereas in corruption it does have a contrary in some way, and it corrupts because it gradually loses existence through parts.[14]

It is noteworthy that Valla here assigns two causes to the diminution of the *virtus impressa*: its self-expending character, which comes from its being more distant from "its principle," apparently the projector; and its weakening from the internal resistance it encounters from the stone's *gravitas* in second act. Apparently both causes operate at once in the case Valla visualizes, and thus it is difficult to ascertain what he would have held in the case of the ball moving on a horizontal, resistance-less plane, where there would be no *virtus motiva* to act against the *virtus impressa.*[15]

Such preliminaries aside, Valla lists two opinions and their proofs before proceeding to his own conclusions. The first opinion is that the projectile is moved by the medium in the way he has already explained, and he attributes this to Burley, Aristotle, Themistius, Philoponus, Averroës, St. Thomas, and others, adding that all of these admit that some *virtus impressa* is also present in the medium. The proofs he offers for this include various texts of Aristotle that speak of the air alone moving the projectile, and then the various arguments already given by Menu against the existence of such a *virtus* in the projectile itself. The second opinion, he says, holds that the projectile is moved by a *virtus impressa*, and this is the teaching of Scaliger, Albert of Saxony, and Buridan. Paul of Venice and Simplicius, he adds, seem to put this *virtus* in the medium, and if it is there, no reason why it is not in the projectile, and so violent motion in a vacuum would be possible. St. Thomas and the *Doctores Parisienses*, he continues, admitted such a *virtus* in the medium, and so did practically all the ancients, including Aristotle himself. The arguments he then offers in support of this opinion are basically those given by Menu against the projectile's being moved by the medium alone. To these he adds only one new argument: if this were so, then a projectile would be projected more easily in water than in air, since water is better able to propel an object than is air, but such is not found to be the case.[16]

Valla's two conclusions, finally, are stated a bit more straight-forwardly than the corresponding conclusions in Menu. The first is that the medium alone is not sufficient to explain the motion of the projectile, and that some other *virtus* must also be admitted.[17] The proofs offered in support of this conclusion are essentially those given by Menu to justify his second conclusion, including the experiment taken from Scaliger, whose source is explicitly acknowledged by Valla. After this the second conclusion follows, without any further proofs, stating that, although the *virtus impressa* is sufficient to cause the motion of projectiles, almost always such a motion is aided by the motion of the medium.[18] In this way, Valla concludes, one can "save" Aristotle and the Peripatetics and all the experiments that have been discussed: the *virtus* is sufficient to move the projectile without the medium, but without the *virtus* the medium can move the projectile either not at all or merely for a very short distance.

There remains only Valla's resolution of the various difficulties that have been raised against his position. Of these one is noteworthy, namely, the way in which he replies to the objection that the motion of the projectile would then be one and continuous, which is against the text of Aristotle. Valla's answer is the following:

> I answer that continuous can be taken in two ways: first, for a motion that is not interrupted by any intermediate point of rest; second, for a uniform and regular motion. If taken in the first way, I concede the inference; nor did Aristotle deny this, nor could he deny it, since experience shows that the stone, when thrown upward, does not come to rest in any way. If taken in the second way, I deny the inference, and this is what Aristotle meant in that place. For he concluded from the fact that the motion of the projectile is not regular and uniform that it could not be perpetual. And such a motion is not regular, both because the *virtus impressa* gradually weakens and because the medium that assists the motion is not always uniform – and it is this second reason of which Aristotle is speaking in that text.[19]

Here again Valla assigns two reasons for the slowing down of the projectile's motion, explaining why it would not be perpetual. More interesting, however, is his admission that the projectile thrown upward does not come to rest "in any way," thus allowing the possibility that the *virtus impressa* remains throughout the entire course of the thrown stone's motion, much as this was understood by the young Galileo.[20]

V. VITELLESCHI AND RUGERIUS

The views of Mutius Vitelleschi and Ludovicus Rugerius regarding *impetus* are less important, for purposes of this essay, than those of Menu and Valla, mainly because Vitelleschi has no formal treatment of the motion of projectiles, as do all the others, and because Rugerius begins to draw away from the *impetus* concept and return to a teaching similar to that of Pererius. Vitelleschi assumes importance for another reason, however, since he raises a question similar to Galileo's about the existence of motions intermediate between the natural and the violent, and whether or not such motions might be perpetual. Rugerius treats this question also, but not with the clarity of Vitelleschi, and therefore the latter's exposition will be the focus in what follows.

Vitelleschi's discussion of natural and violent motions occurs at the end of his commentary on the fifth book of the *Physics*, where he takes up questions relating to various kinds of motion, namely, regular and irregular, simple and compound, fast and slow, and natural and violent.[21] In treating the last distinction he inquires whether there exists any motion intermediate (*motus medius*) between natural motion and violent motion; the question arises, he says, because of the circular motion of fire when it rises to the orb of the moon. Such a motion does not seem to be natural, because fire's natural motion is upward and not circular; nor does it seem to be violent, since once at the orb of the moon it will circulate forever, and no violent motion is perpetual. Arguments can be given from Aristotle in support of both sides of the question, and thus the apparent impasse.

The problem can be solved, Vitelleschi observes, when one realizes that the term violent can be understood in two ways: either (1) for something contrary to the nature of the thing, imposed on it from without, and to which the thing acted upon contributes no force (*nullam vim conferenti passo*), i.e., for something to which the thing has no positive inclination, even though it has no repugnance to it or is not contrary to its principles, but instead is beyond its nature (*praeter illius naturam*); or (2) for something that is positively repugnant to the nature and inclination of the thing. If the violent is taken in the first way, every motion is either natural or violent; if in the second way, it is possible to have a type of intermediate motion that is beyond and even above nature (*praeter naturam . . . super naturam*), but not according to or contrary to nature (*secundum naturam . . . contra naturam*). Understood in this way, the upward motion of the fire would be natural, its forced downward motion would be violent, and its circular motion would be neither natural nor violent but preternatural. The statement that nothing that is *praeter naturam* is perpetual is then to be understood of the violent in the second sense, which is contrary to nature. The reason is that this type of violence takes away the force of nature (*afferat vim naturae*) and sometimes destroys it, so that the motion cannot be perpetual. The other alternative, not explicitly stated by Vitelleschi, is that violence of the first type would not take away the force of nature, and thus a preternatural motion could go on forever.[22]

Rugerius, finally, takes up the motion of projectiles at the end of his questions on the *De caelo* in a tractate devoted entirely to the local motion of heavy and light objects.[23] He incorporates much of the material that is found in Menu and Valla, but he organizes it differently and comes to a different conclusion. This is expressed as a qualified acceptance of *impetus* or *virtus impressa*, with a preferable endorsement of the more traditional Aristotelian view. Rugerius concludes that, "although it is not yet too improbable to posit this *virtus impressa*, it is more philosophical and more in conformity with Peripatetic principles to attribute the motion of projectiles to the medium alone, and not to any *virtus* newly impressed."[24] His explanation of the qualifying clause with which he begins his statement is interesting, for it contains a description of how he conceives *impetus*, even though he is not fully prepared to accept the concept:

> This proposition, with respect to its first part, is proved: because practically all of the difficulties brought against the *virtus impressa* can be solved if one maintains that this *virtus* is different in kind from natural motive qualities, and that it is of only one species and moves up and down and in other ways as the impeller begins the movement. Again, it is corrupted in two ways: one by the natural quality of the subject that is opposed to it; the other by rest. Likewise, it undergoes intension and remission, it is produced in time, and it is received in the entire object that is moved. Nevertheless, the motion that results from this *virtus* is not natural, because there is in the object another *virtus* inclining to an opposite motion, and so the [impressed] *virtus* is not natural to the body but violent. Secondly, it can also be said that this *virtus* is not any really distinct quality that is newly produced, but is a kind of modification of *gravitas* in the heavy element, or in some way mixed with the *gravitas*, and that the natural quality is modified by the one impelling the object so that the *gravitas* inclines the object in the direction intended by the projector. The modification is more or less according as the object is impelled with greater or lesser force, within a certain latitude of *gravitas*. The motion therefore is not natural, because the object is not carried by its *gravitas* according to the modification the *gravitas* has by nature, but rather by that it receives from the projector . . .[25]

This view of the *virtus impressa* is indeed quite original, and seems to be a development of ideas contained in Menu and Valla, while differing from them in significant respects. One point of interest is that Rugerius explicitly states that *impetus*, in his view, corrupts when the body comes to rest, which Maier proposes as the fourteenth-century concept of *impetus* but which is not noted by the other Jesuit professors at the Collegio Romano.

VI. COMPARATIVE ANALYSIS

This, then, completes the exposition of the sixteenth-century ambience in which Galileo likely make his first attempts at composing a treatise on motion. The thought of the professors at the Collegio contains a wide variety of insights into the nature of *impetus* and how it might be used to explain the motion of projected bodies. How much of this was known to Galileo is, of course, problematical, and yet it is quite remarkable that the key ideas contained in his *De motu antiquiora* are not far different from those espoused by one or other of the Jesuits noted above. A few remarks along this line may help the reader to situate Galileo's composition vis-à-vis the contributions of Menu, Valla, and Vitelleschi.

According to Menu's analysis, *impetus* is self-expending, as Galileo first held it to be. It is also conserved in the object at the turning point of the motion. Menu's example is the ball thrown to the ground, for in this case he maintains that the *impetus* changes direction but still continues to move the ball upward after the moment of impact. Galileo's example, of course, is the ball thrown upward, but here the upward *impetus* does not change direction, but it is still conserved and continues to act on the body after the turning point, just as does Menu's.[26] Finally, for both Menu and Galileo, the *virtus impressa* acts against the body's natural *gravitas* and impedes the motion it would normally produce, but it does not actually remove the *gravitas*, so that the body remains heavy throughout the entire motion.

In Valla's exposition, *impetus* is not merely self-expending but is also made to diminish by the contrary action of the body's motive power. Valla introduces, moreover, distinctions of "first act" and "second act" in discussing the interactions between *impetus* and *gravitas* as motive powers; in his view, as in Galileo's, *impetus* can act to deprive a body of the effects of its gravity, and so effectively is a *privatio gravitatis*, although Valla does not use this term. Like Galileo, he is interested in preserving the body's natural *gravitas* even though the latter's action is temporarily suspended owing to the influence of the *impetus*. And he explicitly states that the object thrown upward does not come to rest, and thus in his view, as in Galileo's, the *virtus impressa* continues to act in it throughout its entire motion.[27]

What Valla would have done with the case of the object moved along the frictionless horizontal plane is difficult to say, but Vitelleschi definitely provides materials that contribute to its analysis. Like Galileo, he recognized the possibility of motions that are neither natural nor violent, but are preternatural and intermediate between the two. In his early *De motu antiquiora* Galileo raises the question whether such motion would be perpetual, but postpones his answer; later he inclines to the view that a preternatural motion could go on *ad infinitum*. Vitelleschi gives the answer that is lacking in Galileo's early work: there is no reason why such motion should not be perpetual, since there is nothing to bring it to rest.[28] Since Vitelleschi has no treatment of *impetus*, it is difficult to say whether he would have regarded it as inherently self-expending or as diminished only by the action of some contrary motive force or resistance. Galileo later took the second option, and this was perfectly consistent with Vitelleschi's teaching, even though the particular application did not come under the latter's consideration.

Whatever the actual influence of these Jesuits on Galileo's early thought, however, it seems that Maier was quite right in pointing out significant differences between fourteenth- and sixteenth-century theories of impetus, and in suspecting that Galileo's interpretations of Aristotle and his tradition arose in a different context from that of the medievals with which she was so familiar.

NOTES

1 Anneliese Maier, 'Galilei und die scholastische Impetustheorie,' *Ausgehendes Mittelalter II*, Storia e Letteratura 105, Rome: Edizioni di Storia e Letteratura, 1967.

2 It is noteworthy that at p. 480 Maier interprets Galileo as holding that the projectile actually comes to rest at the turning point of its trajectory, which is contrary to his teaching in another chapter of the *De motu antiquiora* entitled 'In which, in opposition to Aristotle and the general view, it is shown that at the turning point [an interval of] rest does not occur.' For Galileo's view, see the English translation in *Galileo Galilei ON MOTION and ON MECHANICS*, eds. and trs. I. E. Drabkin and S. Drake, Madison: The University of Wisconsin Press, 1960, pp. 94–100; the question discussed by Maier occurs earlier in Drabkin's translation at pp. 76–84. Whereas many Schoolmen held that there was a point of rest at the top of the trajectory that interrupted the flight of the projectile and divided it into two different motions, one up and the other down, Galileo argued that its motion was one and continuous from beginning to end.

3 The most important attempt has been that of Antonio Favaro in his introductions to the various treatises in the National Edition of Galileo's works, *Le Opere di Galileo Galilei*, 20 vols. in 21, Florence: G. Barbera Editore, 1890–1909, reprinted 1968; see especially Vol. 1, pp. 9–13, and Vol. 9, pp. 275–282. Favaro's results have been generally adopted by Ludovico Geymonat, Stillman Drake, and others in their accounts of the development of Galileo's early thought.

4 See the essays in Part III of this volume.

5 See *Galileo's Early Notebooks*, pp. 12–21. It was customary at the Collegio Romano during this period for each professor to teach the whole of philosophy in a three-year cycle, the first year being devoted to logic, the second to natural philosophy, and the third to metaphysics. Because of the large amount of material to be covered in the second year, however, some matter pertaining to natural philosophy was postponed until the beginning of the third year. Thus it happened that treatises on the elements and on the motion of heavy and light bodies were sometimes covered during the second year, sometimes during the third, depending on how quickly the individual professor had progressed through the assigned material.

6 Cod. Uberlingen 138, Leopold-Sophien-Bibliothek, *In philosophiam naturalem, anno 1577.* The MS lacks foliation and thus precise references to its contents cannot be given in what follows; a general description is given in *Galileo's early Notebooks*, pp. 16–17.

7 Caput sextum. De motu violento proiectorum gravium et levium.

8 De hac materia lege Themistium, octavo Physicorum, tex. 83; Simplicium, tex. 82; Scaligerum, exercit. 28; Philoponum, quarto Physicorum in tractatu de vacuo; Paulum Venetum, octavo Physicorum, 82; Buridanum, ibid., q. 12; Magistrum Sotum, q. 3, ad 2um, octavi Physicorum; Albertum de Saxonia, ibid., q. ultima; Gratiadei, lec. 22, q. 2; Albertum Magnum, ibid., trac. 4, cap. 4; P. Benedictum [Pererium], lib. 14.

9 Omissis vero opinionibus, dicendum est primo: Probabile est proiecta moveri a medio, scilicet, aere vel aqua, ut tenet Aristoteles et Peripatetici. . . . Dicendum secundo: Esse satis probabile et forte probabilius proiecta moveri non solum ab aere sed etiam ab aliqua qualitate ut virtute impressa.

10 See J. C. Scaliger, *De subtilitate*, exercitatione 28, Lyons: A. de Harsy, 1615, pp. 100–105.

11 Here it is noteworthy that Maier, loc. cit., p. 477, adopts the position that Aristotle himself placed a *vis* in the medium; apparently this was disputed among sixteenth-century Aristotelians, and thus Menu takes a neutral stand on the matter, attributing the position to Themistius and Simplicius rather than asserting positively that it was Aristotle's own.

12 Archivum Pontificiae Universitatis Gregorianae, Fondo Curia, Cod. 1710, *Commentaria in libros Meteororum Aristotelis.* This MS also lacks foliation, and thus precise references to its contents cannot be given either; for a general description, see *Galileo's Early Notebooks* (note 11 *supra*), pp. 17–18.

13 For details, see the author's commentary on Galileo's tractate *De elementis* in *Galileo's Early Notebooks*, pp. 281–303.

14 . . . respondeo hanc qualitatem corrumpi successive et in tempore, quoniam corrumpitur a se ipsa. Qualitas enim impressa habet partes graduales et caret contrario, et corrumpitur tantum quia paulatim deserit esse et quo magis removetur a suo principio eo magis debilitare, cum habet latitudinem quae causat talem successionem. Corrumpitur enim haec qualitas per hoc quod resistitur actui secundo; illius actus autem secundus est contrarius actui secundo virtutis motivae lapidis, et ideo potest in hac corruptione esse successio. Neque est absurdum aliquid produci in instanti et corrumpi in tempore, quia in productione nullum habet contrarium et producitur post motum; in corruptione autem habet aliquo modo contrarium, et corrumpitur quia paulatim deperit esse per partes. – On Valla's use of the expression *partes graduales* (degree-like parts), see the

parallel discussion in Galileo's question on the parts or degrees of qualities, *Galileo's Early Notebooks*, p. 174, par. 05.

15 From the way in which Valla expresses his conclusion it is not clear whether or not the motive quality would corrupt if there were no resistance present, since the presence of a contrary quality seems essential to his explanation as to why the motive quality corrupts.

16 Valla also changes somewhat Menu's argument F (p. 328, *supra*), proposing that the person be struck by a bat (*a baculo*) rather than by a fist (*a pugno*).

17 Prima conclusio: solum medium non videtur sufficiens ad salvandum motum proiectorum, sed necessario admittenda est aliqua alia virtus.

18 Secunda conclusio: etiamsi sola virtus impressa sufficiens sit ad causandum motum proiectorum, quando tamen proiecta moventur secundum ordinem universi in plano, fere semper talis motus adiuvatur ab aere; tum quia rarefit et condensatur ad motum supra explicatum, ita ut, etiamsi aer solus non sit sufficiens movere proiectum ad eam distantiam et eo impetu quo movetur, possit tamen adiuvare talem motum, sicut fit in motu naturali gravium et levium. Et hoc modo salvamus Aristotelem, Peripateticos, et omnes experientias, si dicamus utrumque concurrere, ita tamen ut virtus sola sufficiens sit movere proiectum sine medio, medium vero sine virtute aut non possit ullo modo aut certe possit tantum ad modicam aliquam distantiam.

19 . . . respondeo continuum sumi posse duobus modis: primo, pro motu qui non est interruptus quiete intermedia; secundo, pro motu uniformi et regulari. Si sumatur primo modo, concedo sequelam; neque hoc negavit aut negare potuit Aristoteles, quia experientia patet lapidem, quando movetur sursum, nullo modo quiescere. Si vero sumatur secundo modo, nego sequelam; et hoc voluit Aristoteles eo loco. Ex eo enim quod motus proiectorum non est regularis et uniformis concludit Aristoteles motum illum non posse esse perpetuum. Motus autem talis non est regularis, tum quia virtus impressa paulatim deficit, tum quia medium quod adiuvat motum non semper eodem modo se habet; et hanc secundam causam assignavit ibi Aristoteles.

20 It is difficult to fathom precisely what Valla had in mind in stating that the stone, when thrown upward, does not come to rest "in any way," if he did not mean by this that there was no rest at the turning point of the motion.

21 Staatsbibliothek Bamberg, Cod. 70 (H.J.VI.21), Lectiones in octo libros Physicorum et quatuor De caelo, anno 1589 et 1590. For more details concerning Vitelleschi's teachings on motion, see Essay 7, *supra*; also *Galileo's Early Notebooks*, pp. 18–19 and passim.

22 Cod. Bamberg. SB 70, fols. 258v–259r.

23 Cod. Bamberg. SB 62.4, Quaestiones in quatuor libros Aristotelis De caelo et mundo, anno 1591, fols. 86r–114v.

24 Quanquam non est usque ideo improbabile ponere hanc virtutem impressam, nihilominus tamen magis philosophicum est et Peripateticis principiis conformius attribuere motum proiectorum soli medio, non ulli virtuti de novo impressae. – *ibid.*, fol. 104r.

25 Haec propositio quo ad primam partem probatur, quia fere omnes difficultates contra virtutem impressam allatae solvi possent, si quis diceret hanc virtutem esse diversae rationis a qualitatibus naturalibus motivis, et esse unam tantum speciem quae moveat sursum et deorsum et aliis modis, sicut expellens moveri inceperit. Itemque corrumpi duobus modis: uno a qualitate naturali subiecti qui illi adversatur; altero per quietem. Item intendi ac remitti, produci in tempore, recipi in toto mobili. Neque tamen

motus qui secundum hanc virtutem fit esse naturalem, quia in mobili est alia virtus inclinans ad motum oppositum; haec autem virtus non est naturalis subiecto sed violenta. Secundo etiam dici posset, hanc virtutem non esse aliquam qualitatem realiter distinctam ac de novo productam, sed esse modificationem quandam gravitatis in elemento gravi, vel quoquo modo commixto gravitati, modificari enim qualitatem illam naturalem ab expellente mobili, ut in eam partem gravitas ipsa inclinet in quam expellens direxerit mobile. Esse autem maiorem vel minorem modificationem prout maiore vel minore vi impulsum fuerit mobile intra certam tamen latitudinem gravitatis. Motum vero non ideo esse naturalem quia non fertur a sua gravitate secundum eam modificationem quam gravitas habet a natura, sed secundum eam quam habet a projiciente . . . – *ibid.*

[26] It is noteworthy, on this account, that neither speaks of the body "coming to rest" at the turning point of the motion.

[27] See text cited at note 19; also notes 2, 20, and 26 *supra.*

[28] Compare Galileo's teaching in the *De motu antiquiora*, Drabkin translation, p. 75, with Vitelleschi's, as cited in note 22 *supra*; see also the corresponding treatment of Rugerius in Cod. Bamberg. SB 62.4, fols. 106r–110r.

16. ERNEST MOODY: GALILEO AND NOMINALISM

The death of Ernest A. Moody in December of 1975 deprived the academic world of one of its foremost medievalists and intellectual historians, a person to be ranked surely with Pierre Duhem and Anneliese Maier for the many difficult texts he made available to scholars and for the novelty of the insights with which he continually stimulated them. Fortunate it was that just six months before his death the University of California Press saw fit to publish his collected papers, together with an autobiographical preface that explained his intellectual odyssey, why and when he wrote what he did from beginning to end, and how he finally evaluated the results of all his labors.[1] This series of papers, together with Moody's three books,[2] stand as a monument to the man's impressive scholarship; they also afford those of us who knew and admired his work the opportunity to reflect on his achievement and to offer a critique of his central theses.

Our own research interests have paralleled Moody's in a remarkable way, although we came to approach our common area from diametrically opposite directions. In our case Thomism provided the initial framework for a deep interest in Aristotle and in the medieval commentaries on the *Physics, De caelo*, etc., that led, by howsoever circuitous a route, to Galileo and his *nuova scienza*. In Moody's case it was Ockham who provided a similar inspiration, and this, oddly enough, precisely because of opposition to him from the Thomist camp. As he tells us,

> What attracted me to Ockham, in the first instance, was the bad publicity given to him by the Thomists and particularly by Gilson, who portrayed him as a diabolical genius who tore down the beautiful edifice of scholastic philosophy and theology erected by Saint Thomas Aquinas. Since it was natural for me to side with the underdog, I felt the urge to find out what Ockham had to say.[3]

This enticed Moody into his doctoral study of Ockham's logic, which in turn led, after years of maturation, to his most famous work, *Truth and Consequence in Medieval Logic*. Logic and methodology then gave way to concern with physical science, and here Moody found in Jean Buridan a congenial figure with whom to continue his Ockhamist interests. The fourteenth century became his focal point for ever more detailed studies, and the more he studied it, the more he saw that century as the one to which our own age is most in debt. As Lynn White points out in his foreword, quoting Moody's overall conclusion,

> ... if the later fourteenth century "has seemed to the historians of philosophy an age of decline, to the historians of science and logic it has seemed an age of rebirth and advance. ... For better or worse, it gave a new character and direction to all later philosophy, of which we have not yet seen the end."[4]

Our first contact with Moody came, predictably, shortly after the appearance of his classic essay, 'Galileo and Avempace: The Mechanics of the Leaning Tower Experiment,'[5] at which time we took issue with the mechanical doctrines he there attributed to St. Thomas and particularly with his attaching the lables "Cartesian" and "Platonist" to Aquinas's thought.[6] As a result of an initial interchange both of us prepared notes for the *Journal of the History of Ideas* and corresponded about them over a considerable period; in the end, however, neither was pleased that he had understood and met the other's objections, and by mutual consent we withdrew our manuscripts. Neither of us returned to the precise matter of the interchange, although we later attempted to set the record straight on Aquinas's contribution to medieval mechanics in our treatment of him in the *Dictionary of Scientific Biography*, without, however, making reference to Moody's interpretation.[7] Fortunately, in the intervening years James A. Weisheipl has written two scholarly articles wherein he makes essentially the same points we had indicated to Moody, without himself being aware of that interchange.[8] Since Aquinas's teaching is thus now well exposed in the literature, a few comments may serve here to relate that teaching to Moody's exposition of it in "Galileo and Avempace."

Paralleling his work in medieval logic, where he was able to translate the discursive Latin texts of the fourteenth century into the symbolic expressions of twentieth-century logic, Moody attempted to formulate a key problem of medieval and early modern dynamics in terms of equations that would be intelligible to twentieth-century physicists. He thus pictured the difference between Aristotle and Galileo over the possibility of motion through a void, a topic discussed in Galileo's Pisan work *De motu*, as captured in the two equations, $V = P/M$ (Aristotle) and $V = P-M$ (Galileo), where V stands for the velocity or speed of motion, P for the motive power urging the body moved, and M for the resistive medium through which the body passes.[9] In a void, of course, since there is no resistance to motion, M takes the value of zero. For Aristotle this has the consequence that the motion becomes instantaneous, which is another way of saying that motion in a void is impossible; for Galileo, on the other hand, motion takes a definite value determined by the motive power alone, and thus motion through a void is possible. Then, searching out the medieval antecedents of these very different conceptions, Moody discovered them quite unexpectedly in the teachings of Averroës and Avempace: Averroës upheld the validity of Aristotle's equation, $V = P/M$, whereas Avempace rejected Aristotle's equation and in its place substituted the equation later to be found in Galileo, $V = P-M$.[10] More than that, Avempace's progressive views were not unappreciated in the Latin West; although some scholastics rejected them, "the outstanding defender of Avempace's theory was St. Thomas Aquinas,"[11] who not only defended that theory but actually adopted "Avempace's 'law of difference' represented by the formula $V = P-M$ . . . "[12] Thus Aquinas, acting as an intermediary for Avempace, played a key role in the development of Galileo's new science.

Flattering as it may be to propose Aquinas as such a precursor of Galileo, Moody's way of doing so does not do justice either to Aquinas's discussion of the possibility of motion through a void or to Aquinas's exegesis of Aristotle's text. As Weisheipl makes clear, Aquinas did not subscribe to the view that the dynamic formula $V = P/M$ represents Aristotle's own teaching, for he regarded the arguments in Aristotle's text on which this formula is based as merely dialectical and not in any way demonstrative.[13] Thus Aquinas had no reason to endorse either that formula or an alternate one such as Avempace's. It is true that fourteenth-century thinkers, following Thomas Bradwardine, became interested in dynamic formulas of various types, and that earlier Averroës (whose views on this matter Aquinas regarded as *omnino frivola*[14]) had championed $V = P/M$ as Aristotle's authentic teaching. But Averroës did this because of his idiosyncratic philosophical understanding of the principle *omne quod movetur ab alio movetur* and how that principle could be justified in the case of falling bodies. In no event did Aquinas agree with Averroës on such matters, although unfortunately Anneliese Maier thought that all scholastics shared common views both on the principle *omne quod movetur* and on the problem of motion through a void – views that in her estimation constituted a fatal barrier to the rise of classical physics.[15] Weisheipl has been at pains, because of Maier's widespread influence, to show how diverse were the teachings of scholastics on these matters, and particularly how nuanced was Aquinas's view, being incapable of ready assimilation into what is fast becoming a standardized exposition among historians of medieval science.[16]

Weisheipl's studies are mentioned here as only a mild corrective to some of Moody's statements in the "Galileo and Avempace" article, for Moody rightly discerned Aquinas's rejection of the more obvious aspects of Averroës teaching, and this was indeed a contribution at the time of his writing. Since that time twenty five years have elapsed, and our own recent researches, mainly in Galileo's early notebooks, have uncovered further connections between Aquinas and Galileo.[17] With regard to Moody's overall thesis these new discoveries work two ways: they serve to ground in an unsuspected fashion Moody's suspicion of Aquinas as an influence on Galileo, and at the same time they tend to diminish Ockham's importance and to highlight Buridan's – not indeed as an Ockhamist, as Moody thought, but as an unlikely transmitter of Aquinas's methodological doctrines to Galileo.

Buridan's importance lies in his explanation of the methodology of *ex suppositione* reasoning, a topic touched on in one of Moody's papers in this collection entitled "Ockham, Buridan, and Nicholas of Autrecourt."[18] In view of Buridan's well-known condemnation, while rector of the University of Paris, of Nicholas's teaching and the suspicion that this condemnation was actually directed against Ockhamism, Moody decided to study the complex relationships between Ockham, Buridan, and Nicholas to ascertain the precise target of the condemnation and whether Ockhamism was *de facto* involved. Moody's conclusion, which comes as no surprise, is that the condemnation was indeed against Nicholas but that it was not anti-Ockhamist, at least not against the type of Ockhamism advocated by either Ockham or Buridan.[19] Moody points to various passages in Buridan's commentaries on Aristotle where he defends, apparently against Nicholas, the validity of causal analysis and man's ability to achieve certain knowledge of nature; such passages, of course, can easily serve to align Buridan with the Thomistic tradition, as Moody was well aware.[20] What is surprising is Moody's attempt to align Buridan with Ockham's position on similar matters. Now Ockham's denial of local motion as a distinct reality and his clear assertion of the inapplicability of causal analysis to this phenomena was certainly *not* accepted by Buridan; had it been, the impetus theory would never have been developed, to say nothing of the subsequent studies in medieval dynamics that make Buridan and his followers so important for the history of science generally.[21] And in the matter of certain, scientific knowledge of the world of nature, Buridan's commitment was much stronger than Ockham's; if it is to be identified with any medieval tradition, it fits more readily with Aquinas's than with that of the Venerable Inceptor.

Ockham, like Aristotle, had a theory of demonstration, but as De Rijk has made clear, for Ockham a demonstration is nothing more than a disguised hypothetical argument and thus is not completely apodictic.[22] Unfortunately

Moody reads Buridan with precisely this Ockhamist bias, and so he interprets Buridan's claim that *scientia naturalis* is capable of attaining truth and certitude in a rather peculiar way. Failing to understand, as we see it, the technique of demonstration *ex suppositione*, which for Aquinas could lead to true and certain results, Moody interprets Buridan's use of the expression *ex suppositione* to mean that Buridan is advocating a type of hypothetico-deductive reasoning as proper to the natural sciences. So he draws the inference that, with Buridan,

> an ineradicable element of *hypothesis* is introduced into the science of nature, and, as its counterpart, the principle that all scientific hypotheses require empirical verification, and retain an element of probability which cannot be completely eliminated.[23]

We do not believe that this is the correct meaning of Buridan's thesis. Its exposition occurs in Bk. 2, q. 1, of Buridan's commentary on Aristotle's *Metaphysics*, which inquires, "Whether it is possible for us to comprehend the truth concerning things?"[24] Buridan answers the question affirmatively through a precise and thorough analysis of the types of evidence on which truth and certitude must ultimately rest. From this he draws an inference that is quite different from the one Moody attributes to him. Buridan's own words read:

> It follows as a corollary that some people do great harm when they attempt to destroy the natural and moral sciences because of the fact that in many of their principles and conclusions there is no evidence *simpliciter*, and so they can be falsified through cases that are supernaturally possible; for evidence *simpliciter* is not required for such sciences, since it suffices for them that they have evidence *secundum quid* or *ex suppositione*. Thus Aristotle speaks well in the second [book] when he says that mathematical certitude is not to be sought in every science. And since it is now apparent that firmness of truth and firmness of assent are possible for us in all the aforementioned modes, we can conclude with regard to our question that the comprehension of truth with certitude is possible for us.[25]

To affirm that "the comprehension of truth with certitude is possible for us" seems quite different from affirming, as Moody does, that "all scientific hypotheses require empirical verification and retain an element of probability which cannot be completely eliminated." The later affirmation would reduce science to dialectics, it would clearly eliminate apodictic certitude from all scientific conclusions, and this is precisely the error Buridan has set himself to refute.

Now it seems more than coincidental that Galileo made many epistemological claims for science and demonstration in the matters with which he worked, and that he, like Aquinas and Buridan, very frequently justified his results by an appeal to reasoning *ex suppositione.*[26] This technique, as explained elsewhere, is implicit in Aristotle's *Physics* and *Posterior Analytics*, and it was explicitly shown by Aquinas to be capable of generating strict demonstration in the contingent subject matters with which natural science is concerned.[27] However, most commentators on Galileo, and most translators of his works, fail to grasp the nuances of this methodology and interpret Galileo, as Moody interprets Buridan, to be advocating and employing the hypothetico-deductive methods used in twentieth-century science. Such methods, of course, could never achieve the results that Galileo claimed, either by demonstrating the truth of the Copernican system or by establishing the *nuova scienza* of local motion of which he was so justly proud. To see Galileo as practicing a method that derives from Aquinas, on the other hand, and perhaps *via* Buridan but surely not *via* Ockham, would be to locate him in a methodological tradition that provided adequate canons for attaining the demonstrative certitude he claimed, however defective he himself might have been in applying such canons to the materials he had at hand.

Moody, moreover, notes of Galileo that his medieval thought context was essentially that of the thirteenth century, and suggests that "the sources of [his] dynamics . . . are to be sought elsewhere than in the tradition of fourteenth century Oxford or Paris, or than in the tradition of fifteenth century Padua."[28] Now our recent work on the sources of Galileo's Pisan notebooks, oddly enough, would appear to confirm the validity of this particular insight. Much yet remains to be done, for the work is still actively in progress, but results to date strongly suggest that the main source of Galileo's early writings on logic and the physical sciences were contemporary Jesuit professors at the Collegio Romano.[29] These Jesuits were all thoroughly trained in the Thomistic tradition, but they were also eager to search through and evaluate the common teachings of the Schools, and their works are replete with references to Averroists, Scotists, nominalists, and others. It would not be surprising if Galileo derived his knowledge of Avempace, for example, from the writings (mainly *reportationes* of lectures) of such Jesuits. And this fact alone would serve to explain why Galileo's discussion continues to focus on issues that were central in thirteenth-century thought, even though they touch tangentially on problems dating from the fourteenth and fifteenth centuries that have been regarded for so long as the seed bed of modern science.

Moody's heroes, by his own admission, were the fourteenth-century thinkers who contributed much to logic and to the mathematical modes of thought that have become popular among philosophers in our own "age of analysis." Like many others, Duhem and Maier included, he did his history of science with an ulterior goal in mind: he thought that careful studies of the type he engaged in would cast light on present-day problems and perhaps point the way to new directions for the future.[30] Having such a goal did not corrupt his historical scholarship: withal he was careful, objective, dogged in his search for truth, and ever willing to pursue that search wherever it might lead. His loss at this time, needless to say, is deeply felt, all the more because of the new research materials that are becoming available on Galileo and his relationships to medieval science. That particular problem engaged much of Moody's effort over a long period of his life, and he was uniquely endowed to give a critical evaluation of the many factors that bear on its solution. Our own reaction to the new materials (again, predictably) is that they connect Galileo's *nuova scienza* much more strongly with the *via antiqua* of Aquinas than they do with the *via moderna* of Ockham. This is not to say, of course, that Ockham and the nominalist movement were unimportant either for Galileo or for the rise of modern science. To be convinced of that all one need do is read these collected papers that summarize Moody's life work so well, and that now stand as such a fitting memorial to his scholarly endeavors.

NOTES

1 Ernest A. Moody, *Studies in Medieval Philosophy, Science, and Logic*. Collected Papers, 1933–1969. Berkeley: University of California Press, 1975, 477 pp.
2 Moody's books include *The Logic of William of Ockham* (New York and London: 1935), *The Medieval Science of Weights* (*Scientia de ponderibus*), coauthored with Marshall Clagett (Madison, Wisconsin: 1952), and *Truth and Consequence in Medieval Logic* (Amsterdam: 1953).
3 Preface, p. xi.
4 Foreword, p. viii; see also pp. 300, 302.
5 Pp. 203–286.
6 P. 244.
7 'Saint Thomas Aquinas,' *DSB* 1: 196–200, esp. p. 198.
8 'The Principle *Omne quod movetur ab alio movetur* in Medieval Physics,' *Isis* 56 (1965), pp. 26–45, and 'Motion in a Void: Aquinas and Averroes', in *St. Thomas Aquinas Commemorative Studies 1274–1974*, 2 vols., Toronto: Pontifical Institute of Mediaeval Studies, 1974, Vol. 1, pp. 469–488.
9 Moody, *Studies*, p. 215.
10 *Ibid.*, p. 227.
11 *Ibid.*, p. 236.
12 *Ibid.*, p. 242.
13 Weisheipl, 'Motion in a Void,' pp. 476, 487.
14 *Ibid.*, p. 480.

15 *Ibid.*, pp. 469–470.
16 *Ibid.*, pp. 487–488.
17 The beginnings of these researches are reported in Essay 9 *supra*; see also notes 26 and 29 below.
18 Moody, *Studies*, pp. 127–160.
19 *Ibid.*, pp. 157–160.
20 *Ibid.*, p. 154.
21 Some references to these teachings are given in my *Causality and Scientific Explanation*, 2 vols., Ann Arbor: The University of Michigan Press, 1972–1974, Vol. 1, pp. 53–55, 104–109.
22 L. M. De Rijk, 'The Development of *Suppositio naturalis* in Mediaeval Logic,' *Vivarium* **11** (1973), pp. 43–79, esp. p. 54.
23 Moody, *Studies*, p. 156.
24 Iohannes Buridanus, *In metaphysicen Aristotelis quaestiones* . . . , Paris: 1518 (reprinted Frankfurt a. M.: 1964), fol. 8r, Utrum de rebus sit nobis possibilis comprehensio veritatis.
25 *Ibid.*, fol. 9r: Ideo conclusum est correlarie quod aliqui valde mali dicunt volentes interimere scientias naturales et morales eo quod in pluribus earum principiis et conclusionibus non est evidentia simplex sed possunt falsificari per casus supernaturaliter possibiles, quia non requiritur ad tales scientias evidentia simpliciter sed sufficiunt predicte evidentie secundum quid sive ex suppositione. Ideo bene dicit Aristoteles secundo huius quod non in omnibus scientiis mathematica acribologia est expetenda. Et quia iam apparuit quod omnibus predictis modis firmitas veritatis et firmitas assensus sunt nobis possibiles, ideo concludendum est quid querebatur, scilicet, quod nobis est possibilis comprehensio veritatis cum certitudine . . . .
26 The specific texts are discussed and analyzed at length in Essay 8 *supra*.
27 For a summary account see the author's 'Aquinas on the Temporal Relation Between Cause and Effect,' *Review of Metaphysics* 27 (1974), pp. 569–584, esp. pp. 572–574.
28 Moody, *Studies*, p. 274.
29 See Essay 10 *supra*.
30 Moody, *Studies*, pp. 287–304 and 305–320.

BIBLIOGRAPHY

Beltrán de Heredia, V., 1960: *Domingo de Soto.* Estudio biográfico documentado, Biblioteca de Teologos Españoles, Salamanca.

Blake, R. M., Ducasse, C. J., and Madden, E. H., 1960: *Theories of Scientific Method.* The Renaissance through the Nineteenth Century, Univ. of Washington Press, Seattle.

Boehner, P., ed., 1957: *William of Ockham: Philosophical Writings*, Thomas Nelson, Edinburgh.

Borgnet, A., ed., 1890–1899: Albertus Magnus, *Opera omnia*, 38 vols., Ludovicus Vivès, Paris.

Boyer, C. B., 1949: *The Concepts of the Calculus*, Columbia Univ. Press, New York.

Boyer, C. B., 1959: *The Rainbow: From Myth to Mathematics*, Thomas Yoseloff, New York.

Busa, R., ed., 1975: *Index Thomisticus.* Sancti Thomae Aquinatis Operum Omnium Indices et Concordantiae, Friedrich Frommann Verlag, Stuttgart.

Butterfield, H., 1965: *The Origins of Modern Science*, rev. ed., The Free Press, New York.

Clagett, M., 1959: *The Science of Mechanics in the Middle Ages*, Univ. of Wisconsin Press, Madison.

Clagett, M., 1968: *Nicole Oresme and the Medieval Geometry of Qualities and Motions*, Univ. of Wisconsin Press, Madison.

Clavelin, M., 1974: *The Natural Philosophy of Galileo*, tr. A. J. Pomerans, The M.I.T. Press, Cambridge, Mass.

Copleston, F., 1972: *A History of Medieval Philosophy*, Methuen & Co., Ltd., London.

Cosentino, G., 1970: 'Le Matematiche nella 'Ratio Studiorum' della Compagnia di Gesù,' *Miscellanea Storica Ligure* **2**, 171–123.

Cosentino, G., 1971: 'L'Insegnamento delle Matematiche nei Collegi Gesuitici nell'Italia settentrionale. Nota Introduttiva,' *Physis* **13**, 205–217.

Crombie, A. C., 1953: *Robert Grosseteste and the Origins of Experimental Science*, Oxford Univ. Press, Oxford.

Crombie, A. C., 1959: *Medieval and Early Modern Science*, 2 vol., rev. ed., Doubleday Anchor, New York.

Crombie, A. C., 1975: 'Sources of Galileo's Early Natural Philosophy,' in *Reason, Experiment, and Mysticism in the Scientific Revolution*, M. L. Righini Bonelli and W. R. Shea, eds., Science History Publications, New York, 157–175, 303–305.

Crosby, H. L., ed., 1955: *Thomas of Bradwardine: His 'Tractatus de Proportionibus,'* Univ. of Wisconsin Press, Madison.

Dales, R. C., 1973: *The Scientific Achievement of the Middle Ages*, Univ. of Pennsylvania Press, Philadelphia.

Dampier, W. C., 1968: *A History of Science and Its Relation to Philosophy and Religion*, 4th ed., Cambridge Univ. Press, Cambridge.

De Rijk, L. M., 1973: 'The Development of *Suppositio naturalis* in Medieval Logic,' *Vivarium* **11**, 43–79.

Dijksterhuis, E. J., 1961: *The Mechanization of the World Picture*, tr. C. Dikshoorn, Oxford Univ. Press, Oxford.

Drake, S., 1957: *Discoveries and Opinions of Galileo*, Doubleday, Garden City.

Drake, S., 1959: 'An Unrecorded Manuscript Copy of Galileo's *Cosmography*,' *Physics* **1**, 294–306.

Drake, S., and Drabkin, I. E., eds., 1960: *Galileo Galilei*: *'On Motion' and 'On Mechanics*,' Univ. of Wisconsin Press, Madison.

Drake, S., and Drabkin, I. E., eds., 1969: *Mechanics in Sixteenth-Century Italy*, Univ. of Wisconsin Press, Madison.

Drake, S., 1970: *Galileo Studies: Personality, Tradition, and Revolution*, Univ. of Michigan Press, Ann Arbor.

Drake, S., 1973a: 'Galileo's Discovery of the Law of Free Fall,' *Scientific American* **228**, 84–92.

Drake, S., 1973b: 'Galileo's Experimental Confirmation of Horizontal Inertia: Unpublished Manuscripts,' *Isis* **64**, 209–305.

Drake, S., ed., 1974: *Galileo Galilei: Two New Sciences*, Univ. of Wisconsin Press, Madison.

Duhem, P., 1906: *Etudes sur Léonard de Vinci*, Vol. 1, Hermann et Fils, Paris.

Duhem, P., 1908: *Sozein ta Phainomena*: Essai sur la notion de théorie physique de Platon à Galilée, Hermann et Fils, Paris.

Duhem, P., 1909: *Etudes sur Léonard de Vinci*, Vol. 2, Hermann et Fils, Paris.

Duhem, P., 1913: *Etudes sur Léonard de Vinci*, Vol. 3, Hermann et Fils, Paris.

Duhem, P., 1954: *The Aim and Structure of Physical Theory*, tr. P. P. Wiener, Princeton Univ. Press, Princeton.

Duhem, P., 1913–1959: *Le Système du monde*, 10 vols., Hermann et Fils, Paris.

Duhem, P., 1969: *To Save the Phenomena*, tr. E. Doland and C. Maschler, Univ. of Chicago Press, Chicago.

Edwards, P., ed., 1967: *Encyclopedia of Philosophy*, 8 vols., Macmillan and Free Press, New York.

Effler, R. R., 1962: *John Duns Scotus and the Principle 'Omne Quod Movetur ab Alio Movetur*,' Franciscan Institute, St. Bonaventure, New York.

Elie, H., 1950–1951: 'Quelques maitres de l'université de Paris vers l'an 1500,' *Archives d'histoire doctrinale et littéraire du moyen âge* **18**, 193–243.

Favaro, A., ed., 1890–1909: *Le Opere di Galileo Galilei*, 20 vols. in 21, G. Barbèra, Florence.

Favaro, A., 1916: 'Léonard de Vinci a-t-il exercé une influence sur Galilée et son école?' *Scientia* **20**, 257–265.

Favaro, A., 1918: 'Galileo Galilei e i Doctores Parisienses,' *Rendiconti della R. Accademia dei Lincei* **27**, 3–14.

Favaro, A., 1968: *Galileo Galilei a Padova*, Editrice Antenore, Padua.

Finnochiaro, M. A., 1972: '*Vires acquirit eundo*: The Passage Where Galileo Renounces Space-Acceleration and Causal Investigation,' *Physics* **14**, 125–145.

Fredette, R., 1972: 'Galileo's *De motu antiquiora*,' *Physics* **14**, 321–348.

Fredette, R., 1975: 'Bringing to Light the Order of Composition of Galileo Galilei's *De motu antiquiora*,' unpublished paper delivered at the Workshop on Galileo at Virginia Polytechnic Institute and State University, Blacksburg, Virginia.

Garin, E., 1969: *Science and Civic Life in the Italian Renaissance*, tr. P. Munz, Doubleday Anchor, New York.

Geymonat, L., 1965: *Galileo Galilei: A Biography and Inquiry into His Philosophy of Science*, tr. S. Drake, McGraw-Hill Book Co., New York.

Gilbert, N. W., 1960: *Renaissance Concepts of Method*, Columbia Univ. Press, New York.

Gillispie, C. C., 1970–1980: *Dictionary of Scientific Biography*, 16 vols., Charles Scribner's Sons, New York.

Gilson, E., 1938: *Reason and Revelation in the Middle Ages*, Charles Scribner's Sons, New York.

Gilson, E., 1955: *History of Christian Philosophy in the Middle Ages*, Random House, New York.

Golino, C. L., ed., 1966: *Galileo Reappraised*, Univ. of California Press, Berkeley.

Grant, E., ed., 1966: *Nicole Oresme: 'De proportionibus proportionum' and 'Ad pauca respicientes,'* Univ. of Wisconsin Press, Madison.

Grant, E., 1971: *Physical Science in the Middle Ages*, John Wiley and Sons, New York.

Grant, E., ed., 1974: *A Source Book in Medieval Science*, Harvard Univ. Press, Cambridge, Mass.

Hales, W., 1830: *A New Analysis of Chronology and Geography, History and Prophecy*, 4 vols., C. J. G. & F. Rivington, London.

Harré, R., 1972: *The Philosophies of Science: An Introductory Survey*, Oxford Univ. Press, Oxford and New York.

Hoskin, M. A., and Molland, A. G., 1966: 'Swineshead on Falling Bodies: An Example of Fourteenth-Century Physics,' *British Journal for the History of Science* **3**, 150–182.

Jammer, M., 1957: *Concepts of Force*, Harvard Univ. Press, Cambridge, Mass.

Knowles, D., 1962: *The Evolution of Medieval Thought*, Helicon Press, Baltimore.

Koertge, N., 1974: 'Galileo and the Problem of Accidents,' paper read to the British Society for the Philosophy of Science; see 1977.

Koertge, N., 1977: 'Galileo and the Accidents,' *Journal of the History of Ideas* **38**, 389–408.

Koyré, A., 1939: *Etudes Galiléenes*, Hermann, Paris, reprinted 1966.

Koyré, A., 1957: *From the Closed World to the Infinite Universe*, The Johns Hopkins Press, Baltimore.

Koyré, A., 1968: *Metaphysics and Measurement. Essays in the Scientific Revolution*, Harvard Univ. Press, Cambridge, Mass.

Koyré, A., and Cohen, I. B., eds., 1972: *Isaac Newton's Philosophiae Naturalis Principia Mathematica*, 2 vols., Harvard Univ. Press, Cambridge, Mass.

Leff, G., 1958: *Medieval Thought: St. Augustine to Ockham*, Penguin Books, Baltimore.

Leff, G., 1968: *Paris and Oxford Universities in the Thirteenth and Fourteenth Centuries: An Institutional and Intellectual History*, John Wiley and Sons, New York.

Leff, G., 1975: *William of Ockham: The Metamorphosis of Scholastic Discourse*, Manchester Univ. Press, Manchester.

Lewis, C. J. T., 1976: 'The Fortunes of Richard Swineshead in the Time of Galileo,' *Annals of Science* **33**, 561–584.

Lindberg, D. C., 1976: *Theories of Vision from al-Kindi to Kepler*, Univ. of Chicago Press, Chicago.

Lindberg, D. C., ed., 1978: *Science in the Middle Ages*, Univ. of Chicago Press, Chicago.

Litt, T., 1963: *Les Corps célestes dans l'univers de saint Thomas d'Aquin*, Publications Universitaires, Louvain.

Lohr, C. H., 1974–1979: 'Renaissance Latin Aristotle Commentaries,' *Studies in the*

Renaissance **21**, 228–289, and *Renaissance Quarterly* **28**, 689–741; **29**, 714–745; **30**, 681–741; and **31**, 532–603.

Lohr, C. H., 1976: 'Jesuit Aristotelianism and Suarez's *Disputationes Metaphysicae*', *Paradosis: Studies in Memory of E. A. Quain*, New York.

Machamer, P., 1978: 'Galileo and the Causes,' in *New Perspectives on Galileo*, R. E. Butts & J. C. Pitt, eds., D. Reidel Publishing Company, Dordrecht and Boston, 161–180.

MacLachlan, J., 1973: 'A Test of an "Imaginary" Experiment of Galileo,' *Isis* **64**, 374–379.

Maier, A., 1949: *Die Vorläufer Galileis im 14. Jahrhundert*, Edizioni di Storia e Letteratura, Rome.

Maier, A., 1952: *An der Grenze von Scholastik und Naturwissenschaft*, Edizioni di Storia e Letteratura, Rome.

Maier, A., 1958: *Zwischen Philosophie und Mechanik*, Edizioni di Storia e Letteratura, Rome.

Maier, A., 1967: *Ausgehendes Mittelalter II*, Edizioni di Storia e Letteratura, Rome.

McDonald, W., ed., 1967: *New Catholic Encyclopedia*, 15 vols., McGraw-Hill Book Co., New York.

McMullin, E., ed., 1967: *Galileo Man of Science*, Basic Books, New York.

McMullin, E., 1978: 'The Conception of Science in Galileo's Work,' in *New Perspectives on Galileo*, R. E. Butts & J. C. Pitt, eds., D. Reidel Publishing Co., Dordrecht and Boston, 209–257.

McTighe, T. P., 1967: 'Galileo's 'Platonism': A Reconsideration,' in *Galileo Man of Science*, E. McMullin, ed., Basic Books, New York, 365–387.

Menut, A. D., and Denomy, A. J., eds., 1968: *Nicole Oresme: Le Livre du ciel et du monde*, Univ. of Wisconsin Press, Madison.

Moody, E. A., 1935: *The Logic of William of Ockham*, Sheed and Ward, New York and London.

Moody, E. A., 1951: 'Galileo and Avempace: The Mechanics of the Leaning Tower Experiment,' *Journal of the History of Ideas* **12**, 163–193, 375–422.

Moody, E. A., and Clagett, M., eds., 1952: *The Medieval Science of Weights* (*Scientia de ponderibus*), Univ. of Wisconsin Press, Madison.

Moody, E. A., 1953: *Truth and Consequence in Medieval Logic*. North-Holland Publishing Company, Amsterdam.

Moody, E. A., 1966: 'Galileo and His Precursors,' in *Galileo Reappraised*, C. L. Golino, ed., Univ. of California Press, Berkeley, 23–43.

Moody, E. A., 1975: *Studies in Medieval Philosophy, Science, and Logic: Collected Papers, 1933–1969*, Univ. of California Press, Berkeley.

Muñoz Delgado, V., 1964: *La Logica Nominalista en la Universidad de Salamanca* (*1510–1530*), Publicaciones del Monasterio de Poyo, Madrid.

Muñoz Delgado, V., 1967: 'La Logica en Salamanca durante la primera mitad del siglo XVI,' *Salmanticensis* **14**, 171–207.

Murdoch, J. E., 1962: *Rationes mathematice: Un Aspect du rapport des mathématiques et de la philosophie au moyen âge*, Univ. of Paris, Paris.

Murdoch, J. E., 1974: 'Philosophy and the Enterprise of Science in the Later Middle Ages,' in *The Interaction Between Science and Philosophy*, Y. Elkana, ed., Humanities Press, Atlantic Highlands, N.J.

Murdoch, J. E., 1975: 'From Social into Intellectual Factors: An Aspect of the Unitary Character of Late Medieval Learning,' in *The Cultural Context of Medieval Learning*, J. E. Murdoch and E. D. Sylla, eds., D. Reidel Publishing Company, Dordrecht and Boston, 271–384.

Murdoch, J. E., 1978: 'The Development of a Critical Temper: New Approaches and Modes of Analysis in Fourteenth-Century Philosophy, Science, and Theology,' *Medieval and Renaissance Studies* (Univ. of North Carolina) 7, 51–79.

Naylor, R. H., 1974: 'Galileo and the Problem of Free Fall,' *The British Journal for the History of Science* 7, 105–134.

Naylor, R. H., 1980: 'Galileo's Theory of Projectile Motion,' *Isis* **71**, 550–570.

Oberman, H. A., 1975: 'Reformation and Revolution: Copernicus's Discovery in an Era of Change,' in *The Cultural Context of Medieval Learning*, J. E. Murdoch and E. D. Sylla, eds., D. Reidel Publishing Company, Dordrecht and Boston, 397–435.

Pederson, O., 1953: 'The Development of Natural Philosophy, 1250–1350,' *Classica et Mediaevalia* **14**, 86–155.

Pirotta, A. M., 1936: *Summa Philosophiae Aristotelico-Thomisticae*, 4 vols., Libraria Marietti, Turin, **2**, 315–329.

Purnell, F., 1972: 'Jacopo Mazzoni and Galileo,' *Physis* **14**, 273–294.

Ramírez, S. M., 1952: 'Hacia una renovación de nuestros estudios filosóficos: Un indice de la producción filosófica de los Dominicos españoles,' *Estudios Filosóficos* **1**, 5–25.

Randall, J. H., Jr., 1961: *The School of Padua and the Emergence of Modern Science*, Editrice Antenore, Padua.

Rashdall, H., 1936: *The Universities of Europe in the Middle Ages*, 3 vols., ed. F. M. Powicke and A. B. Emden, Oxford Univ. Press, Oxford.

Ray Pastor, J., 1926: *Los Mathemáticos españoles del siglo XVI*, Biblioteca Scientia, Toledo.

Righini Bonelli, M. L., and Shea, W. R., 1975: *Reason, Experiment, and Mysticism in the Scientific Revolution*, Science History Publications, New York.

Rose, P. L., and Drake, S., 1971: 'The Pseudo-Aristotelian *Questions of Mechanics* in Renaissance Culture,' *Studies in the Renaissance* **18**, 65–104.

Russell, J. L., 1976: 'Action and Reaction Before Newton,' *The British Journal for the History of Science* **9**, 25–38.

Sarton, G., 1931: *Introduction to the History of Science*, 3 vols., Williams and Wilkins, Baltimore, Vol. 2.

Scaduto, M., 1964: *Storia della Compagnia di Gesù in Italia*, 2 vols., Gregorian University Press, Rome.

Schmitt, C. B., 1967: 'Experimental Evidence for and against a Void: The Sixteenth-Century Arguments,' *Isis* **58**, 325–366.

Schmitt, C. B., 1969: 'Experience and Experiment: A Comparison of Zabarella's View with Galileo's in *De motu*,' *Studies in the Renaissance* **16**, 80–138.

Schmitt, C. B., 1971: *A Critical Survey and Bibliography of Studies on Renaissance Aristotelianism 1958–1969*, Editrice Antenore, Padua.

Schmitt, C. B., 1972: 'The Faculty of Arts at Pisa at the Time of Galileo,' *Physics* **14**, 243–272.

Schmitt, C. B., 1975: 'Philosophy and Science in Sixteenth-Century Universities: Some Preliminary Comments,' in *The Cultural Context of Medieval Learning*, J. E.

Murdoch and E. D. Sylla, eds., D. Reidel Publishing Company, Dordrecht and Boston, 485–537.

Schmitt, C. B., 1976: 'Hieronymus Picus, Renaissance Platonism, and the Calculator,' *International Studies in Philosophy* 8, 57–80.

Settle, T. B., 1961: 'An Experiment in the History of Science,' *Science* **133**, 19–23.

Settle, T. B., 1967: 'Galileo's Use of Experiment as a Tool of Investigation,' in *Galileo Man of Science*, E. McMullin, ed., Basic Books, New York, 315–337.

Shapere, D., 1974: *Galileo: A Philosophical Study*, Univ. of Chicago Press, Chicago and London.

Shapiro, H., 1957: *Motion, Time and Place According to William Ockham*, Franciscan Institute, St. Bonaventure, New York.

Shea, W. R., 1972: *Galileo's Intellectual Revolution. Middle Period 1610–1632*, Science History Publications, New York.

Shea, W. R., and Wolf, N. S., 1975: 'Stillman Drake and the Archimedean Grandfather of Experimental Science,' *Isis* **66**, 397–400.

Smith, D. E., 1908: *Rara Arithmetica*: A Catalogue of the Arithmetics Written before the Year MDCI with a Description of those in the Library of George Arthur Plimpton of New York, Ginn and Company, Boston and London.

Solana, M., 1941: *Historia de la Filosofía Española: Época del Renacimiento (Siglo XVI)*, 3 vols., Real Academia de Ciencias Exactas, Físicas y Naturales, Madrid.

Sommervogel, C., 1891: *Bibliothèque de la Compagnie de Jésus*, Vol. 2, Alphonse Picard, Paris.

Stegmüller, F., 1959: *Filosofia e Teologia nas Universidades de Coimbra e Evora no Seculo XVI*, Univ. of Coimbra, Coimbra.

Stock, B., 1972: *Myth and Science in the Twelfth Century*, Princeton Univ. Press, Princeton.

Taton, R., ed., 1964: *History of Science: The Beginnings of Modern Science from 1450 to 1800*, tr. A. J. Pomerans, Basic Books, New York.

Thorndike, L., 1949: *The 'Sphere' of Sacrobosco and Its Commentators*, Univ. of Chicago Press, Chicago.

Valsanzibio, S. da, 1949: *Vita e dottrina di Gaetano di Thiene, filosofo dello studio di Padova, 1387–1465*, 2d ed., Studio Filosofico dei Fratrum Minorum Cappuccini, Padua.

Van Steenberghen, F., 1955: *Aristotle in the West: The Origins of Latin Aristotelianism*, tr. L. Johnson, E. Nauwelaerts, Louvain.

Vignaux, P., 1959: *Philosophy in the Middle Ages: An Introduction*, tr. E. C. Hall, Meridian Books, New York.

Villoslada, R. G., 1938: *La Universidad de Paris durante los estudios de Francisco de Vitoria, O.P., 1507–1522*, Analecta Gregoriana **14**, Gregorian Univ. Press, Rome.

Villoslada, R. G., 1954: *Storia del Collegio Romano dal suo inzio (1551) alla soppressione della Compagnia di Gesù (1773)*, Analecta Gregoriana **66**, Gregorian Univ. Press, Rome.

Wallace, W. A., 1959: *The Scientific Methodology of Theodoric of Freiberg*, Studia Friburgensia, N.S. **26**, Fribourg Univ. Press, Fribourg.

Wallace, W. A., 1961: 'Gravitational Motion According to Theodoric of Freiberg,' *The Thomist* **24**, 327–352.

Wallace, W. A., 1964: 'The Reality of Elementary Particles,' *Proceedings of the American Catholic Philosophical Association* **38**, 154–166.

Wallace, W. A., 1967: 'The Concept of Motion in the Sixteenth Century,' *Proceedings of the American Catholic Philosophical Association* **41**, 184–195, revised as Essay 4 in this volume.

Wallace, W. A., 1968a: 'The Enigma of Domingo de Soto: *Uniformiter difformis* and Falling Bodies in Late Medieval Physics,' *Isis* **59**, 384–401, revised as Essay 6 in this volume.

Wallace, W. A., 1968b: 'Elementarity and Reality in Particle Physics,' *Boston Studies in the Philosophy of Science*, R. S. Cohen and M. W. Wartofsky, eds., Vol. 3, Humanities Press, New York, 236–271.

Wallace, W. A., 1968c: 'Toward a Definition of the Philosophy of Science,' *Mélanges à la mémoire de Charles De Koninck*, Laval Univ. Press, Quebec, 465–485.

Wallace, W. A., 1969: 'The 'Calculatores' in Early Sixteenth-Century Physics,' *The British Journal for the History of Science* **4**, 221–232, revised as Essay 5 in this volume.

Wallace, W. A., 1971: 'Mechanics from Bradwardine to Galileo,' *Journal of the History of Ideas* **32**, 15–28, revised as Essay 3 in this volume.

Wallace, W. A., 1972: *Causality and Scientific Explanation*, Vol. 1, Medieval and Early Classical Science, Univ. of Michigan Press, Ann Arbor.

Wallace, W. A., 1973: 'Experimental Science and Mechanics in the Middle Ages,' *Dictionary of the History of Ideas*, ed. P. P. Wiener, 4 vols., Charles Scribner's Sons, New York, **2**, 196–205, revised as Essay 2 in this volume.

Wallace, W. A., 1974a: *Causality and Scientific Explanation*, Vol. 2, Classical and Contemporary Science, Univ. of Michigan Press, Ann Arbor.

Wallace, W. A., 1974b: 'Galileo and the Thomists,' *St. Thomas Aquinas Commemorative Studies 1274–1974*, 2 vols., Pontifical Institute of Mediaeval Studies, Toronto, **2**, 293–330, revised as Essay 9 in this volume.

Wallace, W. A., 1974c: 'Theodoric of Freiberg: On the Rainbow,' *A Source Book in Medieval Science*, E. Grant, ed., Harvard Univ. Press, Cambridge, Mass., 435–441.

Wallace, W. A., 1974d: 'Three Classics of Science: Galileo, *Two New Sciences*, etc.' in *The Great Ideas Today 1974*, Encyclopaedia Britannica, Chicago, 211–272.

Wallace, W. A., 1974e: 'Aquinas on the Temporal Relation Between Cause and Effect,' *The Review of Metaphysics* **27**, 569–584.

Wallace, W. A., 1976a: 'Galileo and Reasoning *Ex suppositione*: The Methodology of the *Two New Sciences*,' *Proceedings of the 1974 Biennial Meeting of the Philosophy of Science Association*, ed. R. S. Cohen et al., D. Reidel Publishing Company, Dordrecht and Boston, 79–104, enlarged with an appendix as Essay 8 in this volume.

Wallace, W. A., 1976b: 'Buridan, Ockham, Aquinas: Science in the Middle Ages,' *The Thomist* **40**, 475–483, revised as Essay 16 in this volume.

Wallace, W. A., 1977a: *Galileo's Early Notebooks: The Physical Questions*. A Translation from the Latin, with Historical and Paleographical Commentary, Notre Dame University Press, Notre Dame.

Wallace, W. A., 1977b: *The Elements of Philosophy*. A Compendium for Philosophers and Theologians. Alba House, New York.

Wallace, W. A., 1978a: 'Galileo Galilei and the *Doctores Parisienses*,' *New Perspectives on Galileo*, R. E. Butts and J. C. Pitt, eds., D. Reidel Publishing Company, Dordrecht and Boston, 87–138, enlarged with an appendix as Essay 10 in this volume.

Wallace, W. A., 1978b: 'Causes and Forces in Sixteenth-Century Physics,' *Isis* **69**, 400–412, revised as Essay 7 in this volume.

Wallace, W. A., 1978c: 'The Philosophical Setting of Medieval Science,' *Science in the Middle Ages*, D. Lindberg, ed., Univ. of Chicago Press, Chicago and London, 91–119, revised as Essay 1 in this volume.

Wallace, W. A., 1978d: 'Galileo's Knowledge of the Scotistic Tradition,' *Regnum Hominis et Regnum* Dei, C. Bérubé, ed., 2 vols., Societas Internationalis Scotistica, Rome, 2, 313–320, revised as Essay 11 in this volume.

Wallace, W. A., 1979: 'Medieval and Renaissance Sources of Modern Science: A Revision of Duhem's Continuity Thesis, Based on Galileo's Early Notebooks,' *Proceedings of the Patristic-Medieval-Renaissance Conference 1977*, Augustinian Historical Institute, Villanova, Pennsylvania, 2, 1–17, revised as Essay 14 in this volume.

Wallace, W. A., 1980a: 'Galileo's Citations of Albert the Great,' *Albert the Great: Commemorative Essays*, F. J. Kovach and R. W. Shahan, eds., Univ. of Oklahoma Press, Norman, 261–283, appearing as Essay 12 in this volume.

Wallace, W. A., 1980b: 'Albertus Magnus on Suppositional Necessity in the Natural Sciences,' *Albertus Magnus and the Sciences: Commemorative Essays 1980*, J. A. Weisheipl, ed., Pontifical Institute of Mediaeval Studies, Toronto, 103–128.

Webering, D., 1953: *Theory of Demonstration According to William Ockham*, Franciscan Institute, St. Bonaventure, New York.

Weisheipl, J. A., 1956: *Early Fourteenth-Century Physics of the Merton 'School' with Special Reference to Dumbleton and Heytesbury*, D. Phil. Dissertation, Oxford Univ., Oxford.

Weisheipl, J. A., 1959: *The Development of Physical Theory in the Middle Ages*, Sheed and Ward, New York and London, reprinted 1971, Univ. of Michigan Press, Ann Arbor.

Weisheipl, J. A., 1965: 'The Principle *Omne quod movetur ab alio movetur* in Medieval Physics,' *Isis* 56, 26–45.

Weisheipl, J. A., 1974a: *Friar Thomas d'Aquino. His Life, Thought, and Work*, Doubleday and Company, Garden City, New York.

Weisheipl, J. A., 1974b: 'Motion in a Void: Averroës and Aquinas,' *St. Thomas Aquinas Commemorative Studies 1274–1974*, 2 vols., Pontifical Institute of Mediaeval Studies, Toronto, 1, 467–488.

Weisheipl, J. A., 1976: 'Aristotle's Concept of Nature: Avicenna and Aquinas,' Tenth Annual Conference, *Nature in the Middle Ages*, Center for Medieval and Early Renaissance Studies, State University of New York, Binghamton.

Weisheipl, J. A., ed., 1980: *Albertus Magnus and the Sciences: Commemorative Essays 1980*, Pontifical Institute of Mediaeval Sudies, Toronto.

Westfall, R. S., 1971: *Force in Newton's Physics*, American Elsevier and Macdonald, New York and London.

Whitehead, A. N., 1925: *Science and the Modern World*, The Macmillan Company, New York.

Whittaker, E., 1946: *Space and Spirit*. Theories of the Universe and the Arguments for the Existence of God. Thomas Nelson and Company, Edinburgh.

Wiener, P. P., ed., 1973: *Dictionary of the History of Ideas*, 4 vols., Charles Scribner's Sons, New York.

Wilson, C. A., 1960: *William Heytesbury: Medieval Logic and the Rise of Mathematical Physics*, Univ. of Wisconsin Press, Madison.

Wisan, W. L., 1974: 'The New Science of Motion: A Study of Galileo's *De motu locali*,' *Archive for History of Exact Sciences* **13**, 103–306.

Wisan, W. L., 1978: 'Galileo's Scientific Method: A Reexamination,' *New Perspectives on Galileo*, R. E. Butts and J. C. Pitt, eds., D. Reidel Publishing Company, Dordrecht and Boston, 1–57.

Wolff, M., 1971: *Fallgesetz und Massebegriff: Zwei wissenschafts-historische Untersuchungen zur Kosmologie des Johannes Philoponus*, Walter de Gruyter & Company, Berlin.

Wolter, A., ed., 1962: *Duns Scotus: Philosophical Writings*, Thomas Nelson and Sons, Edinburgh, London, and New York.

Yates, F. A., 1964: *Giordano Bruno and the Hermetic Tradition*, Univ. of Chicago Press, Chicago.

INDEX OF NAMES

Abelard, Peter 6, 20
Abū'l Barakāt 41, 42
Achillini, Alessandro 87, 196, 248, 250, 271, 272
Adam Wodham 65
Ahmad ibn Yusf 37
Albategni 196
Albert the Great 8, 9, 13, 16, 19–23, 31, 132, 157, 162, 171, 184, 196, 247–250, 264–284, 290, 294, 304, 306, 308, 316, 319, 326, 338
Albert of Saxony 24, 25, 40, 42, 45, 54–57, 62, 65–71, 80–88, 93–108, 190, 192, 238, 245–251, 306, 326, 329, 332, 338
Albumasar 197
Alciati, Terentius 212
Alexander of Aphrodisias 45, 162, 164, 180, 184, 196, 267–276, 308
Alexander of Hales 8, 186
Alfonso, King of Castile 196, 221
Alfraganus 196
Algazel 15, 196, 274
Alhazen 34
Alkindi 32
Almainus 173
Alpetragius 196
Alvaro Thomaz 46, 56, 63, 76, 80–90, 100–102, 109
Amat de Graveson, Hyacinth 238, 239
Ambrose, St. 308
Angelo da Fossombrone 98
Anselm of Canterbury, St. 6
Antonius Andreas 196, 248, 253–260
Apollonius of Perga 44
Aquarius 247
Archimedes 32, 44, 134, 141–144, 150, 151, 184, 248–251, 308–313
Aristarchus of Samos 197
Aristotle 4–45, 53, 55, 66, 70, 81–85, 90, 91, 95, 102, 110, 111, 119–123, 132–136, 151–154, 160–169, 177, 185–189, 194–195, 207, 212–219, 226, 244–250, 258, 259, 264–268, 272–280, 287–290, 295–298, 303–317, 320–325, 328–333, 337–348
Arnauld of Villanova 32
Astudillo, Diego de 79, 83–89, 102–105, 109, 187–190
Augustine of Hippo, St. 4–11, 15, 19, 25, 196, 308
Aureolus, Peter 186, 189, 243, 258
Avempace 48, 125, 196, 271, 272, 308, 314, 315, 342–346
Averroës 15–19, 45, 48, 136, 157, 160, 164, 171, 172, 180, 181, 186, 195, 196, 207, 230, 246–250, 259, 268, 271–277, 280, 283, 286–289, 305–308, 332, 343, 344
Avicebron 196, 271–273
Avicenna 15–19, 41, 171, 196, 248, 271–276, 286–289, 297, 298, 308

Balduinus, Hieronymus 197
Baliani, Giovanni Battista 142, 144
Bañez, Domingo 77, 85, 88, 90, 175–177, 185, 187, 190
Basil, St. 308
Bassanus Politus 80, 86, 87
Bede, Venerable 308
Bellarmine, St. Robert 140, 231, 232, 241
Benedetti, Giovanni Battista 47, 58, 59, 310, 312
Biel, Gabriel 196, 213, 243, 251, 255–258
Boethius 5, 6, 9, 20, 37, 82
Boethius of Sweden 16
Bonaventure, St. 8, 12, 16, 19, 185, 186, 196, 214, 243, 271–273

Borri (Borro), Girolamo 116, 248, 250, 310, 313
Bradwardine, Thomas 23, 24, 32, 37–40, 46, 51–54, 59, 82–89, 92, 283, 306, 312, 319, 343
Bruno, Giordano 45
Buccaferreus, Ludovicus 153, 171, 196, 197, 248, 280
Bufalo, Stephanus del 182
Buonamici, Francesco 59, 180, 190, 193, 194, 208, 254, 308, 310, 314
Buridan, Jean 24, 25, 41, 42, 54, 55, 62, 65–74, 82, 122, 152, 192, 230–232, 238, 245, 246, 283, 306, 326, 329, 332, 338, 341, 344–346
Burley, Walter 13, 23, 31, 68–70, 82–87, 180, 187, 196, 213, 243, 246, 248, 251, 258, 259, 326, 332

Cajetan, Tommaso de Vio 136, 160–190, 196, 197, 247, 262, 271, 272, 309
Calcavy, Pierre 142–144
Calculator: *see* Swineshead, Richard
Calculatores 40, 78–86, 91–94, 98, 107, 229, 231, 311–315
Canonicus, John 169, 196, 213, 214, 243, 247, 251, 253–260
Capreolus, Joannes 76, 80, 136, 160–190, 196, 214, 244, 258, 271, 272, 309
Cardano, Girolamo 47, 87, 196–198
Carpentarius, Jacobus 197, 248
Cartarius, Joannes Ludovicus 173, 197
Carugo, Adriano 199, 200, 204, 207, 236, 242, 254, 255, 262, 284
Casiglio, Antonius 212
Caubraith, Roger 80
Celaya, Juan de 46, 47, 56, 57, 63, 65, 71, 76, 79–87, 102, 105, 108, 231
Christina, Grand Duchess 160, 167
Chrysostom, St. John 200, 201, 216, 308
Ciruelo, Pedro 79–88
Clagett, Marshall 52, 74, 241, 304, 318
Clavius, Christopher 136, 138, 180, 197–217, 225–240, 255, 269, 311
Conciliator: *see* Pietro d'Abano
Contarenus, Gaspar 196, 197, 248, 278, 279
Copernicus, Nicolaus 140, 196–198
Coronel, Antonio 76, 79
Coronel, Luis Nuñes 65, 70–75, 79–86, 101
Crinitus, Petrus 197
Crokart of Brussels, Peter 65, 66, 83, 89, 188–190
Crombie, A. C. 136–138, 199, 200, 232, 234, 236, 239, 240, 242, 254, 255, 284, 318

Dampier, W. C. 303, 317
De Angelis, Mutius 283, 285
Democritus 13, 196
Descartes, René 5, 36, 44, 193, 284
Deza, Diego de 76, 89
Diaz, Froilán 90
Diest, Diego 79–88, 102–109
Diophantus 44
Doctores Parisienses xi, 78, 122, 163, 186, 192–251, 303, 306, 311, 329, 332
Dolz del Castellar, Juan 80
Drabkin, I. E. 166
Drake, Stillman ix, 129, 130, 142, 149, 184, 234, 237, 240, 287, 338
Duhem, Pierre ix–xi, 17, 25, 33, 41, 54, 78, 135, 138, 185, 192, 194, 209, 217, 225, 229, 231, 234, 241, 251, 303–316, 341, 347
Dullaert of Ghent, Jean 46, 56, 65–75, 79–84, 95, 99, 100, 104
Dumbleton, John of 38, 52, 60, 61
Duns Scotus, John 18–23, 65, 70, 71, 136, 163, 172–173, 184–189, 196, 197, 214, 230, 244, 248, 251, 253–262, 288, 289, 308
Durán, Tomás 89
Durandus 163, 164, 171, 173, 186, 189, 196, 214, 244, 255–259

Empedocles 164, 196
Encinas, Fernando de 79, 80, 84
Erasmus 44

Espinosa, Pedro de 79, 84, 86, 90
Euclid 34, 37, 44, 240, 311
Eudaemon-Ioannes, Andreas 212
Eudemus 275, 276

Fantoni, Filippo 226
Faventinus, Benedictus Victorius 86–87
Favaro, Antonio 137, 157, 160, 166, 174, 179, 184, 185, 192–194, 200, 204, 208, 209, 217–220, 229, 234, 238, 240, 262, 265, 308, 337
Feribrigge, Robert 93
Fermat, Pierre 142, 143, 157
Ferrariensis (Francesco Silvestri di Ferrara) 136, 168–190, 196, 197, 213, 214, 247, 257, 271, 272, 309
Ficino, Marsilio 44
Finé, Oronce 228
Flandria, Dominicus de 272
Fonseca de Bobadilla, Juan de 76
Foscarini, Paolo Antonio 140, 231, 241
Francis of Marchia 24, 41
Francis of Meyronnes 251
Fracostoro, Girolamo 121
Fredette, Raymond 135, 318

Gaetano da Thiene 45, 55–58, 63, 87, 95–103, 108, 136, 160, 184, 196, 248, 306
Galen 8, 196, 197, 227, 308
Galilei, Galileo ix–xi, 4, 25, 30, 33, 42, 45, 48, 59, 64, 68, 73, 86, 90, 91, 104, 107, 111, 119, 122–124, 129–347 *passim*
Galuzzi, Tarquinio 239
Garin, Eugenio 180
Gentili, Giovanni 250
George of Brussels 68, 70, 74
Gerard of Brussels 32, 92
Gerson, Rabbi 221
Geymonat, Ludovico 337
Gilbert, William 32, 121
Giles of Rome 16, 165–173, 184–188, 196, 247, 258, 271–273, 278, 304–308, 316, 326
Godfrey of Fontaines 187, 258
Grant, Edward 52, 317, 318
Gregoriis, Hieronymus de 181, 182, 212, 216
Gregory of Nyssa, St. 4, 308
Gregory of Rimini 66, 69, 70, 82–85, 163, 171, 173, 187, 213, 243, 248, 251, 257, 258
Grosseteste, Robert 7, 9, 22, 23, 30, 34, 45, 132, 157, 304–306, 316
Guiducci, Mario 239, 240

Hales, William 221, 224
Harvey, William 264
Henry of Ghent 171, 184, 185, 188, 196, 207, 214, 237, 243, 258
Hero of Alexandria 32, 58
Hervaeus Natalis 76, 136, 160–188, 196, 214, 258, 272, 309
Heytesbury, William 23, 24, 38, 45, 52–56, 60–63, 70, 71, 80–87, 92–105, 192, 197, 306
Hipparchus 119, 308, 315
Hippocrates 8, 196
Hugo Senensis 250

Iamblichus 196, 267
Ibn Gebirol 12
Isocrates 217
Isidorus de Isolanis 87

James of Forlì (Forliviensis) 70, 87, 180, 251
Jandun, John of 18, 171, 188, 189, 196, 247–250, 278
Javelli, Chrysostomus 87, 123, 136, 160–190, 196, 197, 247, 309
John of Ascoli (Gratiadei) 188, 246, 247, 326, 338
John of Bacon 253, 255, 258
John of Casali 44, 87
John of Holland 40, 93, 94, 98, 100
John of Mirecourt 21, 22
Jones, Robertus 182, 255, 256, 260, 263
Jordanus Nemorarius 32, 45, 58
Josephus 222, 224

Kepler, Johannes 32, 121, 122, 221

Koertge, Noretta 149, 158
Koyré, Alexandre ix, 59, 76, 91, 129, 130, 303, 317

Lax, Gaspar 46, 79, 84, 87
Leonardo da Vinci 42, 44, 192, 229, 241, 303
Lerma, Cosme de 90
Lince, Domingo 90
Lohr, C. H. 234, 238
Lokart, George 65, 193
Ludovicus, Antonius 121
Lychetus, Franciscus 197, 198, 253, 255, 258

Maestlin, Michael 221
Maier, Anneliese ix–xi, 52, 54, 110, 122, 241, 304, 318, 320–341, 343, 347
Major, John 46, 56, 65, 69–71, 79, 82–84, 89, 99, 101
Margallo, Pedro 79–84
Marliani, Giovanni 197
Marsilius of Inghen 42, 54, 55, 62, 68, 82, 171, 196, 214, 238, 245, 331
Mas, Diego 88, 124
Mazzoni, Jacopo 138, 227
Menu, Antonius 181–184, 243–247, 255, 256, 260–282, 325–339
Mertonians 38, 40, 51–55, 58, 62, 73, 100, 107, 122, 138, 303, 306, 311
Messinus 98
Meygret, Amadeus 83, 89, 272
Mirandulanus, Bernardus Antonius 196, 198, 271, 272
Moody, E. A. ix–xi, 125, 161, 185, 241, 242, 286, 304, 314, 319, 341–347
Moses, Rabbi 271, 272, 274
Moss, J. D. xvi
Murdoch, John 233, 234, 316, 318

Nardo, Francesco di 136
Naylor, R. H. 130, 159
Newton, Sir Isaac 4, 33, 36, 40, 42, 48, 130, 193, 221, 284
Nicholas of Autrecourt 21, 22, 24, 37, 344
Nicholas of Cusa 44
Nifo, Agostino 115, 196, 198, 247–250, 268
Nobili, Flaminio 171, 180, 181, 196, 198
Nuñes, Pedro 229

Ockham, William of 18, 20–23, 31, 36, 53, 54, 61, 64, 66, 69, 82, 136, 163, 171–173, 189, 196, 197, 213, 214, 230, 232, 242, 243, 251, 257–259, 305, 308, 341–347
Oña, Pedro de 88
Oresme, Nicole 24, 25, 39, 42–46, 70, 71, 80–88, 93–100, 306
Ortega, Juan de 89
Ortiz, Diego 90

Paludanus, Peter 65
Pappus 44, 58
Pardo, Geronimo 79, 83, 84
Parra, Jacinto de la 90
Paul of Venice 45, 55, 56, 65–70, 85, 87, 96, 99, 172, 196, 213, 246, 248, 258, 306, 326, 329, 332, 338
Pavesius, Joannes Josephus 197, 198, 254
Pecham, John 7, 82
Pererius, Benedictus 86, 117, 121, 136, 153, 163, 172, 177–191, 196–200, 207–209, 212, 214, 221–248, 255–262, 288, 289, 298, 325, 326, 334, 338
Peter of Mantua 87
Peter Peregrinus of Maricourt 32
Peter of Spain 65
Philoponus, John 41, 110, 111, 123, 136, 171, 184, 185, 196, 197, 207, 237, 243, 246, 249, 271–280, 290–299, 308, 326, 329, 332, 338
Pietro d'Abano 184, 196, 250, 278, 279
Pico della Mirandola, Giovanni 179
Picus, Hieronymus 87
Plato 4, 6, 24, 44, 136, 164, 168, 184, 195–197, 269, 274–277, 303, 308
Plotinus 4, 5, 196
Pomponazzi, Pietro 121, 196–198, 280
Porphyry 5, 6, 196

Proclus 196, 207, 214, 249
Pseudo-Bradwardine 93, 95
Ptolemy 34, 44, 184, 195–197, 228, 246, 250, 251, 308
Puerbach 197
Pythagoras 196

Randall, J. H., Jr. 304, 316
Regiomontanus 196
Reinhold, Erasmus 221
Ricci, Ostilio 310
Richard of Middleton 173
Roger Bacon 7, 32
Rubeyro, Juan de 80, 87
Rugerius, Ludovicus 117, 125, 153, 158, 159, 182–184, 243, 249–251, 260, 261, 266, 267, 271–283, 311–314, 325, 334–336, 340

Sacrobosco, John of 82, 84, 89, 90, 137, 196, 200, 216, 227, 228, 239, 269, 311
Sarton, George 303, 317
Scaliger, Joseph 222, 238, 246
Scaliger, Julius Caesar 196, 198, 248, 271–273, 326–329, 332, 333, 338
Schmitt, C. B. 137, 234, 242
Settle, T. B. 129, 155, 234, 240
Shapere, Dudley 129
Shea, W. R. 130, 138, 139, 143, 306
Siger of Brabant 16, 17
Silíceo, Juan Martinez 46, 79–88
Simplicius 136, 160–164, 171, 180, 196, 197, 246–250, 267–277, 291, 295, 305, 308, 326, 329, 332, 338
Sixtus of Siena 197, 198, 219
Sommervogel, Carlos 237, 238
Soncinas, Paul 136, 160–190, 196, 214, 258, 271, 272, 309
Soto, Domingo de x, 47, 57, 58, 63, 65, 71, 72, 76–109, 117, 124, 135, 136, 160–190, 196, 198, 229, 231, 241, 244, 247, 251, 288, 289, 298, 309, 312, 326, 338
Soto, Francisco de 71, 76, 102
Soto, Pedro de 229
Spinola, Favius Ambrosius 212
Strabo 197
Swineshead, Richard 23, 24, 38–40, 46, 52–62, 70–89, 92, 100–102, 184, 192, 197, 251, 306
Syrianus Magnus 250, 267

Taiapetra, Hieronymus 197, 198
Taisnier, Jean 312, 319
Tartaglia, Niccolò 47, 58, 240, 310
Tateret, Peter 66
Telesio, Bernardino 121
Tempier, Etienne 16, 74, 316
Terminists, Parisian 24, 25, 54, 80, 303
Themistius 196, 246, 248, 250, 326, 329, 332, 338
Themo Judaeus, 192, 193, 238
Theodoric of Freiberg 34–36, 133, 145, 305
Thomas Aquinas 8–24, 31, 65, 70–73, 80, 83, 122, 129, 132, 133, 136, 152, 160–188, 195, 196, 213, 214, 230, 232, 236, 243–250, 256, 258, 262, 264, 268, 271–283, 286, 288, 293, 297, 304–309, 316, 317, 329, 332, 341–346
Thomas de Garbo 248
Titelmans, Franz 85
Toletus, Franciscus 86, 88, 121, 136, 180–190, 247–251, 255, 261, 262, 312, 325
Torni, Bernardo 99
Trombetta, Antonius 247

Valeriola, Franciscus 197, 198
Valla, Paulus 117, 124, 181–184, 212–217, 225, 227, 238, 243–248, 255–263, 266, 274–285, 311–313, 319, 325, 330–339
Valles, Francisco 116, 172, 177, 179, 185, 188–190, 197, 198, 246, 248, 250
Veracruz, Alonso de la 79, 85, 86, 90
Villalpando, Gaspar Cardillo de 189, 190
Villoslada, R. G. 229, 241
Vinta, Belisario 139
Vitelleschi, Mutius 111–125, 181–184,

212–217, 225, 238, 243, 247–249, 255–263, 266–283, 289–298, 311–319, 325, 334–340
Vitoria, Francisco de 83, 104, 109, 189
Viviani, Vincenzio 308

Weisheipl, J. A. 288, 342, 343
Welser, Mark 237
William of Auvergne 8, 16
William of Ralione 258
Wisan, W. L. 130, 150, 154, 159
Witelo 34

Zabarella, Jacopo 45, 113, 119, 250
Zimara, Marc Antonio 196, 198, 248, 278

SUBJECT INDEX

acceleration, uniform 40, 51, 58, 286
act, first and second 116, 120, 124, 322, 336
action and reaction 120
affirming the consequent, fallacy of 131, 154
Alcalá, University of 46, 57, 72, 78–82
alteration 161, 164, 165, 169, 188, 190, 195
  see also intensification
Aristotelianism 3, 7, 8, 121, 136, 316
  heterodox 16, 37
  scholastic 230, 233, 315
astronomy 200, 216, 307
  see also heavens
Augustinianism 7, 99
Averroism 15, 17, 45, 46, 234, 238, 268, 286, 306, 343
  Latin 16, 129

balance 134, 141, 146, 151, 152, 158
Bologna, University of 226

calculations (*calculationes*) 32, 38, 40
calculatory tradition 40, 46, 47, 53, 55, 78, 80–83, 96, 110, 231, 233, 251, 306, 311
  see also sophisms, calculatory
calculus 38, 39, 52, 154
categories 5, 20, 28, 72
causality 22, 53, 117, 149, 231, 286–299, 344
causes 10, 13, 30, 33, 36, 67, 72, 110–126, 132–134, 138
  antecedent 133, 134, 145
  efficient 112–114, 146, 287, 291, 293
  external 293
  final 11, 121
  formal 10, 114, 147, 287
  internal 111, 118, 147, 288, 293
  material 12, 287
  universal 290
Coimbra, University of 82, 121
compounds (*mixta*) 12, 115, 116
Condemnations of 1270 and 1277 16, 17, 22, 36, 283, 303, 305, 315
Collegio Romano 110–126, 129, 138, 181–184, 198, 216, 221, 225, 226, 229, 233, 242, 255, 261, 266, 268, 283, 290, 309, 310, 324, 338, 346
continuity thesis: Duhem's 192–194, 217, 229, 241, 303–319
  qualified 230, 233, 317
creation, chronology of 219–224

definition, mathematical 134, 151–153
demonstration (*demonstratio*) 131, 133, 134, 138, 139, 145–147, 153, 159, 232, 344, 345
  evidence of premises of 151–154, 232
  physico-mathematical 147, 148, 215
distinction, kinds of 28, 72
dynamics 25, 38, 40–46, 51, 53–56, 73, 74, 78, 110, 342, 346

eclecticism 25, 37, 46, 64, 71
effects 67, 138, 230
  see also causes
elements 12, 112, 114, 116, 120, 161, 165, 166, 169, 195, 245, 274–277, 281
  forms of 166, 171, 172, 176, 177, 180, 254, 259, 278, 279
  gravitation and levitation of 248, 250, 313
  movers of 248
  size and shape of 215, 244, 250

ex suppositione 11, 24, 129–159, 232, 233, 262, 344–346
experimentation 10, 29, 31, 34, 47, 52, 78, 129, 149, 155, 159
experiments 33, 51, 56, 110, 124, 130, 144, 146, 154, 313
thought 58, 67, 129

falling bodies 55, 58, 72, 73, 91, 94–96, 119, 121, 144, 154, 321
acceleration of 42, 47, 118, 119, 153, 251, 314
law of 47, 85, 90, 155, 231
falsification 30, 145
fire, circular motion of 270, 283, 334
forces 14, 33, 37, 67, 110–126
impressed 117
internal 111
motive 114–117, 118, 321, 325
occult 120–122
resistive 117–120
forms 11
intension and remission of 169–172, 176, 187, 200, 254, 257–259
latitude of 24, 38, 68
substantial 113, 118, 287

Galileo's early writings: authors cited in 136, 160–191, 195–197, 253–263, 264–285, 308, 309
dating of 217, 225, 226, 240, 262, 265, 281, 324
evidences of copying in 182–184, 206, 211–261, 282
handwriting of 209, 210, 217–219
Latin style of 202–204, 209, 217
sources of 174–184, 216, 254, 255, 282–283, 324
generator 113, 114, 287, 289
geometry 31
analytic 43, 44, 52
God 12
power of 17, 20, 36, 66, 74
gravitation (*gravitatio*) 116, 120, 246
see also elements
gravity (*gravitas*) 112–118, 120, 287, 288, 310, 320, 326
center of 33, 143
degrees of 118, 120
extensive and intensive 115, 312
positional 32
specific 124, 310–312

heavens 161, 175, 176, 255, 288, 289
animation of 165, 253, 257, 258, 286
composition of 164, 165, 168, 169, 200, 253, 257, 258, 270–273, 286
matter of 164, 165, 169, 177, 274
unity of 164
see also spheres, number and order of heavenly
humanism 44, 129
hypothesis 130, 131, 135, 140, 144–150, 151
see also method, hypothetico-deductive; supposition

impediments (*impedimenti*) 132, 142, 145–148, 151–153, 158
impetus 24, 41–43, 54, 100, 114, 117, 124, 245, 246, 303, 306, 310, 314, 320–340
inclined plane 129, 146, 155, 156, 323
inertia 21, 41, 42, 320
circular 124, 271, 284, 313, 314
instruments 115, 118, 287
intensification 92, 170, 195
continuity of 170, 177, 187, 188, 190

kinematics 24, 32, 37–40, 45, 46, 51–56, 73, 74, 78, 92, 98, 100, 110, 306

lever 32, 141, 158
levity (*levitas*) 114, 115
see also elements
light 5, 30
metaphysics of 7, 9, 30, 33

magnet and magnetism 32, 113, 329, 331

mathematics 3, 7, 9, 14, 31, 73, 80, 81, 84, 93, 148, 149, 155, 310
applications of 32, 51, 59, 95, 110, 231, 265, 315, 346
concepts of 25, 59
methods of 23, 29, 67
matter 11
quantity of 40, 42
maxima and minima 86, 172–178, 254–259
mean-speed theorem 33, 39, 43, 47, 85, 91, 95, 105, 106, 108
measurement 10, 27, 44, 47, 52, 56, 149, 152
mechanics 29–48, 51–65, 86, 92, 107, 122
see also motion, science of
medium 117, 119, 310, 326, 332
resistive 94
uniform 97, 108
method: empirical 31, 104
hypothetico-deductive 51, 129, 130, 144, 154, 345, 346
scholastic 6
methodology 138, 139, 157, 231, 341
see also resolution and composition
motion 13, 39, 66, 68, 78, 110, 143, 307, 310
causes of 40, 53, 61, 111, 117
composition of 118
concept of 53, 61, 64–77
decelerated 92, 119
difform 57, 92
difformly difform 58, 92, 101, 105
distinct from terminus attained 72
distinct from thing moved 72, 230
effects of 40, 53
entitative status of 64, 66, 69, 71, 73, 230, 344
examples of types of 93, 97, 98, 102, 109
falling 14, 118, 144, 286
intermediate or neutral 112, 270, 284, 310, 313, 314, 322, 323, 334
natural 116, 153, 271, 288, 289, 334
preternatural 334, 335, 337
reflected 330
rules for comparability of 37, 83, 86, 312, 316
schemata for types of 93–102, 105, 107
science of 64, 71, 78, 265
turning point of 322, 330, 333, 336, 337, 339
uniform 57, 92
uniformly difform 47, 58, 92, 101–106, 303
violent 116, 271, 334
motor causality 19, 21
principle of 61, 67, 72, 287, 321, 322, 343
mover 248
principal and instrumental 287

natural philosophy 9, 22, 23, 26
proper subject of 247
nature 55, 59, 67, 110–115, 120, 122, 147, 286
causality of 286–299
common 19
definition of 286, 288, 291, 294–297
Neoplatonism 4, 7, 29, 45, 130, 288
Neopythagoreanism 31, 130
nominalism 6, 20, 21, 23, 36, 37, 46, 55, 57, 64–67, 70–72, 80–84, 89, 99, 107, 231, 233, 283, 286, 305, 306, 316, 341–347

Ockhamism 87, 343
see also nominalism, terminism
Omne quod movetur ab alio movetur: *see* motor causality, principle of
optics 7, 29–35
Oxford, University of 6, 7, 14, 19, 23, 24, 29–31, 33, 37, 40–42, 45, 53, 54, 121

Padua, University of 45–47, 54, 55, 58, 137, 226, 228
Paris, University of 6, 14, 16, 23, 24, 29, 31, 33, 40, 41, 45, 46, 54, 56, 57, 64, 66, 78–84, 91, 99, 122, 133

physics 9, 229, 310
mathematical 4, 81, 85, 86, 134, 229, 231, 233, 234, 264, 315
see also science, mixed
Pisa, University of 59, 86, 122, 137, 161, 179, 180, 193, 194, 220, 226–228, 253, 288, 307
Platonism 6, 29, 129, 130
potency, kinds of 113
precursors of Galileo 135, 303, 304, 343
principle: accidental 290
active 112, 249, 289
instrumental 113, 287
passive 112
projectile 14, 18, 41, 47, 106, 111, 116, 117, 125, 245, 246, 321, 330, 332
propensity, internal 111, 112

qualities 12, 161, 274, 321
active or alterative 114, 115
motive 112–119, 125
primary 114, 166, 195, 280, 281, 284
quantification 10, 36, 41, 306

rainbow 30, 34, 35, 133, 145
ratio 37, 60, 67, 87, 89
of ratios 38, 84
rationalism 15, 16, 31
realism 24, 46, 53, 55–57, 64–67, 70–72, 80, 99, 107, 233
removens prohibens 113, 114, 287, 288
resistance 14, 33, 37, 116, 117, 119, 120
internal 118
resolution and composition 30, 36, 45

Salamanca, University of 46, 47, 54, 57, 72, 78–85, 91, 102, 136
Saragossa, University of 102
saving the appearances 10, 131, 138, 229, 232, 234, 303, 316
scholasticism 3, 5, 44, 59, 121, 304
early 6, 304
high 22, 304, 305, 309, 315
late 304, 305, 309, 315
second 65
science (*scientia*) 10, 51, 131, 138, 139, 145, 345
continuity of medieval with modern 192, 229, 303, 315, 317
existence of subject of 254, 259, 260
medieval 3, 25, 29–48
mixed (or middle) 9, 10, 26, 151, 232, 233
new (*scienza nuova*) 25, 48, 73, 130, 131, 146, 148, 154, 307, 316, 320, 325
scientia media (*mixta*); *see* science, mixed (or middle)
Scotism 46, 66, 82, 84, 87, 234, 253–263, 286
skepticism 21, 22, 37
sophisms, calculatory (*sophismata calculatoria*) 24, 38, 44, 81, 85, 306
space 97, 98, 100, 108
see also medium
spheres, number and order of heavenly 164, 200, 201, 216, 269
statics 32, 58
supposition (*suppositio*) 36, 132–135, 140, 145, 151, 152, 158, 159
see also hypothesis; *ex suppositione*

terminism 24, 25, 36, 45, 46
theory 33, 51
see also experiment; verification, experimental
Thomism 46, 66, 71, 82, 84, 87, 89, 136, 160–191, 234, 286, 288, 290, 309, 315, 341, 346

uniformiter difformis 73, 91, 95, 99, 102, 105, 107, 108, 192, 251, 306, 313
see also motion, uniformly difform
universals 5, 6, 19, 20, 36
universe 162, 163, 253
age of 219
creation of 163
eternity of 16, 163, 190, 207, 213, 214, 219, 243, 250, 253, 257, 268

Ptolemaic conception of 32
unity and perfection of 175, 256, 268

vacuum (void) 14, 17, 18, 47, 58, 342, 343
Valladolid, University of 102
velocity 92, 100, 103, 115
instantaneous 38–40, 52
verification, experimental 51, 145
via antiqua 21, 53, 60, 347
via moderna 21, 53, 60, 347
virtus impressa 114, 245, 322, 325–327, 330, 332, 336
see also impetus
vis 122
see also force

water vapor, upward motion of 291, 292
weights, science of 58
Works of Aristotle cited: *De caelo et mundo* 9, 85, 136, 161, 162, 164, 194, 195, 216, 226, 249, 266, 274, 305, 307, 311, 324, 325
De generatione et corruptione 9, 83, 85, 136, 162, 194, 195, 226, 274, 307, 311, 324–326
Meteorology 9, 195, 215, 325
Physics 9, 17, 23, 25, 30, 37, 46, 47, 56, 57, 65, 70, 81, 83, 89–91, 99, 114, 132, 136, 152, 164, 195, 230, 267, 289, 305, 307, 311, 341, 346
Posterior Analytics 9, 30, 31, 36, 132, 135, 151, 152, 229, 232, 305, 307, 346
Quaestiones mechanicae 58, 327
Works of Galileo cited: *De motu* (MS Gal. 71) 59, 135, 161, 166, 167, 205, 228, 231, 237, 270, 310, 314, 318, 322–324
Juvenilia (so-called) 135, 161, 193, 194, 220, 225, 228, 240, 265
Logical Questions (MS Gal. 27) 135, 205, 254, 265, 307, 318
Physical Questions (MS Gal. 46) 135–138, 161, 174, 179, 180, 194, 200, 212, 216, 217, 225, 235, 236, 250, 253, 254, 265, 282, 307–309
Two New Sciences 59, 130, 131, 135, 138–144, 146, 149, 150, 157, 158, 232, 234, 252, 307, 310, 311, 318
Two Chief World Systems 139, 142, 232, 307

BOSTON STUDIES IN THE PHILOSOPHY OF SCIENCE

1. Marx W. Wartofsky (ed.), *Proceedings of the Boston Colloquium for the Philosophy of Science 1961-1962.* 1963.
2. Robert S. Cohen and Marx W. Wartofsky (eds.), *In Honor of Philipp Frank.* 1965.
3. Robert S. Cohen and Marx W. Wartofsky (eds.), *Proceedings of the Boston Colloquium for the Philosophy of Science 1964-1966. In Memory of Norwood Russell Hanson.* 1967.
4. Robert S. Cohen and Marx W. Wartofsky (eds.), *Proceedings of the Boston Colloquium for the Philosophy of Science 1966-1968.* 1969.
5. Robert S. Cohen and Marx W. Wartofsky (eds.), *Proceedings of the Boston Colloquium for the Philosophy of Science 1966-1968.* 1969.
6. Robert S. Cohen and Raymond J. Seeger (eds.), *Ernst Mach: Physicist and Philosopher.* 1970.
7. Milic Capek, *Bergson and Modern Physics.* 1971.
8. Roger C. Buck and Robert S. Cohen (eds.), *PSA 1970. In Memory of Rudolf Carnap.* 1971.
9. A. A. Zinov'ev, *Foundations of the Logical Theory of Scientific Knowledge (Complex Logic).* (Revised and enlarged English edition with an appendix by G. A. Smirnov, E. A. Sidorenka, A. M. Fedina, and L. A. Bobrova.) 1973.
10. Ladislav Tondl, *Scientific Procedures.* 1973.
11. R. J. Seeger and Robert S. Cohen (eds.), *Philosophical Foundations of Science.* 1974.
12. Adolf Grünbaum, *Philosophical Problems of Space and Time.* (Second, enlarged edition.) 1973.
13. Robert S. Cohen and Marx W. Wartofsky (eds.), *Logical and Epistemological Studies in Contemporary Physics.* 1973.
14. Robert S. Cohen and Marx W. Wartofsky (eds.), *Methodological and Historical Essays in the Natural and Social Sciences. Proceedings of the Boston Colloquium for the Philosophy of Science 1969-1972.* 1974.
15. Robert S. Cohen, J. J. Stachel and Marx W. Wartofsky (eds.), *For Dirk Struik. Scientific, Historical and Political Essays in Honor of Dirk Struik.* 1974.
16. Norman Geschwind, *Selected Papers on Language and the Brain.* 1974.
18. Peter Mittelstaedt, *Philosophical Problems of Modern Physics.* 1976.
19. Henry Mehlberg, *Time, Causality, and the Quantum Theory* (2 vols.). 1980.
20. Kenneth F. Schaffner and Robert S. Cohen (eds.), *Proceedings of the 1972 Biennial Meeting, Philosophy of Science Association.* 1974.
21. R. S. Cohen and J. J. Stachel (eds.), *Selected Papers of Léon Rosenfeld.* 1978.
22. Milic Capek (ed.), *The Concepts of Space and Time. Their Structure and Their Development.* 1976.
23. Marjorie Grene, *The Understanding of Nature. Essays in the Philosophy of Biology.* 1974.

24. Don Ihde, *Technics and Praxis. A Philosophy of Technology.* 1978.
25. Jaakko Hintikka and Unto Remes, *The Method of Analysis. Its Geometrical Origin and Its General Significance.* 1974.
26. John Emery Murdoch and Edith Dudley Sylla, *The Cultural Context of Medieval Learning.* 1975.
27. Marjorie Grene and Everett Mendelsohn (eds.), *Topics in the Philosophy of Biology.* 1976.
28. Joseph Agassi, *Science in Flux.* 1975.
29. Jerzy J. Wiatr (ed.), *Polish Essays in the Methodology of the Social Sciences.* 1979.
32. R. S. Cohen, C. A. Hooker, A. C. Michalos, and J. W. van Evra (eds.), *PSA 1974: Proceedings of the 1974 Biennial Meeting of the Philosophy of Science Association.* 1976.
33. Gerald Holton and William Blanpied (eds.), *Science and Its Public: The Changing Relationship.* 1976.
34. Mirko D. Grmek (ed.), *On Scientific Discovery.* 1980.
35. Stefan Amsterdamski, *Between Experience and Metaphysics. Philosophical Problems of the Evolution of Science.* 1975.
36. Mihailo Marković and Gajo Petrović (eds.), *Praxis. Yugoslav Essays in the Philosophy and Methodology of the Social Sciences.* 1979.
37. Hermann von Helmholtz: *Epistemological Writings. The Paul Hertz/Moritz Schlick Centenary Edition of 1921 with Notes and Commentary by the Editors.* (Newly translated by Malcolm F. Lowe. Edited, with an Introduction and Bibliography, by Robert S. Cohen and Yehuda Elkana.) 1977.
38. R. M. Martin, *Pragmatics, Truth, and Language.* 1979.
39. R. S. Cohen, P. K. Feyerabend, and M. W. Wartofsky (eds.), *Essays in Memory of Imre Lakatos.* 1976.
42. Humberto R. Maturana and Francisco J. Varela, *Autopoiesis and Cognition. The Realization of the Living.* 1980.
43. A. Kasher (ed.), *Language in Focus: Foundations, Methods and Systems. Essays Dedicated to Yehoshua Bar-Hillel.* 1976.
46. Peter L. Kapitza, *Experiment, Theory, Practice.* 1980.
47. Maria L. Dalla Chiara (ed.), *Italian Studies in the Philosophy of Science.* 1980.
48. Marx W. Wartofsky, *Models: Representation and the Scientific Understanding.* 1979.
50. Yehuda Fried and Joseph Agassi, *Paranoia: A Study in Diagnosis.* 1976.
51. Kurt H. Wolff, *Surrender and Catch: Experience and Inquiry Today.* 1976.
52. Karel Kosík, *Dialectics of the Concrete.* 1976.
53. Nelson Goodman, *The Structure of Appearance.* (Third edition.) 1977.
54. Herbert A. Simon, *Models of Discovery and Other Topics in the Methods of Science.* 1977.
55. Morris Lazerowitz, *The Language of Philosophy. Freud and Wittgenstein.* 1977.
56. Thomas Nickles (ed.), *Scientific Discovery, Logic, and Rationality.* 1980.
57. Joseph Margolis, *Persons and Minds. The Prospects of Nonreductive Materialism.* 1977.
58. Gerard Radnitzky and Gunnar Andersson (eds.), *Progress and Rationality in Science.* 1978.

59. Gerard Radnitzky and Gunnar Andersson (eds.), *The Structure and Development of Science.* 1979.
60. Thomas Nickles (ed.), *Scientific Discovery: Case Studies.* 1980.
61. Maurice A. Finocchiaro, *Galileo and the Art of Reasoning.* 1980.

GPSR Compliance

The European Union's (EU) General Product Safety Regulation (GPSR) is a set of rules that requires consumer products to be safe and our obligations to ensure this.

If you have any concerns about our products, you can contact us on ProductSafety@springernature.com

In case Publisher is established outside the EU, the EU authorized representative is:

Springer Nature Customer Service Center GmbH
Europaplatz 3
69115 Heidelberg, Germany

Batch number: 10761599

Printed by Printforce, the Netherlands